Free From HomeTech!

Send in this registration card and receive valuable products and services absolutely free!

S0-AXV-642

BUSINESS REPLY MAIL
First Class Mail   Permit No. 491   Bethesda, MD 20816

Postage will be paid by addressee

HomeTech Information Systems
5161 RIVER ROAD   SUITE 104
BETHESDA MD   20816-9976

NO POSTAGE
NECESSARY IF
MAILED IN THE
UNITED STATES

HomeTech Information Systems, Inc.
5161 River Road, Bethesda, Maryland 20816
1-800-638-8292

ABBREVIATIONS

THE FOLLOWING ABBREVIATIONS ARE USED IN THIS MANUAL:

CF	PER CUBIC FOOT (ONE FOOT WIDE, ONE FOOT HIGH AND ONE FOOT LONG)
LF	PER LINEAL FOOT (PER RUNNING FOOT)
SF	PER SQUARE FOOT OF AREA
EA	EACH UNIT (A WINDOW UNIT INCLUDES WINDOW, FRAME, TRIM AND LABOR TO INSTALL)
EA PLUS SF	ADD THE AMOUNT OPPOSITE "EA" (EACH) TO THE TOTAL OBTAINED BY MULTIPLYING THE SQUARE FOOTAGE OF AREA BY THE AMOUNT OPPOSITE "SF"
UI	UNITED INCHES

TO CONVERT TO GROSS PROFITS OTHER THAN THE 33-1/3% RATE SHOWN IN THIS MANUAL, USE THE TABLE BELOW:

TO OBTAIN % OF GROSS PROFIT	MULTIPLY BOOK SALES PRICE BY	MULTIPLY BOOK COST BY
50%	1.34	2.00
45%	1.22	1.82
40%	1.12	1.67
35%	1.03	1.54
33-1/3%	- -	1.50
30%	.96	1.43
25%	.89	1.34
20%	.84	1.25

TABLE OF CONTENTS

Preface to 30th

This estimating manual has been prepared for the use of contractors, estimators, supervisors, architects, insurance adjusters, urban renewal specialists, apartment house managers and all other professionals who need up-to-date cost information on which to base estimates and sales proposals.

Job costs include laying out the work, sweeping up and piling debris outside the building, but not removal of debris from premises unless so indicated.

A 33-1/3% gross profit is included in each sales price, which is obtained by adding 50% of the total labor and materials costs for each item. Therefore, 2/3 of the sales price is cost and 1/3 is gross profit.

MATERIALS COSTS

The materials costs in this volume are based on current costs in a metropolitan area including applicable state, provincial and federal sales taxes.

The cost of materials includes everything that goes into the job. For example, drywall material would include the board, drywall nails, corner beads, tape, joint compound, even sandpaper. A normal waste factor is also included.

LABOR COSTS

The labor costs include:
HOURLY WAGES
WORKERS' COMP. INSURANCE
SOCIAL SECURITY TAXES
STATE UNEMPLOYMENT TAX
MINOR DAILY CLEANUP
COFFEE BREAKS
SET UP AND LAYOUT TIME
MINOR MATERIALS PICKUP

Also included are the normal inefficiencies of the remodeling business: time spent in discussions with customers on the job, placing and protecting delivered materials, assisting subcontractors, waiting for customer selections or official building inspections, etc.

Paid holidays, vacation and medical benefits are not included in the labor costs. If your employees receive these fringe benefits, you should calculate what these are as an hourly figure for 2,000 hours per year and add that to the labor cost shown in the manual.

SUBCONTRACTOR COSTS

The costs for plumbing, heating, electrical, built-up roofing, machine excavation, asphalt paving, ceramic tile and blown-in insulation are the amounts subcontractors in the trades would charge a general contractor.

The amounts in the "PRICE" columns throughout the volume are the amounts a general contractor would charge a homeowner or customer for the work. The costs shown, therefore, include the subcontractor's own profit on the job.

While many contractors do not charge a full markup for work done by subcontractors, it is our view that a remodeling contractor cannot make enough gross profit on a job to cover overhead unless everything, including subcontractors, is marked up at least 50%.

Annual Edition

The contractor has spent a good deal of time finding subcontractors who can do a job properly, on time, at a fair price. The contractor is responsible to the customer for quality of work if any problem arises later on. Should the subcontractor be unable or unwilling to correct the problem, the remodeler must do it at his own expense.

OVERHEAD ITEMS

Overhead items must be paid for out of the gross profit which is included in the amount shown in the "PRICE" column.

Overhead items include the following:
SALES COMMISSIONS
OFFICE SALARIES
JOB SUPERVISION
TRUCK EXPENSE
TELEPHONE
ADVERTISING AND MARKETING
OFFICE RENTAL
OFFICE SUPPLIES
LEGAL FEES
ACCOUNTANT FEES
LEASED OFFICE EQUIPMENT
TOOLS AND EQUIPMENT
INTEREST ON LOANS
BAD DEBTS
INSURANCE AND BONDS

JUDGEMENT FACTORS

Remodeling and renovation work is the most difficult to estimate in residential construction. There are more hidden problems resulting in additional costs than normally occur in new home construction. Care should be taken to determine whether structural defects exist inside closed floors, walls and ceilings before submitting a final bid.

Estimators must use judgement in working with this manual. Conditions under which jobs are performed in the building trades are variable, but normal, sunshiny job conditions and easy access to the job have been assumed in estimating the cost of each item. Important factors governing costs and bids received are the season of the year, general business conditions in the area, and the amount of work subcontractors have ahead when bidding your work.

Every effort is made to provide up to date, accurate cost information. Building costs throughout the United States and Canada are monitored, but cost changes occur frequently and are not always consistent with general economic conditions in an area. **No warranty or guarantee is made as to the correctness or sufficiency of the information contained in this book. The editors and publishers assume no responsibility or liability in connection with its use.**

The editors wish to thank all who have generously assisted in the preparation of this volume. We are especially grateful to David Hawbecker, Darrell Lewis and Peter Starbuck for the materials and information which they have supplied. Constructive criticism which will assist us in the preparation of future editions will be welcomed.

Henry Reynolds, Editor
Bethesda, Maryland
December 1, 1994

How to Use the Remodeling and

SPEEDY RECKONER SECTION

Using the Speedy Reckoner, you can estimate many standard jobs quickly and accurately. For example, look at Page 2 to see how a standard shell addition is estimated. First, the specifications are given, clearly and completely, from the foundation through the exterior finish. Next, on Page 3, the pricing information is shown.

The Shell Addition has a base price of $6381, similar to a trip charge including set-up costs and minimum charges that occur regardless of the size of the addition. In addition to the base price there is a per-square-foot price which is $27.47 on a one-story addition. If you are estimating a 200 square foot addition, you would multiply 200 x $27.47 and add the result, $5494, to the $6381 base price for a total price of $11,875.

You will find the prices in the Speedy Reckoner to be accurate so long as the specifications for the job you are estimating are the same as the specifications shown in the manual. For example, the Shell Addition on Page 2 has no door, and one window per 100 SF of living area. If you are estimating a shell addition with more windows and one door, you will need to add the price of the additional windows and a door to the price calculated from the manual.

When used properly, the Speedy Reckoner can produce accurate estimates in as little as five minutes. It is also useful for "ballpark" estimates, and can be invaluable to estimators who do not know how to break down a job into its individual components.

SECTION II (ESTIMATOR)

This section of the manual is broken down into the 25 categories listed in the Table of Contents and on the back cover. These categories are organized in the order that a job is built, from Plans and Permits on through to Clean-Up. Many estimators use these 25 categories as a checklist to be sure they have not forgotten anything. The list of categories on the back cover matches up with black tab markings on the pages of the manual, so that you can easily flip to the category you want in seconds.

Here is an example of how an individual item is set up.

Look at Page 126, at the Exterior Stud Wall. In the SPECIFICATION section is a clear description of what is included in the item. The unit of measure, "SF", stands for square feet of wall (other items may be "LF" for lineal foot, "CF" for cubic foot, "EA" for each item, etc.). There are columns for Materials Cost, Labor Cost, Total Cost, and then Total Price including a 50% markup.

The labor figures in the manual are based on actual labor for thousands of jobs, translated into unit cost amounts. They provide an accurate guide, but since different companies have different labor efficiencies, you should habitually compare your actual labor costs on jobs with the costs in the manual.

TOTAL COST: The Job Cost Total column is the sum of the Materials and Labor Cost columns. In categories such as Plumbing, Electrical, and Heating, you will see that only a Total Job Cost is

Renovation Cost Estimator

given. These trades are usually subcontracted in remodeling, and the subcontractor supplies materials and labor. The Cost column in these categories includes the subcontractor's markup. We recommend that you negotiate prices with subcontractors, using the manual as a guide.

TOTAL PRICE: The Price column marks up the Job Cost Total by 50% for a 33-1/3% gross profit margin. If you want to charge a higher or lower markup than 50%, see the table on Page ii. Home-Tech recommends a minimum markup of 50% over cost to make a profit in remodeling and renovation.

GENERAL INFORMATION

This manual is the best estimating guide available in the industry, and many contractors use it without adjustment. In order to use it properly, you must check it against the realities of your operation and adjust the figures as needed. For example, if the manual shows that drywall costs 75¢ per square foot, and your drywall hanger charges 87¢ per square foot, you should change the figure in the manual to 87¢ to reflect the reality.

Unit cost estimating allows you to estimate standard work very quickly and accurately. But that is only about 90% of a good, detailed bid. You need to apply a judgement analysis to any estimate to reflect special conditions. That analysis should include the following:

PROJECT ANALYSIS: If there are unusual characteristics to the job, such as a tricky roof tie-in or difficult materials matching, increase the job costs.

JOB CONDITIONS: Difficult conditions such as poor access, little or no storage, high risk of theft, etc., should be reflected in increased job costs.

CUSTOMER ANALYSIS: Approximately one out of five customers will be unusually demanding, and can cost you all the profit on a job unless you recognize them for what they are and increase your estimate to allow you to satisfy them and still make a profit.

COMPANY CAPABILITY: Certain kinds of work may be easier or harder than usual for your workers, and your estimate should be adjusted to reflect that. If you will need to hire more people or pay overtime in order to do a particular job, your increased costs should be reflected in your estimate.

If you have specific questions about using this manual, please call us. Use our toll-free number, 1-800-638-8292.

> DATABASE ITEM NUMBERS <

This manual includes the item numbers from HomeTech's computer estimating database. The three digits in the far right column on each page are the extension of the number for that item in the database. The section number of the manual is the prefix of the item number.

For example, on Page 126, the 2x4 stud wall is item number 7.000, while the 2x6 stud wall is 7.001. On Page 226, 3/8" gypsum lath is item number 18.115 and 1/2" gypsum lath is 18.116.

For more information about HomeTech computer estimating, call 1-800-638-8292.

SÁMPLE ESTIMATE

ESTIMATE SHEET

Name	J.M. KEYNES	Date 1 NOV 19 94	Architect JJB
Address	474 MAIN ST., CITY	Telephone 555-3210	Estimator TRH
Job Address	SAME	Job Telephone SAME	Checked By NT

Item	Unit	Unit Quantity	MATERIALS		LABOR & SUB	
			Unit Cost	COST	Unit Cost	COST
PLANS	EA+ #M	13,000	–	–	200.+ 10	200 130
TEAR OUT MASONRY WALL	LF	3	18.	54	46.	138
BLOCK WALL 8x8x16	SF	150	1.84	276	2.47	371
CONC. FOOTINGS 12x24, 36" BELOW GRD.	LF	50	6.53	327	14.	700
WOOD BEAM	LF	12	5.67	68	2.08	25
STEEL COLUMN	EA	1	37.50	38	14.50	15
MUDSILL	LF	50	1.25	63	.65	33
FLOOR JOISTS 2x10, 16" OC	SF	290	1.71	496	.80	232
SUBFLOOR 3/4" T&G PLYWOOD	SF	290	.67	194	.62	180
WALL FRAMING 2x4, 16" OC	SF	384	.76	292	.62	238
SHEATHING 1/2" CDX	SF	432	.45	195	.42	181
FURRING 1x3	SF	192	.17	33	.60	115
ROOF FRAMING 2x8, 4/12	SF	314	1.72	540	1.30	408
ROOF SHEATHING 1/2" CDX	SF	314	.45	141	.37	116
ROOF COVERING #225 FIBERGL.	SF	314	.38	119	.51	160
ROOF COVERING, SMALL JOB	EA		–	–	40.	40
ALUMINUM FLASHING	LF	25	2.00	50	4.00	100
GUTTER & DOWNSPOUT, ALUMINUM	LF	34	1.72	58	1.40	48
FASCIA, 6"	LF	25	.79	20	1.01	25
SOFFIT, 6"	LF	25	.77	19	1.32	33
RAKE & RAKE MOULDING	LF	25	.99	25	1.21	30
SIDING – 1/2" BEVELED CEDAR	SF	432	1.75	756	.97	419
WDH WINDOW 2-8x4-6, DBL. GL.	EA	3	193.	579	59.	177
INTER. STUD WALL 2x4, 16" OC	SF	80	.73	58	.65	52
REAR DOOR, 3 PANEL, 4 LIGHTS	EA	1	317.	317	91.	91
INTER. BIRCH DOOR, FLUSH, 2-6x6-8	EA	1	98.	98	34.	34
BATH – W.C., LAV., TUB	TOTAL			623		1769
ELECTRICAL OUTLETS	EA	12	13.	156	25.	300
SWITCH AND LIGHT	EA	1	24.	24	50.	50
CLEAN UP	EA+ SF	290	–	–	124.+ .26	124 76
			Totals	5,619		6,610

Total Cost	12,229
Add 50 % Markup	6,115
SALES PRICE	18,344

HomeTech Form 128

SAMPLE ESTIMATE

	ESTIMATE SHEET			Sheet 1 of 1				

Name J. M. KEYNES — **Date** 1 NOV 1994 — **Architect** JJB

Address 474 MAIN ST., CITY — **Telephone** 555-3210 — **Estimator** TRH

Job Address SAME — **Job Telephone** SAME — **Checked By** NT

Cat. No.	Specifications	Quantity	Unit Cost	Total			
	AMOUNT FORWARD	---	---				
1	PLANS	13,000	200. + 10.	2	0	0	
				1	3	0	
2	TEAR OUT MASONRY WALL	3 LF	64.	1	9	2	
4	CONC. FOOTINGS 12x24, 36" BELOW GRADE	50 LF	20.53	1 0	2	7	
5	BLOCK WALL 8x8x16	150 SF	4.31	6	4	7	
6	WOOD BEAM	12 LF	7.75		9	3	
	STEEL COLUMN	1	53.		5	3	
	MUDSILL	50 LF	1.90		9	5	
	FLOOR JOISTS 2x10, 16" OC	290 SF	2.51	7	2	8	
	SUBFLOOR 3/4" T & G PLYWOOD	290 SF	1.29	3	7	5	
7	WALL FRAMING 2x4, 16" OC	384 SF	1.38	5	3	0	
	INTER. STUD WALL 2x4, 16" OC	80 SF	1.38	1	1	1	
	SHEATHING 1/2" CDX	432 SF	.87	3	7	6	
	FURRING 1x3	192 SF	.77	1	4	8	
8	ROOF FRAMING 2x8, 4/12	314 SF	3.02	9	4	9	
	ROOF SHEATHING 1/2" CDX	314 SF	.82	2	5	8	
9	ROOF COVERING #225 FIBERGLASS	314 SF	.89	2	8	0	
	ROOF COVERING, SMALL JOB		40.		4	0	
	ALUMINUM FLASHING	25 LF	6.00	1	5	0	
	GUTTER & DOWNSPOUT, ALUMINUM	34 LF	3.12	1	0	6	
10	FASCIA, 6"	25 LF	1.80		4	5	
	SOFFIT, 6"	25 LF	2.09		5	2	
	RAKE & RAKE MOULDING	25 LF	2.20		5	5	
11	SIDING - 1/2" BEVELED CEDAR	432 SF	2.72	1 1	7	5	
12	REAR DOOR, 3 PANEL, 4 LIGHTS	1	408.	4	0	8	
	INTER. BIRCH DOOR, FLUSH, 2-6x6-8	1	132.	1	3	2	
13	WDH WINDOW 2-8x4-6, DBL. GL.	3	252.	7	5	6	
14	BATH - W.C., LAV., TUB	TOTAL		2 3	9	2	
16	ELECTRICAL OUTLETS	12	38.	4	5	6	
	SWITCH & LIGHT	1	74.		7	4	
25	CLEAN UP	290 SF	24. + .26	1	2	4	
					7	6	

General Description of Work

Total Cost	1 2	2	2	9
33 % Gross Profit	6	1	1	5
TOTAL AMOUNT	1 8	3	4	4

Form 135

SAMPLE ESTIMATE

REMODELING SPEEDY RECKONER ESTIMATE

Name	J.M. KEYNES	Telephone 555-3210	Job Telephone SAME
Address	474 MAIN ST., CITY	Estimator TRH	Date 1 NOV 1994
Job Address	SAME	Description of Work	SHELL ADDITION, BATHROOM, ELECTRICAL OUTLETS

BASIC JOB, EXTRAS AND ALLOWANCES

Item	Unit	Unit Quantity	ADD Unit Price	ADD PRICE	DEDUCT Unit Price	DEDUCT PRICE
BASIC PRICE	EACH	---	6381.	6,381		---
SQUARE FOOT AREA	SF	290	25.26	7,325		---
TEAR OUT MASONRY WALL	LF	3	96.	288		
INTERIOR STUD WALL 2x4, 16"O.C.	LF	10	16.56	166		
SUBSTITUTE BEVELED SIDING	SF	432	.86	372		
IN LIEU OF VINYL SIDING						
REAR DOOR 2-8x6-8	EA	1	612.	612		
BATH - W.C., LAV., TUB	TOTAL			3,210		
ELECTRICAL OUTLETS	EA	12	57.	684		
SWITCH & LIGHT	EA	1	111.	111		
INTER. BIRCH DOOR 2-6x6-8	EA	1	198.	198		
	Totals			19,347		
	Less Deductions			—		
	SALES PRICE			19,347		

HomeTech Form 136

HomeTech
Remodeling and Renovation
Cost Estimator

Section I
Speedy Reckoner

How to Use the Speedy Reckoner

The figures used in the Speedy Reckoner are the prices a contractor would charge a customer for the work. Prices include the contractor's 1/3 gross profit (50% markup over direct job costs).

After the basic price has been entered on your Speedy Reckoner estimate, there may be additional items that are not included in the basic job specifications. There may also be items in the basic price specifications that are to be eliminated in the particular job being estimated. Enter these items on your estimate sheet in the "ADD" or "DEDUCT" columns (see the sample Speedy Reckoner Estimate on Page -x-). If the items are not under "Extras, Allowances, Alternate Specifications" in the Speedy Reckoner, look in Section II (starting on Page 52) and use the figures shown in the "PRICE" columns as the basis for computing the additions and deductions.

To Add Items Not Included In Basic Job:

Compute the price based on the actual quantity of the work to be done and set that amount down in the "ADD" column.

To Eliminate Items Included In Basic Job:

If an item specified in a basic job is not to be included in the job being estimated, you may eliminate it on your Speedy Reckoner Estimate as follows: compute what the total price of the item would have been if it had been included as specified and set that amount down in the "DEDUCT" column.

To Substitute One Item For Another In Basic Job:

When a different item is to be substituted for the item specified, refer to the Shell Addition Extras and Allowances on Pages 14 to 16.

For example, the Sample Shell Addition shown on Page -x- calls for beveled wood siding on the sidewalls instead of the vinyl siding shown in the Shell Addition Specifications on Page 2. On Page 16, select the item that substitutes 6" beveled siding, and set down that price in the "ADD" column. (If you were substituting a less expensive siding, such as hardboard lap siding, you would set down the price from Page 16 in the "DEDUCT" column of your estimate.

SHELL ADDITION -- ON BLOCK FOUNDATION WALL

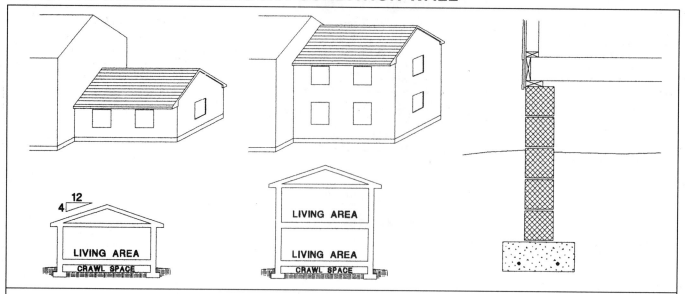

SPECIFICATIONS

PLANS AND PERMIT	PREPARE PLANS AND OBTAIN PERMIT, PERMIT FEE COST **NOT** INCLUDED
CONCRETE AND MASONRY	EXCAVATE BY HAND TO 36" BELOW GRADE, FORM AND POUR CONTINUOUS CONCRETE FOOTINGS WITH REBAR BLOCK FOUNDATION WALL ON FOOTINGS TO 18" ABOVE GRADE WITH 1/2" ANCHOR BOLTS 8'-0" O.C. PARGE, DAMPPROOF, BACKFILL AND ROUGH GRADE AS REQUIRED
FLOOR FRAMING	FLOOR JOISTS AS REQUIRED, 16" O.C. 3/4" T & G FIR PLYWOOD SUBFLOOR, GLUED AND NAILED
WALL FRAMING	2" X 4" STUDDING AND PLATES, 16" O.C. 1/2" CDX PLYWOOD OR FIBERBOARD SHEATHING WITH CORNER BRACING 1" X 2" FURRING, 16" O.C., ON EXISTING HOUSE WALL
ROOF FRAMING	SHED OR GABLE TYPE ROOF, 12" OVERHANG, **NO** OVERLAY OF EXISTING HOUSE ROOF RAFTERS AND CEILING JOISTS AS REQUIRED, 16" O.C. 1/2" CDX PLYWOOD ROOF SHEATHING
ROOFING, GUTTERS, FLASHING	#215 FIBERGLASS STRIP SHINGLES OVER #15 FELT PAPER ALUMINUM GUTTER AND DOWNSPOUT AS REQUIRED ALUMINUM FLASHING AT EXISTING HOUSE
EXTERIOR TRIM	TIGHT KNOT PINE OR FIR FASCIA AND SOFFIT TIGHT KNOT PINE OR FIR RAKE AND RAKE MOULDING
SIDING	8" VINYL SIDING (INSULATED)
WINDOW	**ONE** WOOD DOUBLE HUNG 2-8 X 4-6 WINDOW, DOUBLE GLAZED, FOR EACH 100 SF OF LIVING AREA
CLEAN-UP	REMOVE RUBBISH FROM PREMISES, DUMPING FEE **NOT** INCLUDED
	NO PAINTING

SHELL ADDITION -- ON BLOCK FOUNDATION WALL

SPECIFICATIONS		UNIT	JOB COST			PRICE	LOCAL AREA MODIFICATION			
			MATLS	LABOR	TOTAL		MATLS	LABOR	TOTAL	PRICE
SHELL ADDITION	ONE STORY	EA PLUS	2088.00	2166.00	4,254.00	6,381.00				
	SF = SF LIVING AREA	SF	10.20	6.64	16.84	25.26				
	TWO STORY	EA PLUS	2360.00	2830.00	5,190.00	7,785.00				
	SF = SF LIVING AREA 2 FLOORS	SF	9.90	5.37	15.27	22.91				

EXTRAS AND ALLOWANCES

SPECIFICATIONS		UNIT	JOB COST			PRICE	LOCAL AREA MODIFICATION			
			MATLS	LABOR	TOTAL		MATLS	LABOR	TOTAL	PRICE
DIFFERENT FOOTING DEPTH	***Depth of Bottom of Footing Below Grade***									
	12" **DEDUCT**	LF	3.68	10.28	13.96	20.94				
	24" **DEDUCT**	LF	1.84	5.13	6.97	10.46				
	48" **ADD**	LF	1.83	5.12	6.95	10.43				
	LF = FOOTINGS									
CONCRETE FOUNDATION WALL	SUBSTITUTE CONCRETE FOUNDATION WALL IN PLACE OF BLOCK									
	Depth of Bottom of Footing Below Grade									
	24" **ADD**	LF	.57	8.36	8.93	13.40				
	36" **ADD**	LF	.50	8.05	8.55	12.83				
	48" **ADD**	LF	.44	7.75	8.19	12.29				
	LF = WALL									
FOOTING AND COLUMN	EXCAVATE, FORM, POUR 24" X 24" X 12" CONCRETE FOOTING AND INSTALL 3" STEEL COLUMN UNDER BEAM **ADD**	EA	63.00	41.00	104.00	156.00				
WOOD BEAM	INSTALL WOOD BEAM TO SUPPORT JOISTS									
	6" X 8" **ADD**	LF	5.67	2.08	7.75	11.63				
	6" X 10" **ADD**	LF	9.33	2.63	11.96	17.94				
WINDOWS	ADD OR OMIT ONE WOOD DOUBLE HUNG WINDOW, 2-8 X 4-6, DOUBLE GLAZED **ADD** OR **DEDUCT**	EA	193.00	59.00	252.00	378.00				
	SUBSTITUTE DOUBLE GLAZED 4-0 X 4-0 ALUMINUM SLIDING WINDOW **DEDUCT**	EA	50.00	20.00	70.00	105.00				

NOTE: FOR ADDITIONAL SHELL ADDITION EXTRAS AND ALLOWANCES, TURN TO PAGE 14

SHELL ADDITION -- SLAB ON GRADE

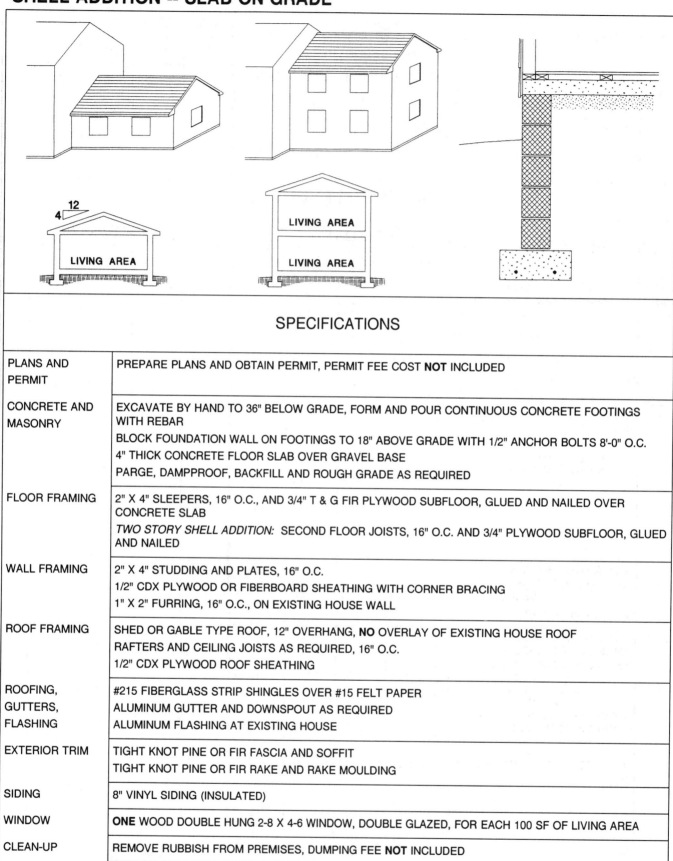

SPECIFICATIONS

PLANS AND PERMIT	PREPARE PLANS AND OBTAIN PERMIT, PERMIT FEE COST **NOT** INCLUDED
CONCRETE AND MASONRY	EXCAVATE BY HAND TO 36" BELOW GRADE, FORM AND POUR CONTINUOUS CONCRETE FOOTINGS WITH REBAR BLOCK FOUNDATION WALL ON FOOTINGS TO 18" ABOVE GRADE WITH 1/2" ANCHOR BOLTS 8'-0" O.C. 4" THICK CONCRETE FLOOR SLAB OVER GRAVEL BASE PARGE, DAMPPROOF, BACKFILL AND ROUGH GRADE AS REQUIRED
FLOOR FRAMING	2" X 4" SLEEPERS, 16" O.C., AND 3/4" T & G FIR PLYWOOD SUBFLOOR, GLUED AND NAILED OVER CONCRETE SLAB *TWO STORY SHELL ADDITION:* SECOND FLOOR JOISTS, 16" O.C. AND 3/4" PLYWOOD SUBFLOOR, GLUED AND NAILED
WALL FRAMING	2" X 4" STUDDING AND PLATES, 16" O.C. 1/2" CDX PLYWOOD OR FIBERBOARD SHEATHING WITH CORNER BRACING 1" X 2" FURRING, 16" O.C., ON EXISTING HOUSE WALL
ROOF FRAMING	SHED OR GABLE TYPE ROOF, 12" OVERHANG, **NO** OVERLAY OF EXISTING HOUSE ROOF RAFTERS AND CEILING JOISTS AS REQUIRED, 16" O.C. 1/2" CDX PLYWOOD ROOF SHEATHING
ROOFING, GUTTERS, FLASHING	#215 FIBERGLASS STRIP SHINGLES OVER #15 FELT PAPER ALUMINUM GUTTER AND DOWNSPOUT AS REQUIRED ALUMINUM FLASHING AT EXISTING HOUSE
EXTERIOR TRIM	TIGHT KNOT PINE OR FIR FASCIA AND SOFFIT TIGHT KNOT PINE OR FIR RAKE AND RAKE MOULDING
SIDING	8" VINYL SIDING (INSULATED)
WINDOW	**ONE** WOOD DOUBLE HUNG 2-8 X 4-6 WINDOW, DOUBLE GLAZED, FOR EACH 100 SF OF LIVING AREA
CLEAN-UP	REMOVE RUBBISH FROM PREMISES, DUMPING FEE **NOT** INCLUDED
	NO PAINTING

SPECIFICATIONS		UNIT	JOB COST			PRICE	LOCAL AREA MODIFICATION				
			MATLS	LABOR	TOTAL		MATLS	LABOR	TOTAL	PRICE	
SHELL ADDITION	ONE STORY	EA PLUS	2088.00	2166.00	4,254.00	6,381.00					
	SF = SF LIVING AREA	SF	9.49	7.12	16.61	24.92					
	TWO STORY	EA PLUS	2360.00	2830.00	5,190.00	7,785.00					
	SF = SF LIVING AREA 2 FLOORS	SF	8.78	4.96	13.74	20.61					

EXTRAS AND ALLOWANCES

SPECIFICATIONS		UNIT	JOB COST			PRICE	LOCAL AREA MODIFICATION				
			MATLS	LABOR	TOTAL		MATLS	LABOR	TOTAL	PRICE	
DIFFERENT FOOTING DEPTH	*Depth of Bottom of Footing Below Grade* 12" **DEDUCT**	LF	3.68	10.28	13.96	20.94					
	24" **DEDUCT**	LF	1.84	5.13	6.97	10.46					
	48" **ADD**	LF	1.83	5.12	6.95	10.43					
	LF = FOOTINGS										
CONCRETE FOUNDA- TION WALL	SUBSTITUTE CONCRETE FOUNDATION WALL IN PLACE OF BLOCK										
	Depth of Bottom of Footing Below Grade 24" **ADD**	LF	.57	8.36	8.93	13.40					
	36" **ADD**	LF	.50	8.05	8.55	12.83					
	48" **ADD**	LF	.44	7.75	8.19	12.29					
	LF = WALL										
OMIT SLEEPERS	OMIT SLEEPERS AND 5/8" PLYWOOD SUBFLOOR **DEDUCT**	SF	1.25	1.29	2.54	3.81					
	SF = SF FLOOR AREA										
WINDOWS	ADD OR OMIT ONE WOOD DOUBLE HUNG WINDOW, 2-8 X 4-6, DOUBLE GLAZED **ADD** OR **DEDUCT**	EA	193.00	59.00	252.00	378.00					
	SUBSTITUTE DOUBLE GLAZED 4-0 X 4-0 ALUMINUM SLIDING WINDOW **DEDUCT**	EA	50.00	20.00	70.00	105.00					

NOTE: FOR ADDITIONAL SHELL ADDITION EXTRAS AND ALLOWANCES, TURN TO PAGE 14

SHELL ADDITION -- ON MONOLITHIC (SINGLE POUR) SLAB

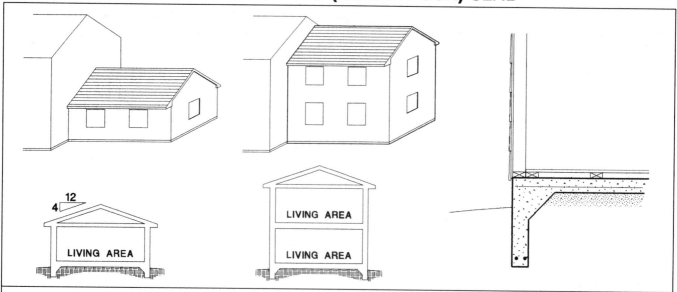

SPECIFICATIONS

PLANS AND PERMIT	PREPARE PLANS AND OBTAIN PERMIT, PERMIT FEE COST **NOT** INCLUDED
CONCRETE	EXCAVATE BY HAND FOR SINGLE POUR COMBINED FLOOR SLAB AND FOOTING FOUNDATIONS; SLAB THICKNESS 4" AND BOTTOM OF SLAB FOOTING 36" BELOW GRADE COVER AREA WITH 4" SAND OR GRAVEL AND PLASTIC VAPOR BARRIER PLACE 1/2" REINFORCING RODS IN FOOTING AREA AND WOVEN WIRE MESH IN SLAB & FOOTING AREA SET FORMS & POUR CONCRETE WITH 1/2" ANCHOR BOLTS 8'-0" O.C., SURFACE 8" ABOVE GRADE BACKFILL AND ROUGH GRADE AS REQUIRED
FLOOR FRAMING	2" X 4" SLEEPERS, 16" O.C., AND 3/4" T & G FIR PLYWOOD SUBFLOOR, GLUED AND NAILED OVER CONCRETE SLAB
WALL FRAMING	2" X 4" STUDDING AND PLATES, 16" O.C. 1/2" CDX PLYWOOD OR FIBERBOARD SHEATHING WITH CORNER BRACING 1" X 2" FURRING, 16" O.C., ON EXISTING HOUSE WALL
ROOF FRAMING	SHED OR GABLE TYPE ROOF, 12" OVERHANG, **NO** OVERLAY OF EXISTING HOUSE ROOF RAFTERS AND CEILING JOISTS AS REQUIRED, 16" O.C. 1/2" CDX PLYWOOD ROOF SHEATHING
ROOFING, GUTTERS, FLASHING	#215 FIBERGLASS STRIP SHINGLES OVER #15 FELT PAPER ALUMINUM GUTTER AND DOWNSPOUT AS REQUIRED ALUMINUM FLASHING AT EXISTING HOUSE
EXTERIOR TRIM	TIGHT KNOT PINE OR FIR FASCIA AND SOFFIT TIGHT KNOT PINE OR FIR RAKE AND RAKE MOULDING
SIDING	8" VINYL SIDING (INSULATED)
WINDOW	**ONE** WOOD DOUBLE HUNG 2-8 X 4-6 WINDOW, DOUBLE GLAZED, FOR EACH 100 SF OF LIVING AREA
CLEAN-UP	REMOVE RUBBISH FROM PREMISES, DUMPING FEE **NOT** INCLUDED
	NO PAINTING

SHELL ADDITION -- ON MONOLITHIC (SINGLE POUR) SLAB

SPECIFICATIONS		UNIT	JOB COST			PRICE	LOCAL AREA MODIFICATION				
			MATLS	LABOR	TOTAL		MATLS	LABOR	TOTAL	PRICE	
SHELL ADDITION	ONE STORY	EA PLUS	2088.00	2166.00	4,254.00	6,381.00					
	SF = SF LIVING AREA	SF	10.21	6.50	16.71	25.07					

EXTRAS AND ALLOWANCES

SPECIFICATIONS		UNIT	JOB COST			PRICE	LOCAL AREA MODIFICATION				
			MATLS	LABOR	TOTAL		MATLS	LABOR	TOTAL	PRICE	
DIFFERENT FOOTING DEPTH	*Depth of Bottom of Footing Below Grade* 12" **DEDUCT**	SF	1.51	1.05	2.56	3.84					
	24" **DEDUCT**	SF	.75	.52	1.27	1.91					
	48" **ADD**	SF	.77	.55	1.32	1.98					
	SF = SHELL ADDITION FIRST FLOOR SF AREA										
OMIT SLEEPERS	OMIT SLEEPERS AND 5/8" PLYWOOD SUBFLOOR **DEDUCT** SF = SF FLOOR AREA	SF	1.25	1.29	2.54	3.81					
WINDOWS	ADD OR OMIT ONE WOOD DOUBLE HUNG WINDOW, 2-8 X 4-6, DOUBLE GLAZED **ADD** OR **DEDUCT**	EA	193.00	59.00	252.00	378.00					
	SUBSTITUTE DOUBLE GLAZED 4-0 X 4-0 ALUMINUM SLIDING WINDOW **DEDUCT**	EA	50.00	20.00	70.00	105.00					

NOTE: FOR ADDITIONAL SHELL ADDITION EXTRAS AND ALLOWANCES, TURN TO PAGE 14

SHELL ADDITION -- ON BRICK/BLOCK OR ROUND CONCRETE PIERS

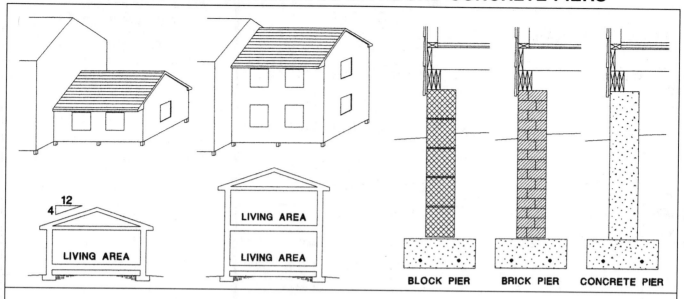

LIVING AREA

LIVING AREA

LIVING AREA

LIVING AREA

BLOCK PIER   **BRICK PIER**   **CONCRETE PIER**

SPECIFICATIONS

PLANS AND PERMIT	PREPARE PLANS AND OBTAIN PERMIT, PERMIT FEE COST **NOT** INCLUDED
CONCRETE AND MASONRY	EXCAVATE BY HAND TO 36" BELOW GRADE, FORM AND POUR 12" X 24" X 24" CONCRETE PIER FOOTINGS BLOCK, BRICK OR CONCRETE IN SONOTUBE PIERS 8'-0" O.C. TO 18" ABOVE GRADE, WITH 1/2" ANCHOR BOLT IN EACH PIER
FLOOR FRAMING	2" X 8" SILL BEAM ON PIERS FASTENED TO ANCHOR BOLTS FLOOR JOISTS AS REQUIRED, 16" O.C. 3/4" T & G FIR PLYWOOD SUBFLOOR, GLUED AND NAILED
WALL FRAMING	2" X 4" STUDDING AND PLATES, 16" O.C. 1/2" CDX PLYWOOD OR FIBERBOARD SHEATHING WITH CORNER BRACING 1" X 2" FURRING, 16" O.C., ON EXISTING HOUSE WALL
ROOF FRAMING	SHED OR GABLE TYPE ROOF, 12" OVERHANG, **NO** OVERLAY OF EXISTING HOUSE ROOF RAFTERS AND CEILING JOISTS AS REQUIRED, 16" O.C. 1/2" CDX PLYWOOD ROOF SHEATHING
ROOFING, GUTTERS, FLASHING	#215 FIBERGLASS STRIP SHINGLES OVER #15 FELT PAPER ALUMINUM GUTTER AND DOWNSPOUT AS REQUIRED ALUMINUM FLASHING AT EXISTING HOUSE
EXTERIOR TRIM	TIGHT KNOT PINE OR FIR 6" FASCIA AND 12" SOFFIT TIGHT KNOT PINE OR FIR RAKE AND RAKE MOULDING
SIDING	8" VINYL SIDING (INSULATED)
WINDOW	**ONE** WOOD DOUBLE HUNG 2-8 X 4-6 WINDOW, DOUBLE GLAZED, FOR EACH 100 SF OF LIVING AREA
CLEAN-UP	REMOVE RUBBISH FROM PREMISES, DUMPING FEE **NOT** INCLUDED
	NO PAINTING

SHELL ADDITION -- ON BRICK/BLOCK OR ROUND CONCRETE PIERS

SPECIFICATIONS		UNIT	JOB COST			PRICE	LOCAL AREA MODIFICATION			
			MATLS	LABOR	TOTAL		MATLS	LABOR	TOTAL	PRICE
SHELL ADDITION ON BRICK PIERS	ONE STORY	EA PLUS	1900.00	2000.00	3,900.00	5,850.00				
		SF	8.69	4.88	13.57	20.36				
	TWO STORY	EA PLUS	1580.00	1670.00	3,250.00	4,875.00				
		SF	9.65	5.30	14.95	22.43				
SHELL ADDITION ON BLOCK PIERS	ONE STORY	EA PLUS	1900.00	2000.00	3,900.00	5,850.00				
		SF	8.58	4.72	13.30	19.95				
	TWO STORY	EA PLUS	1580.00	1670.00	3,250.00	4,875.00				
		SF	9.59	5.22	14.81	22.22				
SHELL ADDITION ON ROUND CONCRETE PIERS	ONE STORY	EA PLUS	1900.00	2000.00	3,900.00	5,850.00				
	SF = SF LIVING AREA	SF	8.68	4.58	13.26	19.89				
	TWO STORY	EA PLUS	1580.00	1670.00	3,250.00	4,875.00				
	SF = TOTAL SQUARE FOOT LIVING AREA, ONE OR TWO FLOORS	SF	9.64	5.15	14.79	22.19				

EXTRAS AND ALLOWANCES

SPECIFICATIONS		UNIT	JOB COST			PRICE	LOCAL AREA MODIFICATION			
			MATLS	LABOR	TOTAL		MATLS	LABOR	TOTAL	PRICE
DIFFERENT FOOTING DEPTH	*Depth of Bottom of Footing Below Grade*									
	12" **DEDUCT**	SF	.10	.70	.80	1.20				
	24" **DEDUCT**	SF	.20	.40	.60	.90				
	48" **ADD**	SF	.20	.40	.60	.90				
	SF = SHELL ADDITION FIRST FLOOR SF AREA									
FOOTING AND COLUMN	EXCAVATE, FORM, POUR 24" X 24" X 12" CONCRETE FOOTING & INSTALL 3" STEEL COLUMN UNDER BEAM **ADD**	EA	63.00	41.00	104.00	156.00				
WOOD BEAM	INSTALL WOOD BEAM TO SUPPORT JOISTS									
	6" X 8" **ADD**	LF	5.67	2.08	7.75	11.63				
	6" X 10" **ADD**	LF	9.33	2.63	11.96	17.94				
WINDOWS	ADD OR OMIT ONE WOOD DOUBLE HUNG WINDOW, 2-8 X 4-6, DOUBLE GLAZED **ADD** OR **DEDUCT**	EA	193.00	59.00	252.00	378.00				
	SUBSTITUTE DOUBLE GLAZED 4-0 X 4-0 ALUMINUM SLIDING WINDOW **DEDUCT**	EA	50.00	20.00	70.00	105.00				

NOTE: FOR ADDITIONAL SHELL ADDITION EXTRAS AND ALLOWANCES, TURN TO PAGE 14

SHELL ADDITION -- OVER FULL BASEMENT

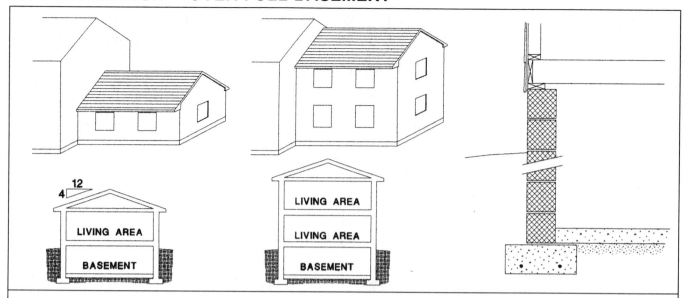

SPECIFICATIONS

PLANS AND PERMIT	PREPARE PLANS AND OBTAIN PERMIT, PERMIT FEE COST **NOT** INCLUDED
CONCRETE AND MASONRY	EXCAVATE FULL BASEMENT WITH FRONT END LOADER OR BACKHOE UNDER LIVING AREA, AND INSTALL CONTINUOUS CONCRETE FOOTINGS WITH REBAR BLOCK FOUNDATION WALL ON FOOTINGS TO 18" ABOVE GRADE WITH 1/2" ANCHOR BOLTS 8'-0" O.C. 4" CONCRETE BASEMENT FLOOR OVER GRAVEL BASE PARGE, DAMPPROOF, BACKFILL AND ROUGH GRADE AS REQUIRED
FLOOR FRAMING	FLOOR JOISTS AS REQUIRED, 16" O.C. 3/4" T & G FIR PLYWOOD SUBFLOOR, GLUED AND NAILED
WALL FRAMING	2" X 4" STUDDING AND PLATES, 16" O.C. 1/2" CDX PLYWOOD OR FIBERBOARD SHEATHING WITH CORNER BRACING 1" X 2" FURRING, 16" O.C., ON EXISTING HOUSE WALL
ROOF FRAMING	SHED OR GABLE TYPE ROOF, 12" OVERHANG, **NO** OVERLAY OF EXISTING HOUSE ROOF RAFTERS AND CEILING JOISTS AS REQUIRED, 16" O.C. 1/2" CDX PLYWOOD ROOF SHEATHING
ROOFING, GUTTERS, FLASHING	#215 FIBERGLASS STRIP SHINGLES OVER #15 FELT PAPER ALUMINUM GUTTER AND DOWNSPOUT AS REQUIRED ALUMINUM FLASHING AT EXISTING HOUSE
EXTERIOR TRIM	TIGHT KNOT PINE OR FIR FASCIA AND SOFFIT TIGHT KNOT PINE OR FIR RAKE AND RAKE MOULDING
SIDING	8" VINYL SIDING (INSULATED)
WINDOW	**ONE** WOOD DOUBLE HUNG 2-8 X 4-6 WINDOW, DOUBLE GLAZED, FOR EACH 100 SF OF LIVING AREA **ONE** WOOD FRAMED SINGLE GLAZED BASEMENT WINDOW FOR EACH 150 SF OF BASEMENT FLOOR AREA
CLEAN-UP	REMOVE RUBBISH FROM PREMISES, DUMPING FEE **NOT** INCLUDED
	NO PAINTING

SPECIFICATIONS		UNIT	JOB COST			PRICE	LOCAL AREA MODIFICATION			
			MATLS	LABOR	TOTAL		MATLS	LABOR	TOTAL	PRICE
SHELL ADDITION	ONE STORY	EA PLUS	2088.00	2166.00	4,254.00	6,381.00				
	SF = SF LIVING AREA	SF	16.19	13.01	29.20	43.80				
	TWO STORY	EA PLUS	2360.00	2830.00	5,190.00	7,785.00				
	SF = SF LIVING AREA 2 FLOORS	SF	12.20	7.84	20.04	30.06				

EXTRAS AND ALLOWANCES

SPECIFICATIONS		UNIT	JOB COST			PRICE	LOCAL AREA MODIFICATION			
			MATLS	LABOR	TOTAL		MATLS	LABOR	TOTAL	PRICE
CONCRETE	SUBSTITUTE CONCRETE FOUNDATION WALL IN PLACE OF BLOCK **ADD** LF = WALL	LF	--	21.50	21.50	32.25				
DOORWAY TO EXISTING BASEMENT	BREAK THROUGH EXISTING HOUSE FOUNDATION WALL TO CONNECT WITH BASEMENT OF SHELL ADDITION AND TRIM OPENING WITH JAMBS AND 2-1/4" CASINGS									
	BLOCK WALL **ADD**	EA	60.00	60.00	120.00	180.00				
	CONCRETE WALL **ADD**	EA	60.00	126.00	186.00	279.00				
	EA = TOTAL COST EACH OPENING									
FOOTING AND COLUMN	EXCAVATE, FORM, POUR 24" X 24" X 12" CONCRETE FOOTING AND INSTALL 3" STEEL COLUMN UNDER BEAM **ADD**	EA	63.00	41.00	104.00	156.00				
WOOD BEAM	INSTALL WOOD BEAM TO SUPPORT JOISTS									
	6" X 8" **ADD**	LF	5.67	2.08	7.75	11.63				
	6" X 10" **ADD**	LF	9.33	2.63	11.96	17.94				
WINDOWS	ADD OR OMIT ONE WOOD DOUBLE HUNG WINDOW, 2-8 X 4-6, DOUBLE GLAZED **ADD** OR **DEDUCT**	EA	193.00	59.00	252.00	378.00				
	SUBSTITUTE DOUBLE GLAZED 4-0 X 4-0 ALUMINUM SLIDING WINDOW **DEDUCT**	EA	50.00	20.00	70.00	105.00				
	ADD OR OMIT ONE WOOD FRAMED BASEMENT WINDOW **ADD** OR **DEDUCT**	EA	70.00	15.00	85.00	127.50				

NOTE: FOR ADDITIONAL SHELL ADDITION EXTRAS AND ALLOWANCES, TURN TO PAGE 14

2nd STORY ADDITION (OVER EXISTING 1st FLOOR)

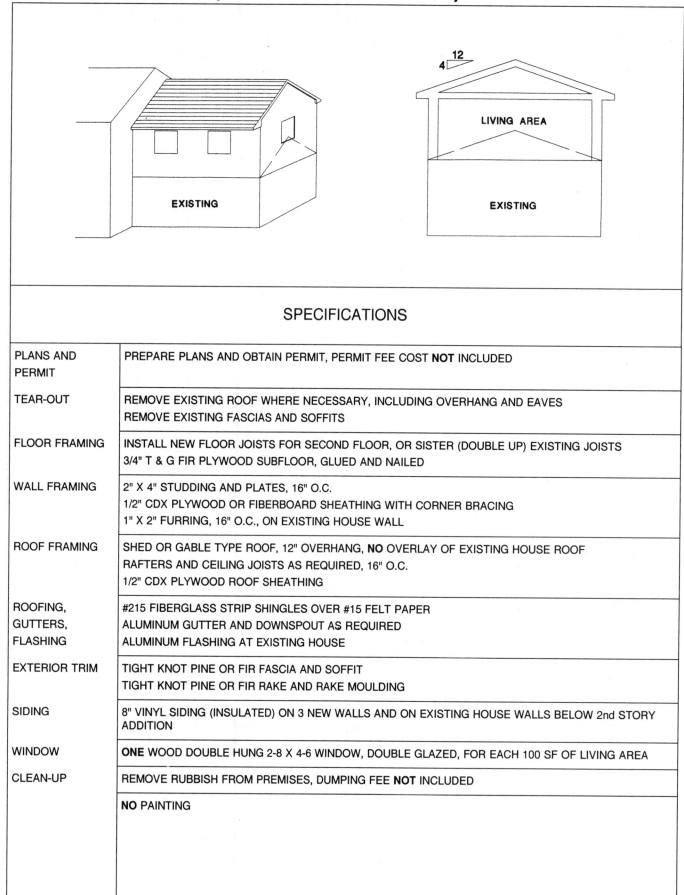

SPECIFICATIONS

PLANS AND PERMIT	PREPARE PLANS AND OBTAIN PERMIT, PERMIT FEE COST **NOT** INCLUDED
TEAR-OUT	REMOVE EXISTING ROOF WHERE NECESSARY, INCLUDING OVERHANG AND EAVES REMOVE EXISTING FASCIAS AND SOFFITS
FLOOR FRAMING	INSTALL NEW FLOOR JOISTS FOR SECOND FLOOR, OR SISTER (DOUBLE UP) EXISTING JOISTS 3/4" T & G FIR PLYWOOD SUBFLOOR, GLUED AND NAILED
WALL FRAMING	2" X 4" STUDDING AND PLATES, 16" O.C. 1/2" CDX PLYWOOD OR FIBERBOARD SHEATHING WITH CORNER BRACING 1" X 2" FURRING, 16" O.C., ON EXISTING HOUSE WALL
ROOF FRAMING	SHED OR GABLE TYPE ROOF, 12" OVERHANG, **NO** OVERLAY OF EXISTING HOUSE ROOF RAFTERS AND CEILING JOISTS AS REQUIRED, 16" O.C. 1/2" CDX PLYWOOD ROOF SHEATHING
ROOFING, GUTTERS, FLASHING	#215 FIBERGLASS STRIP SHINGLES OVER #15 FELT PAPER ALUMINUM GUTTER AND DOWNSPOUT AS REQUIRED ALUMINUM FLASHING AT EXISTING HOUSE
EXTERIOR TRIM	TIGHT KNOT PINE OR FIR FASCIA AND SOFFIT TIGHT KNOT PINE OR FIR RAKE AND RAKE MOULDING
SIDING	8" VINYL SIDING (INSULATED) ON 3 NEW WALLS AND ON EXISTING HOUSE WALLS BELOW 2nd STORY ADDITION
WINDOW	**ONE** WOOD DOUBLE HUNG 2-8 X 4-6 WINDOW, DOUBLE GLAZED, FOR EACH 100 SF OF LIVING AREA
CLEAN-UP	REMOVE RUBBISH FROM PREMISES, DUMPING FEE **NOT** INCLUDED
	NO PAINTING

2nd STORY ADDITION (OVER EXISTING 1st FLOOR)

SPECIFICATIONS		UNIT	JOB COST			PRICE	LOCAL AREA MODIFICATION			
			MATLS	LABOR	TOTAL		MATLS	LABOR	TOTAL	PRICE
SHELL ADDITION	SECOND STORY ADDITION OVER EXISTING FIRST FLOOR	EA PLUS	1900.00	2000.00	3,900.00	5,850.00				
	SF = SF LIVING AREA, 2nd FLOOR ONLY	SF	12.05	7.85	19.90	29.85				

EXTRAS AND ALLOWANCES

SPECIFICATIONS		UNIT	JOB COST			PRICE	LOCAL AREA MODIFICATION			
			MATLS	LABOR	TOTAL		MATLS	LABOR	TOTAL	PRICE
OMIT JOISTS	OMIT INSTALLATION OF NEW JOISTS OR SISTERING EXISTING JOISTS									
	2" X 8" **DEDUCT**	SF	1.32	2.10	3.42	5.13				
	2" X 10" **DEDUCT**	SF	1.71	2.10	3.81	5.72				
	2" X 12" **DEDUCT**	SF	2.04	2.10	4.14	6.21				
SUBFLOOR	REMOVE EXISTING SUBFLOOR									
	1" X 6" OR 1" X 8" **ADD**	SF	--	.52	.52	.78				
	NAILED PLYWOOD **ADD**	SF	--	.45	.45	.68				
	SF = FLOOR									
	INSTALL 3/4" T & G FIR PLYWOOD SUBFLOOR **ADD**	SF	.67	.62	1.29	1.94				
	SF = FLOOR									
SIDING	OMIT SIDING ON EXISTING HOUSE WALL BELOW 2nd STORY ADDITION **DEDUCT**	SF	1.78	.97	2.75	4.13				
	SF = SF EXISTING HOUSE WALL									
WINDOWS	ADD OR OMIT ONE WOOD DOUBLE HUNG WINDOW, 2-8 X 4-6, DOUBLE GLAZED **ADD** OR **DEDUCT**	EA	193.00	59.00	252.00	378.00				
	SUBSTITUTE DOUBLE GLAZED 4-0 X 4-0 ALUMINUM SLIDING WINDOW **DEDUCT**	EA	50.00	20.00	70.00	105.00				

NOTE: FOR ADDITIONAL SHELL ADDITION EXTRAS AND ALLOWANCES, TURN TO PAGE 14

SHELL ADDITION -- EXTRAS AND ALLOWANCES

SPECIFICATIONS		UNIT	JOB COST			PRICE	LOCAL AREA MODIFICATION			
			MATLS	LABOR	TOTAL		MATLS	LABOR	TOTAL	PRICE
SUBSTITUTE GABLE/SHED ROOF SLOPE AND/OR COVERING	SF = SF FLOOR AREA OF TOP FLOOR OF ADDITION ON WHICH THE RAFTERS ARE SUPPORTED. INCLUDES ADDITIONAL GABLE END FRAMING AND SIDING.									
	Same Framing, Same Slope									
	PENNSYLVANIA SLATE **ADD**	SF	4.31	1.31	5.62	8.43				
	16" CEDAR SHINGLES **ADD**	SF	2.09	.91	3.00	4.50				
	CONCRETE TILES **ADD**	SF	.89	.65	1.54	2.31				
	6/12 Slope									
	ASPHALT/FIBERGLASS **ADD**	SF	.48	.33	.81	1.22				
	PENNSYLVANIA SLATE **ADD**	SF	.89	.45	1.34	2.01				
	16" CEDAR SHINGLES **ADD**	SF	.68	.40	1.08	1.62				
	CONCRETE TILES **ADD**	SF	.56	.39	.95	1.43				
	8/12 Slope									
	ASPHALT/FIBERGLASS **ADD**	SF	.61	.51	1.12	1.68				
	PENNSYLVANIA SLATE **ADD**	SF	1.22	.70	1.92	2.88				
	16" CEDAR SHINGLES **ADD**	SF	.91	.62	1.53	2.30				
	CONCRETE TILES **ADD**	SF	.74	.61	1.35	2.03				
	10/12 Slope									
	ASPHALT/FIBERGLASS **ADD**	SF	.99	.82	1.81	2.72				
	PENNSYLVANIA SLATE **ADD**	SF	2.00	1.13	3.13	4.70				
	16" CEDAR SHINGLES **ADD**	SF	1.48	1.01	2.49	3.74				
	CONCRETE TILES **ADD**	SF	1.20	.97	2.17	3.26				
	12/12 Slope									
	ASPHALT/FIBERGLASS **ADD**	SF	1.35	1.12	2.47	3.71				
	PENNSYLVANIA SLATE **ADD**	SF	2.78	1.55	4.33	6.50				
	16" CEDAR SHINGLES **ADD**	SF	2.04	1.38	3.42	5.13				
	CONCRETE TILES **ADD**	SF	1.64	1.33	2.97	4.46				
SUBSTITUTE HIP ROOF FRAMING & ROOF COVERING	SF = SF FLOOR AREA OF TOP FLOOR OF ADDITION ON WHICH THE RAFTERS ARE SUPPORTED. INCLUDES OMITTING GABLE END FRAMING AND SIDING									
	SAME SLOPE **ADD**	SF	.17	--	.17	.26				
	6/12 SLOPE **ADD**	SF	.21	.03	.24	.36				
	8/12 SLOPE **ADD**	SF	.24	.04	.28	.42				
	10/12 SLOPE **ADD**	SF	.28	.06	.34	.51				
	12/12 SLOPE **ADD**	SF	.32	.08	.40	.60				
SUBSTITUTE FLAT ROOF FRAMING & ROOF COVERING	SF = SF FLOOR AREA OF TOP FLOOR OF ADDITION ON WHICH THE RAFTERS (JOISTS) ARE SUPPORTED. INCLUDES OMITTING GABLE END FRAMING AND SIDING									
	1/12 Slope									
	4-PLY BUILT UP **ADD**	SF	.10	1.64	1.74	2.61				
	SINGLE PLY MEMBRANE **ADD**	SF	.10	1.32	1.42	2.13				
	16 OZ. COPPER **ADD**	SF	4.41	5.18	9.59	14.39				
	GALV. SHEET METAL **ADD**	SF	.62	1.60	2.22	3.33				

SPECIFICATIONS		UNIT	JOB COST			PRICE	LOCAL AREA MODIFICATION				
			MATLS	LABOR	TOTAL		MATLS	LABOR	TOTAL	PRICE	
OVERLAY EXISTING ROOF WITH FRAMING AND ROOFING	ASPHALT/FIBERGLASS **ADD**	SF	3.86	4.36	8.22	12.33					
	PENNSYLVANIA SLATE **ADD**	SF	9.41	5.84	15.25	22.88					
	16" CEDAR SHINGLES **ADD**	SF	6.25	4.86	11.11	16.67					
	CONCRETE TILES **ADD**	SF	4.54	4.58	9.12	13.68					
	SF = SF AREA OF ROOF OVERLAY										
TEAR-OUT	REMOVE WOOD OR ALUMINUM SIDING FROM EXISTING HOUSE WALL **ADD**	SF	--	.34	.34	.51					
	SF = SF OF HOUSE WALL										
CONCRETE SLAB AND STEPS	EXCAVATE, FORM, POUR AND FINISH CONCRETE SLAB, 1 OR 2 STEPS TO ADDITION, SLAB 6'-0" X 4'-0"										
	Depth of Bottom of Footing Below Grade										
	12" **ADD**	EA	160.00	236.00	396.00	594.00					
	24" **ADD**	EA	200.00	258.00	458.00	687.00					
	36" **ADD**	EA	253.00	285.00	538.00	807.00					
	48" **ADD**	EA	305.00	315.00	620.00	930.00					
	EA = TOTAL AMOUNT										
2" X 6" STUDS	SUBSTITUTE 2" X 6" STUDS FOR SPECIFIED 2" X 4" **ADD**	LF	.33	.05	.38	.57					
	LF = LENGTH OF WALL										
PARTITION	FRAME OUT PARTITION IN ROOM WITH STUDS AND PLATES										
	2" X 4" **ADD**	LF	6.08	4.96	11.04	16.56					
	2" X 6" **ADD**	LF	8.72	5.36	14.08	21.12					
OMIT FURRING	OMIT FURRING OF EXISTING HOUSE WALL **DEDUCT**	SF	.17	.44	.61	.92					
	SF = EXISTING HOUSE WALL										
SUBSTITUTE OVERHANG AND SOFFIT	ADDITIONAL, OR LESS, ROOF OVERHANG AND SOFFIT, INCLUDING ADDITIONAL OR LESS ROOF FRAMING AND ROOF COVERING										
	6" **DEDUCT**	LF	2.00	1.50	3.50	5.25					
	18" **ADD**	LF	6.20	4.60	10.80	16.20					
	24" **ADD**	LF	8.25	6.20	14.45	21.68					
	LF = LF OF SOFFIT										

SHELL ADDITION -- EXTRAS AND ALLOWANCES

SPECIFICATIONS		UNIT	JOB COST			PRICE	LOCAL AREA MODIFICATION				
			MATLS	LABOR	TOTAL		MATLS	LABOR	TOTAL	PRICE	
SUBSTITUTE SIDING	6" BEVELED SIDING **ADD**	SF	.44	.13	.57	.86					
	BRICK VENEER **ADD**	SF	.95	3.60	4.55	6.83					
	STUCCO **ADD**	SF	--	3.23	3.23	4.85					
	18" CEDAR SHINGLES **ADD**	SF	--	.36	.36	.54					
	3/8" FIR TEXTURE 1-11 **DEDUCT**	SF	.76	.29	1.05	1.58					
	HARDBOARD LAP SIDING **DEDUCT**	SF	.55	.25	.80	1.20					
	SF = SF OF SIDING										
FRAME DOORWAY	REMOVE EXISTING WINDOW AND WALL UNDER AND FRAME OUT FOR DOORWAY FROM HOUSE TO ADDITION **ADD**	EA	16.00	128.00	144.00	216.00					
	BREAK THROUGH EXISTING HOUSE WALL AND FRAME OUT FOR DOORWAY FROM HOUSE TO ADDITION										
	FRAME WALL **ADD**	EA	40.00	178.00	218.00	327.00					
	MASONRY WALL **ADD**	EA	40.00	238.00	278.00	417.00					

SPECIFICATIONS

WINDOWS	TRIM WINDOWS (ONE FOR EACH 100 SQUARE FEET OF LIVING AREA) WITH 2-1/4" CASING, STOOLS, STOPS AND SASH FASTENER
HEATING	EXTEND HEAT AND RETURN DUCTS FROM EXISTING SERVICE
ELECTRIC	DUPLEX WALL OUTLETS TO CODE ON EXISTING SERVICE
INSULA-TION	3-1/2" FIBERGLASS BLANKET INSULATION ON ALL SIDEWALLS 6" FIBERGLASS BLANKET INSULATION IN CEILING
INTERIOR WALL COVERING	1/2" GYPSUM BOARD (SHEETROCK) ON NEW WALLS AND ON FURRING STRIPS OVER EXISTING HOUSE WALL, TAPED AND FINISHED (**NO** PAINTING)
CEILING COVERING	1/2" GYPSUM BOARD (SHEETROCK) TAPED AND FINISHED (**NO** PAINTING)
INTERIOR TRIM	3-1/2" BASE AND SHOE MOULDING ON ALL FOUR WALLS
FLOOR COVERING	2-1/4" SELECT OAK FLOORING, SANDED AND FINISHED
CLEAN-UP	REMOVE TRASH FROM PREMISES, DUMPING FEE **NOT** INCLUDED
	NO PAINTING

SPECIFICATIONS		UNIT	JOB COST			PRICE	LOCAL AREA MODIFICATION				
			MATLS	LABOR	TOTAL		MATLS	LABOR	TOTAL	PRICE	
INTERIOR TRIM AND FINISHING	AS SPECIFIED	EA PLUS	400.00	470.00	870.00	1,305.00					
	SF = TOTAL SQUARE FOOT LIVING AREA TO BE FINISHED	SF	6.27	7.06	13.33	20.00					

ADDITION -- TRIM AND FINISHING EXTRAS AND ALLOWANCES

SPECIFICATIONS		UNIT	JOB COST			PRICE	LOCAL AREA MODIFICATION			
			MATLS	LABOR	TOTAL		MATLS	LABOR	TOTAL	PRICE
INSTALL DOOR FROM ADDITION TO EXISTING HOUSE	IN EXISTING FRAMED OPENING, INSTALL 2-6 X 6-8 X 1-3/8" FLUSH HOLLOW CORE BIRCH DOOR, INCLUDING JAMBS, STOPS, TRIM, HARDWARE & PATCHING **ADD**	EA	95.00	93.00	188.00	282.00				
INSTALL DOOR FROM ADDITION TO EXTERIOR	IN EXISTING FRAMED OPENING, INSTALL DOOR TO OUTSIDE — 2-8 X 6-8 X 1-3/4" 3 PANEL, 4 LIGHT FIR DOOR **ADD**	EA	317.00	91.00	408.00	612.00				
	GLIDING PATIO DOOR, ALUMINUM FRAMED, WITH INSULATED GLASS AND INSECT SCREEN, 6-0 X 6-8 **ADD**	EA	402.00	88.00	490.00	735.00				
DOOR	HANG DOOR IN FRAMED OUT PARTITION, 2-6 X 6-8, BIRCH HOLLOW CORE, INCLUDING JAMBS, STOPS, CASINGS AND HARDWARE **ADD**	EA	95.00	63.00	158.00	237.00				
	BI-FOLD DOORS, BIRCH FLUSH, INCLUDING ALL TRIM AND HARDWARE 2 DOORS 3-0 X 6-8 **ADD** 6-0 X 6-8 **ADD** 4 DOORS 4-0 X 6-8 **ADD** 6-0 X 6-8 **ADD**	 SET SET SET SET	 85.00 111.00 108.00 122.00	 59.00 65.00 62.00 68.00	 144.00 176.00 170.00 190.00	 216.00 264.00 255.00 285.00				
WINDOW	TRIM ADDITIONAL WINDOW WITH 2-1/4" CASINGS AND APRON, STOOL AND STOPS **ADD**	EA	21.00	29.00	50.00	75.00				
HEATING	OMIT EXTENDING HEAT TO ADDITION **DEDUCT** SF = SF LIVING AREA OF ADDITION	SF	.40	1.50	1.90	2.85				
INSULATION	INSULATE FLOOR JOISTS WITH FIBERGLASS BLANKET INSULATION 6" **ADD** 9" **ADD** SF = FLOOR JOISTS	 SF SF	 .41 .58	 .20 .21	 .61 .79	 .92 1.19				

SPECIFICATIONS		UNIT	JOB COST			PRICE	LOCAL AREA MODIFICATION			
			MATLS	LABOR	TOTAL		MATLS	LABOR	TOTAL	PRICE
WALL COVERING	OMIT DRYWALL ON EXISTING HOUSE WALL **DEDUCT** SF = EXISTING HOUSE WALL	SF	.19	.68	.87	1.31				
INTERIOR PARTITION	TRIM AND FINISH STUD PARTITION TWO SIDES OF WALL • DUPLEX OUTLETS TO CODE • 1/2" DRYWALL, TAPED AND FINISHED • 3-1/2" BASE AND SHOE MOULD **ADD** LF = LENGTH OF 8-FOOT HIGH PARTITION	LF	11.00	20.00	31.00	46.50				
CLOSET TRIM	TRIM FRAMED CLOSET WITH BASE AND SHOE, HOOK-STRIP, SHELF, CLOTHES POLE AND CLOTHES POLE SOCKET	LF	9.18	6.60	15.78	23.67				
CEILING MOULDING	CEILING MOULDING AT WALL INTERSECTIONS 3/4" **ADD**	LF	.42	.75	1.17	1.76				
	1-5/8" **ADD**	LF	.73	.79	1.52	2.28				
	3-5/8" **ADD**	LF	1.55	.97	2.52	3.78				
SUBSTITUTE FLOOR COVERING	SUBSTITUTE CARPET AND PAD OVER SUBFLOOR CARPET @ $15/YD **DEDUCT**	SF	2.51	2.23	4.74	7.11				
	@ $25/YD **DEDUCT**	SF	1.04	2.19	3.23	4.85				
	SUBSTITUTE VINYL TILE OVER SUBFLOOR **DEDUCT**	SF	2.71	.98	3.69	5.54				

3 AND 4 FIXTURE BATHROOMS

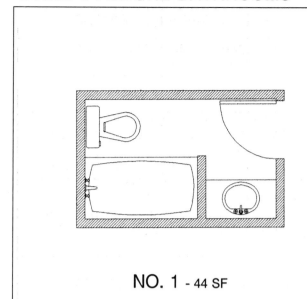

NO. 1 - 44 SF

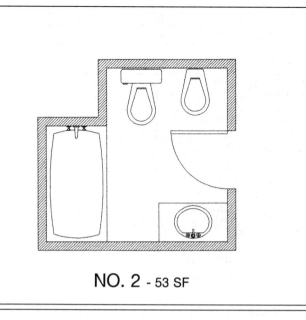

NO. 2 - 53 SF

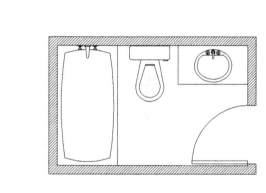

NO. 3 - 43 SF

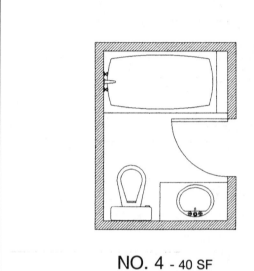

NO. 4 - 40 SF

NO. 5 - 52 SF

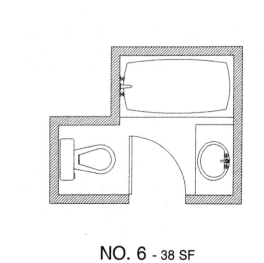

NO. 6 - 38 SF

REMODEL EXISTING BATHROOM | INSTALL NEW BATHROOM

	REMODEL EXISTING BATHROOM	INSTALL NEW BATHROOM
	PREPARE PLANS AND OBTAIN PERMITS AS REQUIRED, PERMIT FEE COST **NOT** INCLUDED	PREPARE PLANS AND OBTAIN PERMITS AS REQUIRED, PERMIT FEE COST **NOT** INCLUDED
	TEAR OUT EXISTING BATHROOM FIXTURES AND NON-CERAMIC FLOOR, WALL AND CEILING COVERINGS	
	RE-ROUGH PLUMBING FOR BATHROOM, NEW COPPER SUPPLY AND NEW PLASTIC WASTE LINES TO MAIN STACK WITH CUTOFFS AT EACH FIXTURE	ROUGH IN PLUMBING FOR BATHROOM, COPPER SUPPLY AND PLASTIC WASTE LINES TO MAIN STACK WITHIN 5 FEET, WITH CUTOFFS AT EACH FIXTURE
	INSTALL BUILDER GRADE WHITE FIXTURES, INCLUDING ALL FITTINGS, CAST IRON TUB, BIDET (BATHROOMS 2 AND 5), AND TWO-PIECE TOILET	INSTALL BUILDER GRADE WHITE FIXTURES, INCLUDING ALL FITTINGS, CAST IRON TUB, BIDET (BATHROOMS 2 AND 5), AND TWO-PIECE TOILET
	20" X 18" LAVATORY SET IN TWO-DOOR BUILDER GRADE PREFINISHED VANITY BASE WITH CULTURED MARBLE TOP AND SPLASH	20" X 18" LAVATORY SET IN TWO-DOOR 36" BUILDER GRADE PREFINISHED VANITY BASE WITH CULTURED MARBLE TOP AND SPLASH
	3/4" T & G PLYWOOD SUBFLOOR	
		2-6 X 6-8 BIRCH FLUSH HOLLOW CORE DOOR WITH BATHROOM STYLE LOCKSET
	SWITCH AND LIGHT OVER LAVATORY	SWITCH AND LIGHT OVER LAVATORY
	GROUND FAULT DUPLEX OUTLET NEXT TO LAVATORY	GROUND FAULT DUPLEX OUTLET NEXT TO LAVATORY
	EXHAUST FAN AND SWITCH VENTED TO OUTSIDE	EXHAUST FAN VENTED TO OUTSIDE
	NEW 1/2" MOISTURE RESISTANT DRYWALL, TAPED AND FINISHED, OVER ALL WALLS AND CEILINGS EXCEPT CERAMIC TILE WALL AREAS	1/2" MOISTURE RESISTANT DRYWALL, TAPED AND FINISHED, OVER ALL WALLS AND CEILINGS EXCEPT CERAMIC TILE WALL AREAS
	NEW CERAMIC TILE ON WALLS OF TUB AREA 6 FEET ABOVE FLOOR, INCLUDING CERAMIC ACCESSORIES	CERAMIC TILE ON WALLS OF TUB AREA 6 FEET ABOVE FLOOR, INCLUDING CERAMIC ACCESSORIES
	3/8" PLYWOOD UNDERLAYMENT AND NEW CERAMIC TILE FLOOR WITH 4" TILE BASE	3/8" PLYWOOD UNDERLAYMENT AND NEW CERAMIC TILE FLOOR WITH 4" TILE BASE
	RECESSED MEDICINE CABINET WITH OVERHEAD FLUORESCENT FIXTURE, 2 MIRROR DOORS, OVERALL SIZE 24" X 16"	RECESSED MEDICINE CABINET WITH OVERHEAD FLUORESCENT FIXTURE, 2 MIRROR DOORS, OVERALL SIZE 24" X 16"
	PRIME ROOM AS NECESSARY AND PAINT WITH 2 COATS TOP QUALITY SEMI-GLOSS ENAMEL	PRIME ENTIRE ROOM AND PAINT WITH 2 COATS TOP QUALITY SEMI-GLOSS ENAMEL
	REMOVE OLD FIXTURES AND TRASH FROM PREMISES AND FINAL CLEAN-UP, DUMPING FEE **NOT** INCLUDED	REMOVE TRASH FROM PREMISES AND FINAL CLEAN-UP, DUMPING FEE **NOT** INCLUDED

SPECIFICATIONS			UNIT	JOB COST MATLS	LABOR	TOTAL	PRICE	LOCAL AREA MODIFICATION MATLS	LABOR	TOTAL	PRICE
REMODEL BATHROOM	REMODEL EXISTING BATHROOM	NO. 1	EA	1538.00	2668.00	4,206.00	6,309.00				
		NO. 2	EA	2138.00	3118.00	5,256.00	7,884.00				
		NO. 3	EA	1626.00	2714.00	4,340.00	6,510.00				
		NO. 4	EA	1663.00	2702.00	4,365.00	6,547.50				
		NO. 5	EA	2216.00	3438.00	5,654.00	8,481.00				
		NO. 6	EA	1472.00	2778.00	4,250.00	6,375.00				
INSTALL NEW BATHROOM	NEW BATHROOM	NO. 1	EA	1890.00	3575.00	5,465.00	8,197.50				
		NO. 2	EA	2447.00	4252.00	6,699.00	10,048.50				
		NO. 3	EA	1807.00	3548.00	5,355.00	8,032.50				
		NO. 4	EA	1842.00	3531.00	5,373.00	8,059.50				
		NO. 5	EA	2450.00	4252.00	6,702.00	10,053.00				
		NO. 6	EA	1853.00	3711.00	5,564.00	8,346.00				

BATHROOMS WITH STALL SHOWER

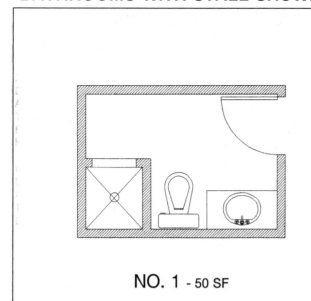

NO. 1 - 50 SF

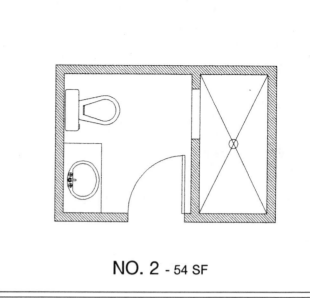

NO. 2 - 54 SF

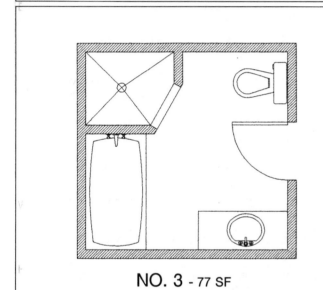

NO. 3 - 77 SF

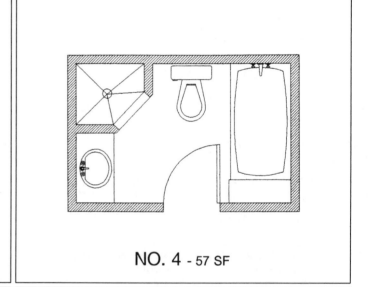

NO. 4 - 57 SF

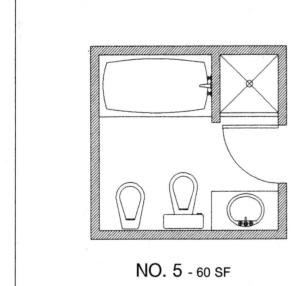

NO. 5 - 60 SF

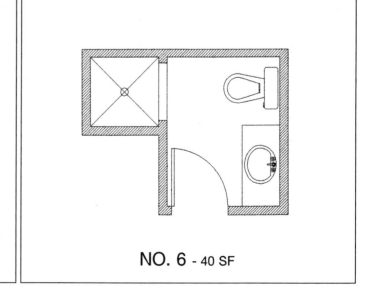

NO. 6 - 40 SF

REMODEL EXISTING BATHROOM	INSTALL NEW BATHROOM
PREPARE PLANS AND OBTAIN PERMITS AS REQUIRED, PERMIT FEE COST **NOT** INCLUDED	PREPARE PLANS AND OBTAIN PERMITS AS REQUIRED, PERMIT FEE COST **NOT** INCLUDED
TEAR OUT EXISTING BATHROOM FIXTURES AND NON-CERAMIC FLOOR, WALL AND CEILING COVERINGS	
RE-ROUGH PLUMBING FOR BATHROOM, NEW COPPER SUPPLY AND NEW PLASTIC WASTE LINES TO MAIN STACK WITH CUTOFFS AT EACH FIXTURE	ROUGH IN PLUMBING FOR BATHROOM, COPPER SUPPLY AND PLASTIC WASTE LINES TO MAIN STACK WITHIN 5 FEET, WITH CUTOFFS AT EACH FIXTURE
INSTALL BUILDER GRADE WHITE FIXTURES, INCLUDING ALL FITTINGS, CAST IRON TUB, BIDET (BATHROOM 5), TWO-PIECE TOILET AND NEW STALL SHOWER PLUMBING AND FITTINGS, INCLUDING VINYL OR RUBBER PAN	INSTALL BUILDER GRADE WHITE FIXTURES, INCLUDING ALL FITTINGS, CAST IRON TUB, BIDET (BATHROOM 5), TWO-PIECE TOILET AND NEW STALL SHOWER PLUMBING AND FITTINGS, INCLUDING VINYL OR RUBBER PAN
ONE OR TWO 20" X 18" LAVATORY BOWLS SET IN BUILDER GRADE PREFINISHED VANITY BASE WITH CULTURED MARBLE TOP AND SPLASH	ONE OR TWO 20" X 18" LAVATORY BOWLS SET IN BUILDER GRADE PREFINISHED VANITY BASE WITH CULTURED MARBLE TOP AND SPLASH
3/4" T & G PLYWOOD SUBFLOOR	
	2-6 X 6-8 BIRCH FLUSH HOLLOW CORE DOOR WITH BATHROOM STYLE LOCKSET
SWITCH AND LIGHT OVER LAVATORY	SWITCH AND LIGHT OVER LAVATORY
GROUND FAULT DUPLEX OUTLET NEXT TO LAVATORY	GROUND FAULT DUPLEX OUTLET NEXT TO LAVATORY
EXHAUST FAN AND SWITCH VENTED TO OUTSIDE	EXHAUST FAN AND SWITCH VENTED TO OUTSIDE
NEW 1/2" MOISTURE RESISTANT DRYWALL, TAPED AND FINISHED, OVER ALL WALLS AND CEILINGS EXCEPT CERAMIC TILE WALL AREAS	1/2" MOISTURE RESISTANT DRYWALL, TAPED AND FINISHED, OVER ALL WALLS AND CEILINGS EXCEPT CERAMIC TILE WALL AREAS
NEW CERAMIC TILE ON WALLS OF TUB AND SHOWER STALL 6 FEET ABOVE FLOOR, INCLUDING CERAMIC ACCESSORIES	CERAMIC TILE ON WALLS OF TUB AND SHOWER STALL 6 FEET ABOVE FLOOR, INCLUDING CERAMIC ACCESSORIES
3/8" PLYWOOD UNDERLAYMENT AND NEW CERAMIC TILE FLOOR WITH 4" TILE BASE	3/8" PLYWOOD UNDERLAYMENT AND CERAMIC TILE FLOOR WITH 4" TILE BASE
RECESSED MEDICINE CABINET WITH OVERHEAD FLUORESCENT FIXTURE, 2 MIRROR DOORS, OVERALL SIZE 24" X 16"	RECESSED MEDICINE CABINET WITH OVERHEAD FLUORESCENT FIXTURE, 2 MIRROR DOORS, OVERALL SIZE 24" X 16"
PRIME ROOM AS NECESSARY AND PAINT WITH 2 COATS TOP QUALITY SEMI-GLOSS ENAMEL	PRIME ENTIRE ROOM AND PAINT WITH 2 COATS TOP QUALITY SEMI-GLOSS ENAMEL
REMOVE OLD FIXTURES AND TRASH FROM PREMISES AND FINAL CLEAN-UP, DUMPING FEE **NOT** INCLUDED	REMOVE TRASH FROM PREMISES AND FINAL CLEAN-UP, DUMPING FEE **NOT** INCLUDED

SPECIFICATIONS			UNIT	JOB COST			PRICE	LOCAL AREA MODIFICATION			
				MATLS	LABOR	TOTAL		MATLS	LABOR	TOTAL	PRICE
REMODEL BATHROOM	REMODEL EXISTING BATHROOM	NO. 1	EA	1298.00	2485.00	3,783.00	5,674.50				
		NO. 2	EA	1666.00	2901.00	4,567.00	6,850.50				
		NO. 3	EA	1958.00	3555.00	5,513.00	8,269.50				
		NO. 4	EA	2094.00	3656.00	5,750.00	8,625.00				
		NO. 5	EA	2668.00	4054.00	6,722.00	10,083.00				
		NO. 6	EA	1580.00	2752.00	4,332.00	6,498.00				
INSTALL NEW BATHROOM	NEW BATHROOM	NO. 1	EA	1753.00	3784.00	5,537.00	8,305.50				
		NO. 2	EA	2017.00	4236.00	6,253.00	9,379.50				
		NO. 3	EA	2635.00	5625.00	8,260.00	12,390.00				
		NO. 4	EA	2338.00	5223.00	7,561.00	11,341.50				
		NO. 5	EA	3125.00	5952.00	9,077.00	13,615.50				
		NO. 6	EA	1904.00	4036.00	5,940.00	8,910.00				

BATHROOMS WITH COMPARTMENTS

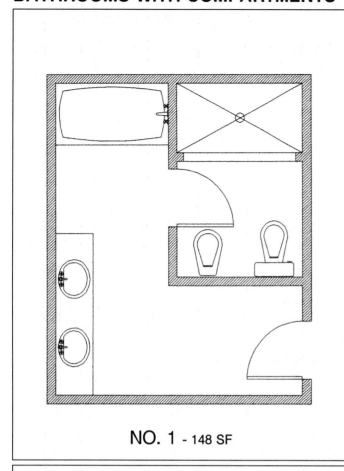

NO. 1 - 148 SF

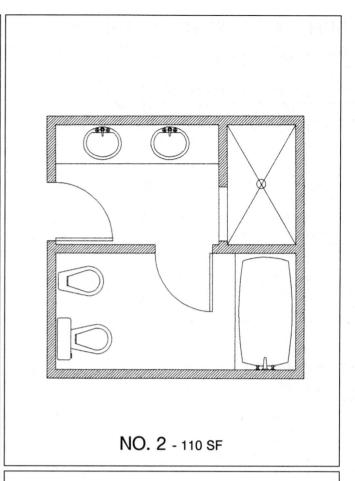

NO. 2 - 110 SF

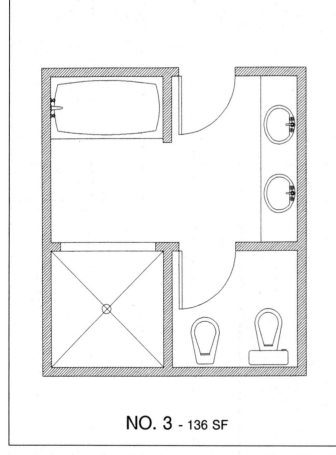

NO. 3 - 136 SF

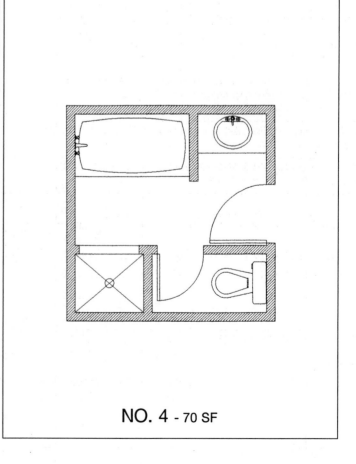

NO. 4 - 70 SF

BATHROOMS WITH COMPARTMENTS

REMODEL EXISTING BATHROOM	INSTALL NEW BATHROOM
PREPARE PLANS AND OBTAIN PERMITS AS REQUIRED, PERMIT FEE COST **NOT** INCLUDED	PREPARE PLANS AND OBTAIN PERMITS AS REQUIRED, PERMIT FEE COST **NOT** INCLUDED
TEAR OUT EXISTING BATHROOM FIXTURES AND NON-CERAMIC FLOOR, WALL AND CEILING COVERINGS	
3/4" T & G PLYWOOD SUBFLOOR	FRAME OUT PARTITIONS WITH 2" X 6", 16" O.C. 3/4" T & G PLYWOOD SUBFLOOR
RE-ROUGH PLUMBING FOR BATHROOM, NEW COPPER SUPPLY AND NEW PLASTIC WASTE LINES TO MAIN STACK WITH CUTOFFS AT EACH FIXTURE	ROUGH IN PLUMBING FOR BATHROOM, COPPER SUPPLY AND PLASTIC WASTE LINES TO MAIN STACK WITHIN 5 FEET, WITH CUTOFFS AT EACH FIXTURE
INSTALL BUILDER GRADE WHITE FIXTURES, INCLUDING ALL FITTINGS, CAST IRON TUB, BIDET (BATHROOMS 1, 2 AND 3), TWO-PIECE TOILET AND NEW STALL SHOWER PLUMBING AND FITTINGS, INCLUDING VINYL OR RUBBER PAN	INSTALL BUILDER GRADE WHITE FIXTURES, INCLUDING ALL FITTINGS, CAST IRON TUB, BIDET (BATHROOMS 1, 2 AND 3), TWO-PIECE TOILET AND NEW STALL SHOWER PLUMBING AND FITTINGS, INCLUDING VINYL OR RUBBER PAN
ONE OR TWO 20" X 18" LAVATORY BOWLS SET IN BUILDER GRADE PREFINISHED VANITY BASE WITH CULTURED MARBLE TOP AND SPLASH	ONE OR TWO 20" X 18" LAVATORY BOWLS SET IN BUILDER GRADE PREFINISHED VANITY BASE WITH CULTURED MARBLE TOP AND SPLASH
	2-6 X 6-8 BIRCH FLUSH HOLLOW CORE DOOR WITH BATHROOM STYLE LOCKSET
SWITCH AND LIGHT OVER LAVATORY	SWITCH AND LIGHT OVER LAVATORY
GROUND FAULT DUPLEX OUTLET NEXT TO EACH LAVATORY BOWL	GROUND FAULT DUPLEX OUTLET NEXT TO EACH LAVATORY BOWL
EXHAUST FAN AND SWITCH VENTED TO OUTSIDE	EXHAUST FAN AND SWITCH VENTED TO OUTSIDE
NEW 1/2" MOISTURE RESISTANT DRYWALL, TAPED AND FINISHED, OVER ALL WALLS AND CEILINGS EXCEPT CERAMIC TILE WALL AREAS	1/2" MOISTURE RESISTANT DRYWALL, TAPED AND FINISHED, OVER ALL WALLS AND CEILINGS EXCEPT CERAMIC TILE WALL AREAS
NEW CERAMIC TILE ON WALLS OF TUB AND SHOWER STALL 6 FEET ABOVE FLOOR, INCLUDING CERAMIC ACCESSORIES	CERAMIC TILE ON WALLS OF TUB AND SHOWER STALL 6 FEET ABOVE FLOOR, INCLUDING CERAMIC ACCESSORIES
3/8" PLYWOOD UNDERLAYMENT AND NEW CERAMIC TILE FLOOR WITH 4" TILE BASE	3/8" PLYWOOD UNDERLAYMENT AND CERAMIC TILE FLOOR WITH 4" TILE BASE
RECESSED MEDICINE CABINET WITH OVERHEAD FLUORESCENT FIXTURE, 2 MIRROR DOORS, OVERALL SIZE 24" X 16", OVER EACH LAVATORY BOWL	RECESSED MEDICINE CABINET WITH OVERHEAD FLUORESCENT FIXTURE, 2 MIRROR DOORS, OVERALL SIZE 24" X 16", OVER EACH LAVATORY BOWL
PRIME ROOM AS NECESSARY AND PAINT WITH 2 COATS TOP QUALITY SEMI-GLOSS ENAMEL	PRIME ENTIRE ROOM AND PAINT WITH 2 COATS TOP QUALITY SEMI-GLOSS ENAMEL
REMOVE OLD FIXTURES AND TRASH FROM PREMISES AND FINAL CLEAN-UP, DUMPING FEE **NOT** INCLUDED	REMOVE TRASH FROM PREMISES AND FINAL CLEAN-UP, DUMPING FEE **NOT** INCLUDED

SPECIFICATIONS			UNIT	JOB COST			PRICE	LOCAL AREA MODIFICATION			
				MATLS	LABOR	TOTAL		MATLS	LABOR	TOTAL	PRICE
REMODEL BATHROOM	REMODEL EXISTING BATHROOM	NO. 1	EA	4102.00	6085.00	10,187.00	15,280.50				
		NO. 2	EA	3817.00	5319.00	9,136.00	13,704.00				
		NO. 3	EA	4220.00	6285.00	10,505.00	15,757.50				
		NO. 4	EA	2377.00	4226.00	6,603.00	9,904.50				
INSTALL NEW BATHROOM	NEW BATHROOM	NO. 1	EA	4805.00	8643.00	13,448.00	20,172.00				
		NO. 2	EA	4489.00	7739.00	12,228.00	18,342.00				
		NO. 3	EA	5106.00	8653.00	13,759.00	20,638.50				
		NO. 4	EA	3023.00	5926.00	8,949.00	13,423.50				

POWDER ROOMS

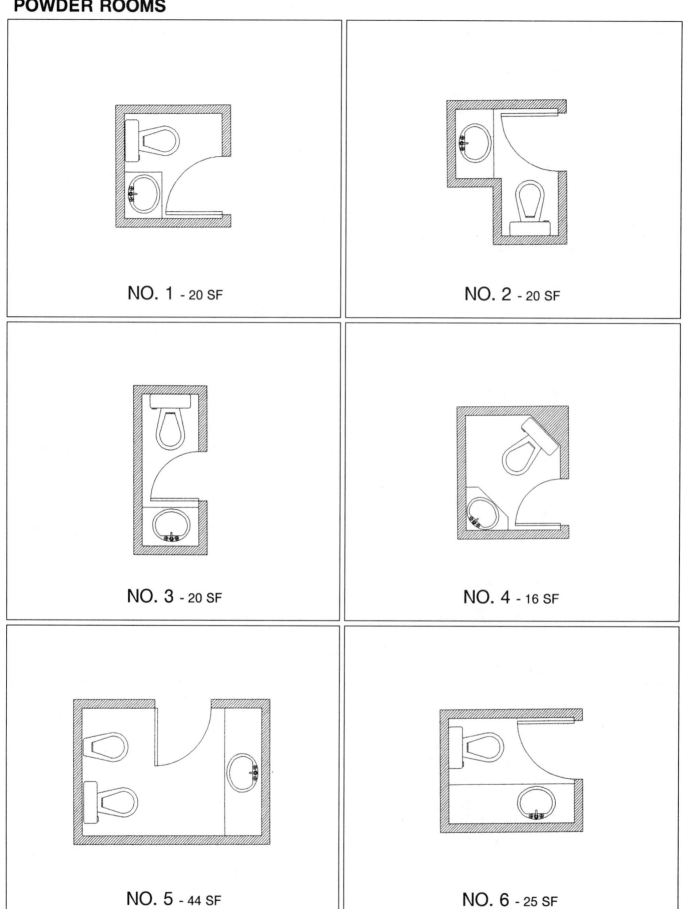

NO. 1 - 20 SF

NO. 2 - 20 SF

NO. 3 - 20 SF

NO. 4 - 16 SF

NO. 5 - 44 SF

NO. 6 - 25 SF

REMODEL EXISTING POWDER ROOM	INSTALL NEW POWDER ROOM
PREPARE PLANS AND OBTAIN PERMITS AS REQUIRED, PERMIT FEE COST **NOT** INCLUDED	PREPARE PLANS AND OBTAIN PERMITS AS REQUIRED, PERMIT FEE COST **NOT** INCLUDED
TEAR OUT EXISTING POWDER ROOM FIXTURES AND NON-CERAMIC FLOOR, WALL AND CEILING COVERINGS	
RE-ROUGH PLUMBING FOR POWDER ROOM, NEW COPPER SUPPLY AND NEW PLASTIC WASTE LINES TO MAIN STACK WITH CUTOFFS AT EACH FIXTURE	ROUGH IN PLUMBING FOR POWDER ROOM, COPPER SUPPLY AND PLASTIC WASTE LINES TO MAIN STACK WITHIN 5 FEET, WITH CUTOFFS AT EACH FIXTURE
INSTALL BUILDER GRADE WHITE FIXTURES, INCLUDING ALL FITTINGS, BIDET (POWDER ROOM 5), AND TWO-PIECE TOILET	INSTALL BUILDER GRADE WHITE FIXTURES, INCLUDING ALL FITTINGS, BIDET (POWDER ROOM 5), AND TWO-PIECE TOILET
20" X 18" LAVATORY BOWL SET IN PREFINISHED VANITY BASE WITH IMITATION MARBLE TOP AND SPLASH	20" X 18" LAVATORY BOWL SET IN PREFINISHED VANITY BASE WITH IMITATION MARBLE TOP AND SPLASH
3/4" T & G PLYWOOD SUBFLOOR	
	2-6 X 6-8 BIRCH FLUSH HOLLOW CORE DOOR WITH BATHROOM STYLE LOCKSET
SWITCH AND LIGHT OVER LAVATORY	SWITCH AND LIGHT OVER LAVATORY
GROUND FAULT DUPLEX OUTLET NEXT TO LAVATORY	GROUND FAULT DUPLEX OUTLET NEXT TO LAVATORY
EXHAUST FAN AND SWITCH VENTED TO OUTSIDE	EXHAUST FAN AND SWITCH VENTED TO OUTSIDE
NEW 1/2" DRYWALL (GYPSUM BOARD), TAPED AND FINISHED, OVER ALL WALLS AND CEILING	1/2" DRYWALL (GYPSUM BOARD), TAPED AND FINISHED, OVER ALL WALLS AND CEILING
3/8" PLYWOOD UNDERLAYMENT AND NEW CERAMIC TILE FLOOR WITH 4" TILE BASE	3/8" PLYWOOD UNDERLAYMENT AND CERAMIC TILE FLOOR WITH 4" TILE BASE
RECESSED MEDICINE CABINET WITH OVERHEAD FLUORESCENT FIXTURE, 2 MIRROR DOORS, OVERALL SIZE 24" X 16"	RECESSED MEDICINE CABINET WITH OVERHEAD FLUORESCENT FIXTURE, 2 MIRROR DOORS, OVERALL SIZE 24" X 16"
PRIME ROOM AS NECESSARY AND PAINT WITH 2 COATS TOP QUALITY SEMI-GLOSS ENAMEL	PRIME ENTIRE ROOM AND PAINT WITH 2 COATS TOP QUALITY SEMI-GLOSS ENAMEL
REMOVE OLD FIXTURES AND TRASH FROM PREMISES AND FINAL CLEAN-UP, DUMPING FEE **NOT** INCLUDED	REMOVE TRASH FROM PREMISES AND FINAL CLEAN-UP, DUMPING FEE **NOT** INCLUDED

SPECIFICATIONS			UNIT	MATLS	LABOR	TOTAL	PRICE	MATLS	LABOR	TOTAL	PRICE	
REMODEL POWDER ROOM	REMODEL EXISTING POWDER ROOM	NO. 1	EA	862.00	1083.00	1,945.00	2,917.50					
		NO. 2	EA	806.00	1041.00	1,847.00	2,770.50					
		NO. 3	EA	868.00	1076.00	1,944.00	2,916.00					
		NO. 4	EA	940.00	1028.00	1,968.00	2,952.00					
		NO. 5	EA	2144.00	1928.00	4,072.00	6,108.00					
		NO. 6	EA	1468.00	1248.00	2,716.00	4,074.00					
INSTALL NEW POWDER ROOM	NEW POWDER ROOM	NO. 1	EA	979.00	2022.00	3,001.00	4,501.50					
		NO. 2	EA	993.00	1985.00	2,978.00	4,467.00					
		NO. 3	EA	1017.00	2073.00	3,090.00	4,635.00					
		NO. 4	EA	1087.00	2003.00	3,090.00	4,635.00					
		NO. 5	EA	2445.00	3329.00	5,774.00	8,661.00					
		NO. 6	EA	1622.00	2192.00	3,814.00	5,721.00					

BATHROOMS AND POWDER ROOMS -- EXTRAS AND ALLOWANCES

SPECIFICATIONS		UNIT	JOB COST			PRICE	LOCAL AREA MODIFICATION				
			MATLS	LABOR	TOTAL		MATLS	LABOR	TOTAL	PRICE	
ADDI-TIONAL AREA	ADDITIONAL FLOOR, WALL AND CEILING AREA TO BATH-ROOM OR POWDER ROOM (ADDITIONAL FIXTURES MAY REQUIRE ADDITIONAL FLOOR SPACE) SF = FLOOR AREA OF ADDITIONAL SPACE	SF	7.57	11.19	18.76	28.14					
FIXTURES	SUBSTITUTE ONE-PIECE LUXOR OR EQUAL W.C. WITH VENTAWAY FEATURE, WHITE **ADD**	EA	204.00	96.00	300.00	450.00					
	SUBSTITUTE STEEL BATH-TUB, FOR SPECIFIED CAST IRON TUB **DEDUCT**	EA	100.00	106.00	206.00	309.00					
	FIBERGLASS WHIRLPOOL TUB, 60" X 30" X 18" **ADD**	EA	1091.00	205.00	1,296.00	1,944.00					
	SUBSTITUTE PEDESTAL SINK, VITREOUS CHINA, 24" X 19" FOR ANY 36" VANITY BASE WITH LAVATORY SET IN **DEDUCT**	EA	216.00	--	216.00	324.00					
	ADD OR OMIT BUILDER GRADE BIDET **ADD** OR **DEDUCT**	EA	600.00	370.00	970.00	1,455.00					
	SUBSTITUTE 36" VANITY BASE AND ONE LAVATORY BOWL FOR SPECIFIED 84" VANITY BASE AND TWO LAV-ATORY BOWLS **DEDUCT**	EA	746.00	164.00	910.00	1,365.00					
	GOING AWAY FROM STACK ON ANY PLUMBING FIXTURE INSTALLATION **ADD** LF = EACH FIXTURE, TO-TAL DISTANCE FROM EXISTING STACK	LF	16.00	20.00	36.00	54.00					

BATHROOMS AND POWDER ROOMS -- EXTRAS AND ALLOWANCES

SPECIFICATIONS		UNIT	JOB COST			PRICE	LOCAL AREA MODIFICATION			
			MATLS	LABOR	TOTAL		MATLS	LABOR	TOTAL	PRICE
ELECTRI-CAL	ADDITIONAL DUPLEX OUT-LETS IN OPEN FRAMING **ADD**	EA	13.00	25.00	38.00	57.00				
	FISHED IN FRAME WALL **ADD**	EA	13.00	51.00	64.00	96.00				
	FISHED IN MASONRY WALL **ADD**	EA	13.00	75.00	88.00	132.00				
	ADDITIONAL EXHAUST FAN IN COMPARTMENT BATH **ADD**	EA	46.00	90.00	136.00	204.00				
	HEAT LIGHT, INFRA-RED **ADD**	EA	126.00	90.00	216.00	324.00				
	ADDITIONAL GFI OUTLET **ADD**	EA	33.00	45.00	78.00	117.00				
	TELEPHONE JACK **ADD**	EA	24.00	80.00	104.00	156.00				
FLOOR COVERING	SUBSTITUTE SHEET VINYL FLOOR COVERING, INCLUDING VINYL BASE AND COR-NERS, FOR SPECIFIED CE-RAMIC TILE FLOOR **DEDUCT** SF = FLOOR	SF	.26	6.30	6.56	9.84				

BREEZEWAY BETWEEN EXISTING HOUSE AND GARAGE

SPECIFICATIONS	
PLANS AND PERMIT	PREPARE PLANS AND OBTAIN PERMITS AS REQUIRED, PERMIT FEE COST **NOT** INCLUDED
CONCRETE AND MASONRY	8" X 16" CONTINUOUS CONCRETE FOOTINGS, 36" BELOW GRADE
	8" X 8" X 16" BLOCK BUILDUP TO 12" ABOVE GRADE
	4" CONCRETE FLOOR SLAB OVER GRAVEL WITH NO. 4 (1/2") REBAR INTO EXISTING HOUSE FOUNDATION WALL, 24" O.C.
FRAMING	4" X 4" PRESSURE TREATED PINE OR FIR POSTS, 6'-0" O.C.
	4" X 10" PRESSURE TREATED HEADERS
	GABLE TYPE ROOF WITH 12" OVERHANG, 4/12 SLOPE, INCLUDING CEILING JOISTS
	1/2" CDX PLYWOOD ROOF SHEATHING
ROOF COVERING, FLASHING, GUTTERS	#215 FIBERGLASS STRIP SHINGLES
	ALUMINUM STEP FLASHING TO EXISTING HOUSE AND GARAGE OR CARPORT WALLS
	ALUMINUM GUTTERS AND DOWNSPOUTS
EXTERIOR TRIM	6" TIGHT KNOT PINE OR FIR FASCIA
	12" TIGHT KNOT PINE OR FIR SOFFIT
	5/8" X 4" FIR BEADED CEILING
	3/4" CEILING COVE MOULDING AT HEADER INTERSECTIONS
CLEAN-UP	REMOVE TRASH FROM PREMISES, DUMPING FEE NOT INCLUDED
	NO PAINTING

SPECIFICATIONS		UNIT	JOB COST			PRICE	LOCAL AREA MODIFICATION			
			MATLS	LABOR	TOTAL		MATLS	LABOR	TOTAL	PRICE
BREEZE-WAY	AS SPECIFIED	EA PLUS	114.00	544.00	658.00	987.00				
		SF	8.96	8.06	17.02	25.53				
DIFFERENT FOOTING DEPTH	*Depth of Bottom of Footing Below Grade*									
	12" **DEDUCT**	LF	3.68	10.28	13.96	20.94				
	24" **DEDUCT**	LF	1.84	5.13	6.97	10.46				
	48" **ADD**	LF	1.83	5.12	6.95	10.43				
	LF = FOOTINGS									

SPECIFICATIONS

PLANS AND PERMIT	PREPARE PLANS AND OBTAIN PERMITS AS REQUIRED, PERMIT FEE COST **NOT** INCLUDED
CONCRETE	MONOLITHIC COMBINATION FOOTING AND 4" CONCRETE FLOOR SLAB, REINFORCED, 36" BELOW GRADE, WITH NO. 4 (1/2") REBAR INTO EXISTING HOUSE FOUNDATION WALL, 24" O.C.
FRAMING	PRESSURE TREATED 4" X 4" PINE OR FIR POSTS, 8'-0" O.C., 4" X 10" PRESSURE TREATED HEADERS GABLE TYPE ROOF WITH 12" OVERHANG, 4/12 SLOPE, **NO** OVERLAY, RAFTERS AND CEILING JOISTS 16" O.C. 1/2" PLYWOOD ROOF SHEATHING
ROOFING, GUTTERS, FLASHING	#215 FIBERGLASS SHINGLES OVER #15 FELT PAPER TWO ALUMINUM GUTTERS AND DOWNSPOUTS ALUMINUM FLASHING AT EXISTING HOUSE
EXTERIOR TRIM	TRIM HEADERS 3 SIDES WITH TIGHT KNOT PINE OR FIR TIGHT KNOT PINE OR FIR FASCIA AND SOFFIT RAKE AND RAKE MOULDING ON GABLE END 5/8" X 4" FIR BEADED CEILING 3/4" CEILING COVE MOULDING AT HEADER INTERSECTIONS
SIDING	8" HORIZONTAL VINYL SIDING, UNINSULATED, ON GABLE END(S)
CLEAN-UP	REMOVE TRASH FROM PREMISES, DUMPING FEE **NOT** INCLUDED
	NO PAINTING

SPECIFICATIONS		UNIT	JOB COST			PRICE	LOCAL AREA MODIFICATION			
			MATLS	LABOR	TOTAL		MATLS	LABOR	TOTAL	PRICE
ATTACHED CARPORT	AS SPECIFIED	EA PLUS	760.00	1020.00	1,780.00	2,670.00				
		SF	8.00	6.50	14.50	21.75				
EXTRAS & ALLOW-ANCES	*Depth of Bottom of Footing Below Grade*									
	12" **DEDUCT**	LF	1.51	1.05	2.56	3.84				
	24" **DEDUCT**	LF	.75	.52	1.27	1.91				
	48" **ADD**	LF	.77	.53	1.30	1.95				
	LF = FOOTINGS									

DECK

SPECIFICATIONS

PLANS AND PERMIT	PREPARE PLANS AND OBTAIN PERMITS AS REQUIRED, PERMIT FEE COST **NOT** INCLUDED
CONCRETE	DIG OUT 12" X 12" HOLES TO 36" BELOW GRADE, 8'-0" O.C.
	BUILD FORMS WITH DIMENSION LUMBER FROM GRADE TO 6" ABOVE GRADE
	POUR CONCRETE INTO EXCAVATION TO TOP OF FORMS AND SET METAL POST ANCHORS IN CONCRETE
FRAMING & DECK	4" X 4" WOOD POSTS, 8'-0" O.C.
	2" X 8" OR 2" X 10" LEDGER BOARD BOLTED TO HOUSE WITH NO. 4 (1/2") LAG BOLTS, 16" O.C.
	(2) 2" X 10" WOOD BEAM AND 2" X 8" OR 2" X 10" JOISTS 16" O.C., CANTILEVERED 24", 2" X 8" OR 2" X 10" BAND
	2" X 6" DECK LAID FLAT WITH RUST RESISTANT NAILS SPACED 1/4" APART
CLEAN-UP	REMOVE TRASH FROM PREMISES, DUMPING FEE COST **NOT** INCLUDED
	NO PAINTING OR STAINING

SPECIFICATIONS		UNIT	JOB COST			PRICE	LOCAL AREA MODIFICATION			
			MATLS	LABOR	TOTAL		MATLS	LABOR	TOTAL	PRICE
WOOD DECK	BUILT WITH PRESSURE TREATED FIR OR PINE FRAMING AND 2" X 6" FIR OR PINE DECK SURFACE	SF	4.25	2.85	7.10	10.65				
	BUILT WITH PRESSURE TREATED FIR OR PINE FRAMING, 1" X 10" CEDAR OR REDWOOD BAND AND 2" X 6" CEDAR OR REDWOOD DECK SURFACE	SF	6.10	3.90	10.00	15.00				
	BUILT WITH PRESSURE TREATED FIR OR PINE FRAMING, 1" X 10" CEDAR OR REDWOOD BAND AND 5/4" X 6" CEDAR OR REDWOOD DECK SURFACE	SF	5.60	2.87	8.47	12.71				
ALTERNATE DECK FOOTINGS SYSTEM	DIG OUT 20" X 20" HOLES TO 36" BELOW GRADE, 8'-0" O.C. AND POUR 8" THICK CONCRETE FOOTINGS. BUILD BLOCK PIER TO 8" ABOVE GRADE AND FILL WITH CONCRETE WITH METAL POST ANCHORS SET IN CONCRETE **ADD**	EA	6.50	12.50	19.00	28.50				

SPECIFICATIONS

DECK RAILING NO. 1	ATTACH 4" X 4" END AND INTERMEDIATE POSTS TO BAND EXTENDING TO 42" ABOVE SURFACE OF DECK, 4'-0" O.C.
	2" X 6" CAP LAID FLAT ON POSTS, 2" X 6" TOP RAIL UNDER CAP AND 2" X 6" MIDDLE AND BOTTOM RAILS ATTACHED TO THE OUTSIDE OF POSTS ON THE DECK SIDE
DECK RAILING NO. 2	ATTACH 4" X 4" END AND INTERMEDIATE POSTS TO BAND EXTENDING TO 42" ABOVE SURFACE OF DECK, 4'-0" O.C.
	2" X 6" CAP LAID FLAT ON POSTS, 2" X 6" TOP RAIL UNDER CAP AND 2" X 2" PICKETS 7-1/2" O.C. NAILED TO OUTSIDE TOP RAIL AND BAND
STEPS	2" X 12" STRINGERS
	TWO 2" X 6" PER TREAD, OPEN RISERS, CONCRETE BOTTOM TREAD ON GROUND
	STEP RAILING ON 2 SIDES WITH CAP LAID FLAT ON POSTS, 2" X 6" TOP RAIL UNDER CAP AND 2" X 6" MIDDLE AND BOTTOM RAILS ATTACHED TO THE OUTSIDE OF POSTS ON STEP SIDE

SPECIFICATIONS		UNIT	JOB COST			PRICE	LOCAL AREA MODIFICATION			
			MATLS	LABOR	TOTAL		MATLS	LABOR	TOTAL	PRICE
DECK RAILING NO. 1	BUILT WITH PRESSURE TREATED FIR OR PINE	LF	3.75	4.45	8.20	12.30				
	BUILT WITH CEDAR OR RED-WOOD	LF	6.72	4.45	11.17	16.76				
DECK RAILING NO. 2	BUILT WITH PRESSURE TREATED FIR OR PINE	LF	3.80	4.60	8.40	12.60				
	BUILT WITH CEDAR OR RED-WOOD	LF	6.85	4.60	11.45	17.18				
STEPS	36" WIDE STEPS BUILT WITH PRESSURE TREATED FIR OR PINE	PER STEP	14.50	12.50	27.00	40.50				
	36" WIDE STEPS BUILT WITH CEDAR OR REDWOOD	PER STEP	30.50	12.50	43.00	64.50				
	60" WIDE STEPS BUILT WITH PRESSURE TREATED FIR OR PINE	PER STEP	22.50	16.50	39.00	58.50				
	60" WIDE STEPS BUILT WITH CEDAR OR REDWOOD	PER STEP	45.50	16.50	62.00	93.00				

SHELL DORMER

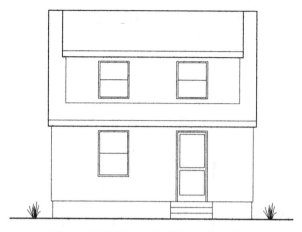

NO.1   STEPPED IN FRONT AND SIDES

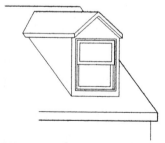

NO.2   STEPPED IN SIDES, FLUSH FRONT

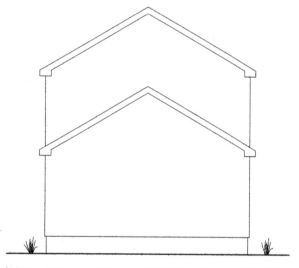

NO.3   ADD A LEVEL -- STEPPED IN ALL AROUND

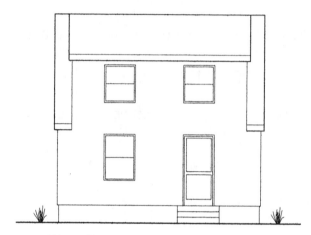

NO.4   ADD A LEVEL -- FLUSH ALL AROUND

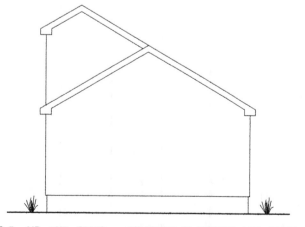

NO.5   UP AND OVER -- STEPPED IN FRONT AND SIDES

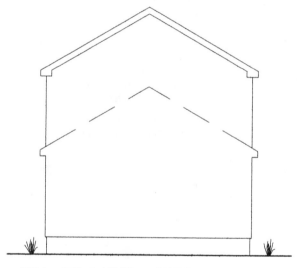

NO.6   SMALL GABLE (DOGHOUSE) DORMER

SPECIFICATIONS

PLANS AND PERMITS	PREPARE PLANS AND OBTAIN PERMITS AS REQUIRED, PERMIT FEE COST **NOT** INCLUDED
TEAR-OUT	REMOVE EXISTING ROOF COVERING, RAFTERS AND SHEATHING WHERE NECESSARY
WALL FRAMING	2" X 4" STUDS, 16" O.C. 1/2" PLYWOOD SHEATHING
ROOF FRAMING	SHED TYPE ROOF WITH 12" OVERHANG RAFTERS AND CEILING JOISTS, 16" O.C., FRAMED INTO EXISTING RIDGE BOARD 1/2" PLYWOOD SHEATHING
ROOFING, GUTTERS, FLASHING	#215 FIBERGLASS STRIP SHINGLES OVER #15 FELT PAPER ALUMINUM GUTTERS AND DOWNSPOUTS ALUMINUM FLASHING WHERE DORMER JOINS EXISTING ROOF
EXTERIOR TRIM	#2 PINE FASCIA AND SOFFIT #2 PINE RAKE WITH RAKE MOULDING SOFFIT VENTS OR GABLE VENTS AS REQUIRED
SIDING	VINYL SIDING, INSULATED, 8" HORIZONTAL
WINDOWS	*DORMERS 1, 2 AND 5:* 3 WOOD DOUBLE HUNG 2-8 X 3-10 WINDOWS, DOUBLE GLAZED *DORMERS 3 AND 4:* 6 WOOD DOUBLE HUNG 2-8 X 3-10 WINDOWS, DOUBLE GLAZED *DORMER 6:* 1 WOOD DOUBLE HUNG 2-8 X 3-10 WINDOW, DOUBLE GLAZED
CLEAN-UP	REMOVE TRASH FROM PREMISES, DUMPING FEE **NOT** INCLUDED
	NO PAINTING

SPECIFICATIONS		UNIT	JOB COST			PRICE	LOCAL AREA MODIFICATION			
			MATLS	LABOR	TOTAL		MATLS	LABOR	TOTAL	PRICE
SHELL DORMER	SF = SHELL DORMER DIMENSIONS (SEE PAGES 34 AND 38)									
	NO. 1 STEPPED IN FRONT AND SIDES	EA PLUS SF	1107.00 4.41	953.00 6.55	2,060.00 10.96	3,090.00 16.44				
	NO. 2 STEPPED IN SIDES, FLUSH FRONT	EA PLUS SF	1107.00 5.09	953.00 6.89	2,060.00 11.98	3,090.00 17.97				
	NO. 3 ADD A LEVEL — STEPPED IN ALL AROUND	EA PLUS SF	2709.00 4.64	1841.00 5.89	4,550.00 10.53	6,825.00 15.80				
	NO. 4 ADD A LEVEL — FLUSH ALL AROUND	EA PLUS SF	2709.00 5.83	1841.00 6.75	4,550.00 12.58	6,825.00 18.87				
	NO. 5 UP AND OVER — STEPPED IN FRONT AND SIDES	EA PLUS SF	455.00 6.64	691.00 4.07	1,146.00 10.71	1,719.00 16.07				
	NO. 6 GABLE (DOGHOUSE) DORMER (**NO** GUTTER AND DOWN-SPOUT)	EA	467.00	553.00	1,020.00	1,530.00				

SHELL DORMER -- EXTRAS AND ALLOWANCES

SPECIFICATIONS		UNIT	JOB COST			PRICE	LOCAL AREA MODIFICATION				
			MATLS	LABOR	TOTAL		MATLS	LABOR	TOTAL	PRICE	
THIRD LEVEL DORMER	DORMER INSTALLED ON 3rd STORY ABOVE GRADE **ADD** SF = DORMER DIMENSION	SF	--	2.30	2.30	3.45					
DORMER ON SLATE ROOF	CUT INTO EXISTING SLATE ROOF AND INSTALL DORMER **ADD** SF = DORMER DIMENSION	SF	--	.80	.80	1.20					
ROOF FRAMING	SUBSTITUTE GABLE STYLE ROOF FOR SPECIFIED SHED TYPE ROOF **ADD** SF = DORMER ROOF	SF	.60	.60	1.20	1.80					
ROOF COVERING	SUBSTITUTE ROOF COVERINGS FOR SPECIFIED FIBERGLASS SHINGLES CONCRETE ROOFING TILES **ADD**	SF	.89	.65	1.54	2.31					
	18" CEDAR SHINGLES **ADD** SF = DORMER ROOF	SF	2.09	.91	3.00	4.50					
TEAR-OUT	REMOVE EXISTING NON-BEARING PARTITION	LF	--	6.64	6.64	9.96					
	REMOVE EXISTING INSULATION AND CEILING COVERING FROM RAFTERS SF = SF RAFTERS	SF	--	.38	.38	.57					
WINDOWS	ADD OR OMIT ONE WOOD DOUBLE HUNG WINDOW, 2-8 X 3-10, DOUBLE GLAZED **ADD** OR **DEDUCT**	EA	173.00	59.00	232.00	348.00					
	SUBSTITUTE DOUBLE GLAZED 4-0 X 4-0 ALUMINUM SLIDING WINDOW **ADD** OR **DEDUCT**	EA	50.00	20.00	70.00	105.00					
BREAK THROUGH AND INSTALL WINDOW	BREAK THROUGH EXISTING GABLE END, INSTALL HEADER AND/OR LINTEL AS REQUIRED AND INSTALL WINDOW, INCLUDING SASH, FRAME AND ALL TRIM, WOOD DOUBLE HUNG, DOUBLE GLAZED, 2-8 X 3-10 FRAME WALL	EA	240.00	290.00	530.00	795.00					
	MASONRY WALL	EA	240.00	320.00	560.00	840.00					

SPECIFICATIONS		UNIT	JOB COST			PRICE	LOCAL AREA MODIFICATION			
			MATLS	LABOR	TOTAL		MATLS	LABOR	TOTAL	PRICE
FLOOR FRAMING	SISTER (DOUBLE UP) EXIST-ING 2nd FLOOR JOISTS									
	2" X 6" **ADD**	SF	.58	1.80	2.38	3.57				
	2" X 8" **ADD**	SF	.93	1.80	2.73	4.10				
	SF = FLOOR JOISTS									
	REMOVE EXISTING FLOOR JOISTS, SUBFLOOR AND CEILING BELOW AND INSTALL HEADER FOR STAIRS	EA	91.00	195.00	286.00	429.00				
KNEEWALL	INTERIOR KNEEWALL, 4 TO 6 FEET HIGH, 2" X 4" 16" O.C.	LF	3.69	4.60	8.29	12.44				
FURRING	FURRING APPLIED OVER FRAME WALLS OR STRAIGHT MASONRY WALL	SF	.17	.44	.61	.92				
	FURRING APPLIED OVER CROOKED MASONRY WALLS	SF	.17	.60	.77	1.16				
FRAME PARTITION	FRAME OUT 2" X 4" WALL 8'-0" HIGH	LF	6.10	5.00	11.10	16.65				
EXTEND CHIMNEY	EXTEND EXISTING CHIMNEY ABOVE ROOF									
	1 FLUE	LF UP	25.00	47.00	72.00	108.00				
	2 FLUES	LF UP	37.00	53.00	90.00	135.00				
SIDING	SUBSTITUTE OTHER SIDINGS FOR SPECIFIED VINYL SIDING									
	BEVELED CEDAR **ADD**	SF	.44	.13	.57	.86				
	18" CEDAR SHINGLES **ADD**	SF	--	.36	.36	.54				
	SF = DORMER WALLS									
RIDGE BOARD	DOUBLE UP RIDGE BOARD									
	2" X 10" **ADD**	LF	1.18	.87	2.05	3.08				
	2" X 12" **ADD**	LF	1.70	.90	2.60	3.90				

DORMER INTERIOR TRIM AND FINISHING

THE DIMENSIONS FOR INTERIOR TRIM AND FINISHING WILL USUALLY BE MORE THAN THE SHELL DORMER DIMENSION AS THE NEW LIVING AREA WILL PROBABLY EXTEND PAST THE EXISTING RIDGEBOARD AS FAR AS ADEQUATE HEADROOM CAN BE OBTAINED.

BE SURE TO INCLUDE ANY LIVING AREA AT EACH SIDE OF THE DORMER INTERIOR IF THE INSTALLATION IS STEPPED IN SIDES.

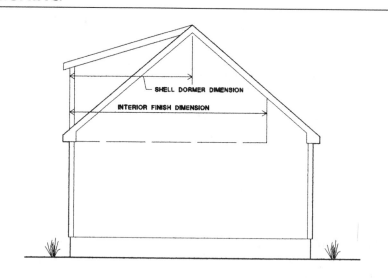

SPECIFICATIONS

WINDOWS	TRIM WINDOWS WITH 2-1/4" CASING, STOOLS, STOPS AND SASH FASTENERS
HEATING	EXTEND HEAT AND RETURN DUCTS FROM EXISTING SERVICE
ELECTRIC	DUPLEX WALL OUTLETS TO CODE ON EXISTING SERVICE
INSULATION	3-1/2" FIBERGLASS BLANKET INSULATION ON ALL SIDEWALLS
	6" FIBERGLASS BLANKET INSULATION IN CEILING
INTERIOR WALL COVERING	1/2" GYPSUM BOARD (SHEETROCK) ON THREE NEW WALLS AND ON FURRING STRIPS OVER EXISTING HOUSE WALL, TAPED AND FINISHED (**NO** PAINTING)
CEILING COVERING	1/2" GYPSUM BOARD (SHEETROCK) TAPED AND FINISHED (**NO** PAINTING)
INTERIOR TRIM	3-1/2" BASE AND SHOE MOULDING ON ALL FOUR WALLS
FLOOR COVERING	2-1/4" SELECT OAK FLOORING, SANDED AND FINISHED
CLEAN-UP	REMOVE RUBBISH FROM PREMISES, DUMPING FEE **NOT** INCLUDED
	NO PAINTING

SPECIFICATIONS		UNIT	JOB COST			PRICE	LOCAL AREA MODIFICATION			
			MATLS	LABOR	TOTAL		MATLS	LABOR	TOTAL	PRICE
INTERIOR TRIM AND FINISHING	AS SPECIFIED	EA PLUS	520.00	500.00	1,020.00	1,530.00				
	SF = TOTAL SQUARE FOOT LIVING AREA TO BE FINISHED	SF	6.00	5.60	11.60	17.40				

DORMER INTERIOR TRIM AND FINISHING -- EXTRAS AND ALLOWANCES

SPECIFICATIONS		UNIT	JOB COST			PRICE	LOCAL AREA MODIFICATION			
			MATLS	LABOR	TOTAL		MATLS	LABOR	TOTAL	PRICE
PARTITION	TRIM AND FINISH FRAMED OUT PARTITION: • DUPLEX OUTLETS TO CODE BOTH SIDES OF WALL • 1/2" DRYWALL BOTH SIDES OF WALL, FINISHED BUT **NOT** PAINTED • 3-1/2" BASE AND SHOE BOTH SIDES OF WALL LF = LENGTH OF PARTITION	LF	10.00	18.00	28.00	42.00				
DOOR	HANG DOOR IN FRAMED OUT PARTITION, 2-6 X 6-8, BIRCH HOLLOW CORE, INCLUDING JAMBS, STOPS, CASINGS AND HARDWARE **ADD**	EA	105.00	63.00	168.00	252.00				
	BI-FOLD DOORS, BIRCH FLUSH, INCLUDING ALL TRIM AND HARDWARE 2 DOORS 3-0 X 6-8 **ADD**	SET	85.00	59.00	144.00	216.00				
	6-0 X 6-8 **ADD**	SET	111.00	65.00	176.00	264.00				
	4 DOORS 4-0 X 6-8 **ADD**	SET	108.00	62.00	170.00	255.00				
	6-0 X 6-8 **ADD**	SET	122.00	68.00	190.00	285.00				
WINDOW	TRIM ADDITIONAL WINDOW OR OMIT TRIMMING WINDOW **ADD** OR **DEDUCT**	EA	21.00	29.00	50.00	75.00				
CLOSET TRIM	TRIM FRAMED CLOSET WITH BASE AND SHOE, HOOK-STRIP, SHELF, CLOTHES POLE AND CLOTHES POLE SOCKET	LF	9.18	6.60	15.78	23.67				
CEILING MOULDING	CEILING MOULDING AT WALL INTERSECTIONS 3/4" **ADD**	LF	.42	.75	1.17	1.76				
	1-5/8" **ADD**	LF	.73	.79	1.52	2.28				
	3-5/8" **ADD**	LF	1.55	.97	2.52	3.78				
STAIRS	CLOSED STAIRWAY TO 2nd FLOOR DORMER ROOM(S), OAK TREADS AND RISERS WITH BIRCH HANDRAIL, SHOP BUILT AND INSTALLED ON JOB BY STAIRBUILDER EA = EACH STAIRWAY	EA	880.00	290.00	1,170.00	1,755.00				
FLOORING	SUBSTITUTE OTHER FLOOR COVERINGS FOR SPECIFIED SELECT OAK FLOORING CARPET AND PAD @ $20/YD **DEDUCT**	SF	1.78	2.21	3.99	5.99				
	UNDERLAYMENT AND VINYL TILE OR SHEET VINYL **DEDUCT**	SF	2.71	.98	3.69	5.54				
	SF = FLOOR									

GARAGE

DETACHED GARAGE	ATTACHED GARAGE
PREPARE PLANS AND OBTAIN PERMITS AS REQUIRED, PERMIT FEE COST **NOT** INCLUDED	PREPARE PLANS AND OBTAIN PERMITS AS REQUIRED, PERMIT FEE COST **NOT** INCLUDED
8" X 16" CONTINUOUS CONCRETE FOOTINGS, 36" BELOW GRADE WITH 8" X 8" X 16" BLOCK BUILDUP TO 12" ABOVE GRADE	8" X 16" CONTINUOUS CONCRETE FOOTINGS, 36" BELOW GRADE WITH 8" X 8" X 16" BLOCK BUILDUP TO 12" ABOVE GRADE
4" CONCRETE FLOOR SLAB OVER GRAVEL	4" CONCRETE FLOOR SLAB OVER GRAVEL WITH NO. 4 (1/2") REBAR INTO EXISTING HOUSE FOUNDATION WALL, 24" O.C.
2" X 4" STUD WALLS, 16" O.C.	2" X 4" STUD WALLS, 16" O.C.
1/2" PLYWOOD SHEATHING	1/2" PLYWOOD SHEATHING
GABLE TYPE ROOF WITH 12" OVERHANG, 4/12 SLOPE, CEILING JOISTS	GABLE TYPE ROOF WITH 12" OVERHANG, 4/12 SLOPE, CEILING JOISTS
1/2" PLYWOOD ROOF SHEATHING	1/2" PLYWOOD ROOF SHEATHING
#215 FIBERGLASS STRIP SHINGLES	#215 FIBERGLASS STRIP SHINGLES
ALUMINUM GUTTERS AND DOWNSPOUTS AS REQUIRED	ALUMINUM STEP FLASHING TO EXISTING HOUSE WALL
	ALUMINUM GUTTERS AND DOWNSPOUTS AS REQUIRED
TIGHT KNOT PINE OR FIR 6" FASCIA	TIGHT KNOT PINE OR FIR 6" FASCIA
TIGHT KNOT PINE OR FIR 12" SOFFIT	TIGHT KNOT PINE OR FIR 12" SOFFIT
RAKE AND RAKE MOULDING	RAKE AND RAKE MOULDING
8" HORIZONTAL VINYL SIDING, UNINSULATED	8" HORIZONTAL VINYL SIDING, UNINSULATED
ONE 8-0 X 7-0 FOUR-SECTION OVERHEAD GARAGE DOOR, INCLUDING EXTERIOR BRICK MOULDING, JAMBS, STOPS AND HARDWARE	ONE 8-0 X 7-0 FOUR-SECTION OVERHEAD GARAGE DOOR, INCLUDING EXTERIOR BRICK MOULDING, JAMBS, STOPS AND HARDWARE
TWO 2-8 X 4-2 WOOD DOUBLE HUNG WINDOWS, SINGLE GLAZED	TWO 2-8 X 4-2 WOOD DOUBLE HUNG WINDOWS, SINGLE GLAZED, INCLUDING INTERIOR CASINGS, STOOL AND STOOL CAP
	5/8" FIRECODE DRYWALL, TAPED AND FINISHED, ON EXISTING HOUSE WALL
REMOVE TRASH FROM PREMISES, DUMPING FEE **NOT** INCLUDED	REMOVE TRASH FROM PREMISES, DUMPING FEE **NOT** INCLUDED
NO PAINTING	**NO** PAINTING

SPECIFICATIONS		UNIT	JOB COST			PRICE	LOCAL AREA MODIFICATION			
			MATLS	LABOR	TOTAL		MATLS	LABOR	TOTAL	PRICE
DETACHED GARAGE	AS SPECIFIED	EA PLUS	2758.00	2510.00	5,268.00	7,902.00				
		SF	6.54	6.15	12.69	19.04				
ATTACHED GARAGE	AS SPECIFIED	EA PLUS	2758.00	2510.00	5,268.00	7,902.00				
		SF	6.38	5.93	12.31	18.47				

SPECIFICATIONS		UNIT	JOB COST			PRICE	LOCAL AREA MODIFICATION				
			MATLS	LABOR	TOTAL		MATLS	LABOR	TOTAL	PRICE	
DIFFERENT FOOTING DEPTH	*Depth of Bottom of Footing Below Grade*										
	12" **DEDUCT**	SF	3.68	10.28	13.96	20.94					
	24" **DEDUCT**	SF	1.84	5.13	6.97	10.46					
	48" **ADD**	SF	1.83	5.12	6.95	10.43					
	LF = FOOTINGS										
ALTERNATE WALL SYSTEM	SUBSTITUTE MASONRY WALL FROM TOP OF SLAB TO START OF GARAGE ROOF										
	8" BLOCK **DEDUCT**	SF	.76	--	.76	1.14					
	BRICK & BLOCK **ADD**	SF	1.76	4.05	5.81	8.72					
	SF = SF GARAGE WALLS										
FURRING	FURR EXISTING HOUSE WALL WITH 1" X 3"										
	STRAIGHT WALL **ADD**	SF	.17	.44	.61	.92					
	CROOKED WALL **ADD**	SF	.17	.60	.77	1.16					
ROOF	OVERLAY EXISTING ROOF **ADD**	SF	3.86	4.36	8.22	12.33					
	SF = ROOF OVERLAY										
SIDING	REMOVE EXISTING WOOD, PLYWOOD OR ALUMINUM SIDING FROM EXISTING HOUSE WALL	SF	--	.34	.34	.51					
	SF = HOUSE WALL										
DOOR	SUBSTITUTE 16-0 X 7-0 OVERHEAD GARAGE DOOR **ADD**	EA	296.00	38.00	334.00	501.00					
	BREAK THROUGH EXISTING HOUSE WALL & INSTALL 3-0 X 6-8 X 3/4" FLUSH SOLID CORE DOOR TRIMMED BOTH SIDES										
	FRAME WALL **ADD**	EA	316.00	480.00	796.00	1,194.00					
	BRICK VENEER **ADD**	EA	316.00	518.00	834.00	1,251.00					
	BRICK AND BLOCK **ADD**	EA	316.00	540.00	856.00	1,284.00					
	REMOVE EXISTING WINDOW AND WALL UNDER AND INSTALL 3-0 X 6-8 X 3/4" FLUSH SOLID CORE DOOR TRIMMED BOTH SIDES	EA	291.00	409.00	700.00	1,050.00					
	INSTALL 3-0 X 6-8 THREE PANEL, 4-LIGHT DOOR, FULLY TRIMMED, FROM GARAGE TO OUTSIDE	EA	317.00	91.00	408.00	612.00					
WINDOWS	ADD OR OMIT ONE WOOD DOUBLE HUNG WINDOW, 2-8 X 4-2, SINGLE GLAZED **ADD OR DEDUCT**	EA	93.00	55.00	148.00	222.00					
ELECTRICAL	EXTEND ELECTRICAL CIRCUIT 50 LF TO DETACHED GARAGE **ADD**	EA	12.00	70.00	82.00	123.00					
	SWITCH, LIGHT AND GFIC OUTLET **ADD**	EA	60.00	100.00	160.00	240.00					
	EA = TOTAL AMOUNT										

SMALL KITCHENS

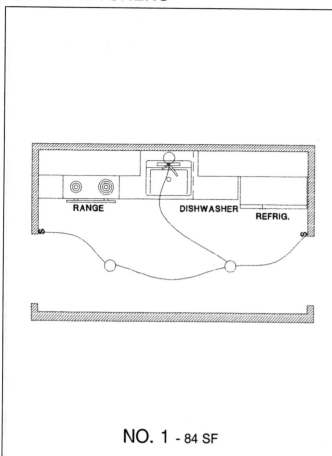

NO. 1 - 84 SF

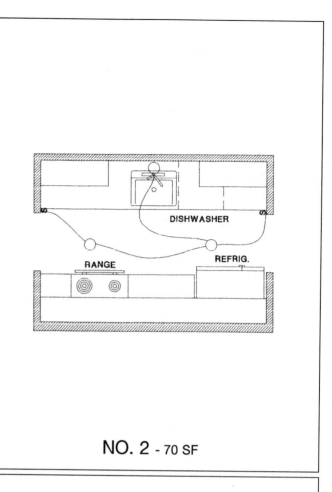

NO. 2 - 70 SF

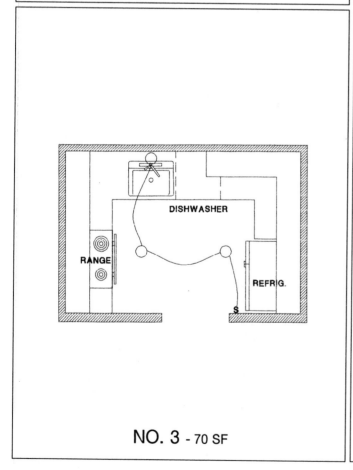

NO. 3 - 70 SF

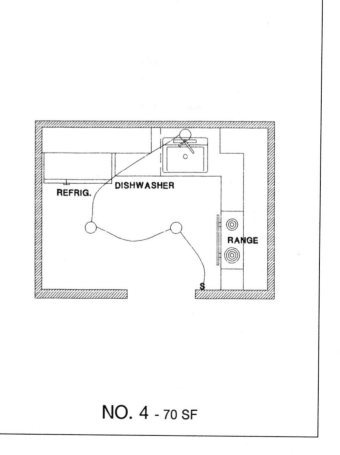

NO. 4 - 70 SF

REMODEL EXISTING KITCHEN

INSTALL NEW KITCHEN

REMODEL EXISTING KITCHEN	INSTALL NEW KITCHEN
PREPARE PLANS AND OBTAIN PERMIT, PERMIT FEE COST **NOT** INCLUDED	PREPARE PLANS AND OBTAIN PERMIT, PERMIT FEE COST **NOT** INCLUDED
TEAR OUT EXISTING KITCHEN CABINETS, COUNTERTOPS, SINK, FLOOR COVERING, STOVE AND REFRIGERATOR AND REMOVE TO OUTSIDE REPAIR MINOR DRYWALL DAMAGE	INSTALL 1/2" DRYWALL ON WALLS AND CEILING, TAPED AND FINISHED
FURNISH AND INSTALL WALL AND BASE CABINETS WITH DRAWER AND DOOR HARDWARE, INCLUDING 36" SINK BASE LAMINATE COUNTERTOPS WITH 4" BACKSPLASH AND SINK CUTOUT 24" X 21" SINGLE BOWL STAINLESS STEEL SINK WITH SINGLE LEVER FAUCET, FURNISHED AND INSTALLED IN EXISTING SINK LOCATION	FURNISH AND INSTALL WALL AND BASE CABINETS WITH DRAWER AND DOOR HARDWARE, INCLUDING 36" SINK BASE LAMINATE COUNTERTOPS WITH 4" BACKSPLASH AND SINK CUTOUT 24" X 21" SINGLE BOWL STAINLESS STEEL SINK WITH SINGLE LEVER FAUCET, FURNISHED AND INSTALLED WITHIN 5 FEET OF EXISTING STACK
REPLACE SWITCHES AND LIGHTS IN EXISTING LOCATIONS AS NECESSARY	SWITCHES AND LIGHTS AS SHOWN DUPLEX OUTLETS TO CODE APPLIANCE OUTLET 220-V RANGE OUTLET
FURNISH AND INSTALL ELECTRIC RANGE IN SAME LOCATION, 30" WHITE, BUILDER GRADE WITH DUCTLESS RANGE HOOD 1/2 HP DISPOSER ON EXISTING CIRCUIT AND PLUMBING DISHWASHER ON EXISTING CIRCUIT AND PLUMBING 16 CF REFRIGERATOR, DOUBLE DOOR, TOP FREEZER, FROST FREE, WHITE, BUILDER GRADE	FURNISH AND INSTALL ELECTRIC RANGE, INCLUDING DEDICATED CIRCUIT AS NECESSARY, 30" WHITE, BUILDER GRADE WITH DUCTLESS RANGE HOOD 1/2 HP DISPOSER ON NEW CIRCUIT AND PLUMBING DISHWASHER ON NEW CIRCUIT AND PLUMBING 16 CF REFRIGERATOR, DOUBLE DOOR, TOP FREEZER, FROST FREE, WHITE, BUILDER GRADE
INSTALL 3/8" PLYWOOD UNDERLAYMENT AND SHEET VINYL LAID IN ADHESIVE, INCLUDING 4" VINYL BASE AND CORNERS	INSTALL 3/8" PLYWOOD UNDERLAYMENT AND SHEET VINYL LAID IN ADHESIVE, INCLUDING 4" VINYL BASE AND CORNERS
PRIME ROOM AS NECESSARY AND PAINT WITH 2 COATS TOP QUALITY SEMI-GLOSS ENAMEL	PRIME ENTIRE ROOM AND PAINT WITH 2 COATS TOP QUALITY SEMI-GLOSS ENAMEL
REMOVE OLD CABINETS, FIXTURES, APPLIANCES AND TRASH FROM PREMISES AND FINAL CLEAN-UP, DUMPING FEE **NOT** INCLUDED	REMOVE TRASH FROM PREMISES AND FINAL CLEAN-UP, DUMPING FEE **NOT** INCLUDED

SPECIFICATIONS			UNIT	JOB COST			PRICE	LOCAL AREA MODIFICATION			
				MATLS	LABOR	TOTAL		MATLS	LABOR	TOTAL	PRICE
REMODEL KITCHEN	REMODEL BUILDER QUALITY KITCHEN	NO. 1	EA	3495.00	1597.00	5,092.00	7,638.00				
		NO. 2	EA	4787.00	2017.00	6,804.00	10,206.00				
		NO. 3	EA	4816.00	2074.00	6,890.00	10,335.00				
		NO. 4	EA	4125.00	1851.00	5,976.00	8,964.00				
INSTALL NEW KITCHEN	INSTALL NEW BUILDER QUALITY KITCHEN	NO. 1	EA	3770.00	2516.00	6,286.00	9,429.00				
		NO. 2	EA	5070.00	2882.00	7,952.00	11,928.00				
		NO. 3	EA	5079.00	2905.00	7,984.00	11,976.00				
		NO. 4	EA	4371.00	2737.00	7,108.00	10,662.00				

LARGE KITCHENS

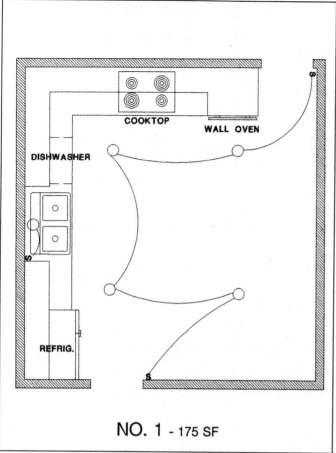

NO. 1 - 175 SF

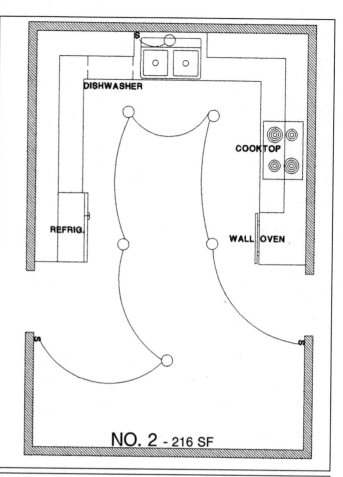

NO. 2 - 216 SF

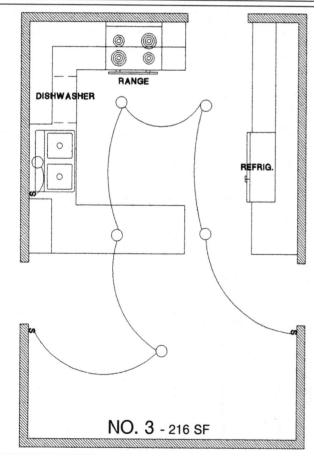

NO. 3 - 216 SF

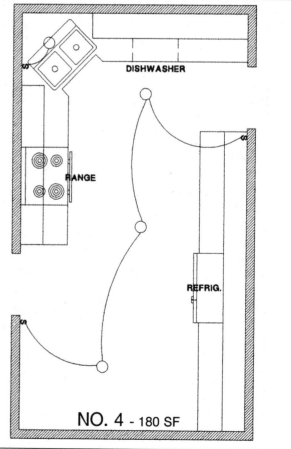

NO. 4 - 180 SF

REMODEL EXISTING KITCHEN	INSTALL NEW KITCHEN
PREPARE PLANS AND OBTAIN PERMIT, PERMIT FEE COST **NOT** INCLUDED	PREPARE PLANS AND OBTAIN PERMIT, PERMIT FEE COST **NOT** INCLUDED
TEAR OUT EXISTING KITCHEN CABINETS, COUNTERTOPS, SINK, FLOOR COVERING, STOVE AND REFRIGERATOR AND REMOVE TO OUTSIDE REPAIR MINOR DRYWALL DAMAGE	INSTALL 1/2" DRYWALL ON WALLS AND CEILING, TAPED AND FINISHED
FURNISH AND INSTALL WALL AND BASE CABINETS WITH DRAWER AND DOOR HARDWARE, INCLUDING 36" SINK BASE LAMINATE COUNTERTOPS WITH 4" BACKSPLASH AND SINK CUTOUT 24" X 21" SINGLE BOWL STAINLESS STEEL SINK WITH SINGLE LEVER FAUCET, FURNISHED AND INSTALLED IN EXISTING SINK LOCATION	FURNISH AND INSTALL WALL AND BASE CABINETS WITH DRAWER AND DOOR HARDWARE, INCLUDING 36" SINK BASE LAMINATE COUNTERTOPS WITH 4" BACKSPLASH AND SINK CUTOUT 24" X 21" SINGLE BOWL STAINLESS STEEL SINK WITH SINGLE LEVER FAUCET, FURNISHED AND INSTALLED WITHIN 5 FEET OF EXISTING STACK
REPLACE SWITCHES AND LIGHTS IN EXISTING LOCATIONS AS NECESSARY	SWITCHES AND LIGHTS AS SHOWN DUPLEX OUTLETS TO CODE APPLIANCE OUTLET 220-V RANGE OUTLET
FURNISH AND INSTALL ELECTRIC RANGE IN SAME LOCATION, 30" WHITE, BUILDER GRADE WITH DUCTLESS RANGE HOOD 1/2 HP DISPOSER ON EXISTING CIRCUIT AND PLUMBING DISHWASHER ON EXISTING CIRCUIT AND PLUMBING 16 CF REFRIGERATOR, DOUBLE DOOR, TOP FREEZER, FROST FREE, WHITE, BUILDER GRADE	FURNISH AND INSTALL ELECTRIC RANGE, INCLUDING DEDICATED CIRCUIT AS NECESSARY, 30" WHITE, BUILDER GRADE WITH DUCTLESS RANGE HOOD 1/2 HP DISPOSER ON NEW CIRCUIT AND PLUMBING DISHWASHER ON NEW CIRCUIT AND PLUMBING 16 CF REFRIGERATOR, DOUBLE DOOR, TOP FREEZER, FROST FREE, WHITE, BUILDER GRADE
INSTALL 3/8" PLYWOOD UNDERLAYMENT AND SHEET VINYL LAID IN ADHESIVE, INCLUDING 4" VINYL BASE AND CORNERS	INSTALL 3/8" PLYWOOD UNDERLAYMENT AND SHEET VINYL LAID IN ADHESIVE, INCLUDING 4" VINYL BASE AND CORNERS
PRIME ROOM AS NECESSARY AND PAINT WITH 2 COATS TOP QUALITY SEMI-GLOSS ENAMEL	PRIME ENTIRE ROOM AND PAINT WITH 2 COATS TOP QUALITY SEMI-GLOSS ENAMEL
REMOVE OLD CABINETS, FIXTURES, APPLIANCES AND TRASH FROM PREMISES AND FINAL CLEAN-UP, DUMPING FEE **NOT** INCLUDED	REMOVE TRASH FROM PREMISES AND FINAL CLEAN-UP, DUMPING FEE **NOT** INCLUDED

SPECIFICATIONS			UNIT	JOB COST			PRICE	LOCAL AREA MODIFICATION			
				MATLS	LABOR	TOTAL		MATLS	LABOR	TOTAL	PRICE
REMODEL KITCHEN	REMODEL BUILDER QUALITY KITCHEN	NO. 1	EA	6039.00	2895.00	8,934.00	13,401.00				
		NO. 2	EA	7425.00	3351.00	10,776.00	16,164.00				
		NO. 3	EA	6029.00	2929.00	8,958.00	13,437.00				
		NO. 4	EA	6482.00	3062.00	9,544.00	14,316.00				
INSTALL NEW KITCHEN	INSTALL NEW BUILDER QUALITY KITCHEN	NO. 1	EA	6144.00	2692.00	8,836.00	13,254.00				
		NO. 2	EA	7946.00	3052.00	10,998.00	16,497.00				
		NO. 3	EA	6219.00	2841.00	9,060.00	13,590.00				
		NO. 4	EA	7159.00	2759.00	9,918.00	14,877.00				

ISLAND KITCHENS

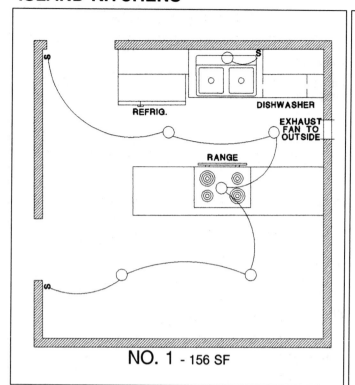

NO. 1 - 156 SF

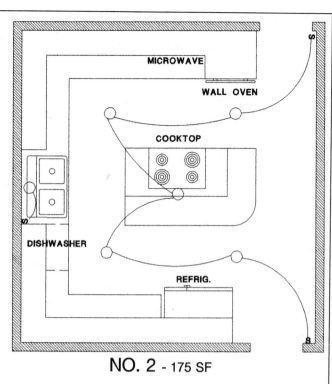

NO. 2 - 175 SF

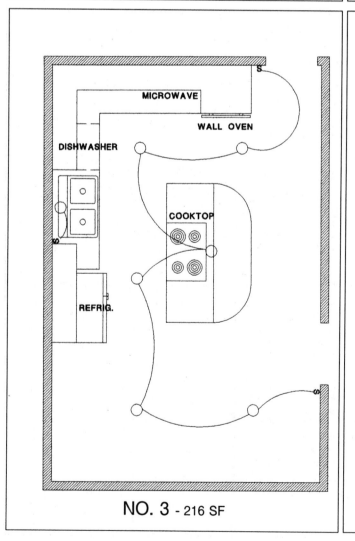

NO. 3 - 216 SF

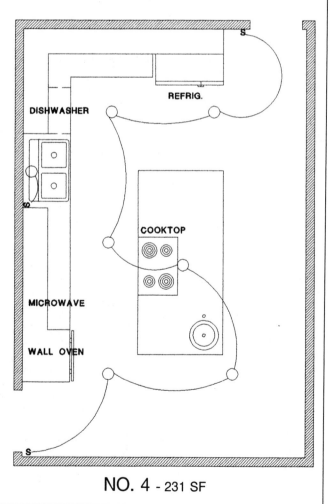

NO. 4 - 231 SF

REMODEL EXISTING KITCHEN	INSTALL NEW KITCHEN
PREPARE PLANS AND OBTAIN PERMIT, PERMIT FEE COST **NOT** INCLUDED	PREPARE PLANS AND OBTAIN PERMIT, PERMIT FEE COST **NOT** INCLUDED
TEAR OUT EXISTING KITCHEN CABINETS, COUNTERTOPS, SINK, FLOOR COVERING, STOVE AND REFRIGERATOR AND REMOVE TO OUTSIDE REPAIR MINOR DRYWALL DAMAGE	INSTALL 1/2" DRYWALL ON WALLS AND CEILING, TAPED AND FINISHED
FURNISH AND INSTALL WALL AND BASE CABINETS, INCLUDING ISLAND BASE AND 36" SINK BASE WITH DRAWER AND DOOR HARDWARE LAMINATE COUNTERTOPS WITH 4" BACKSPLASH AND SINK CUTOUTS 32" X 21" DOUBLE BOWL STAINLESS STEEL SINK WITH SINGLE LEVER FAUCET 15" X 15" BAR SINK WITH BAR FAUCET (KITCHEN NO. 4) FURNISHED AND INSTALLED IN EXISTING SINK LOCATION	FURNISH AND INSTALL WALL AND BASE CABINETS, INCLUDING ISLAND BASE AND 36" SINK BASE WITH DRAWER AND DOOR HARDWARE LAMINATE COUNTERTOPS WITH 4" BACKSPLASH AND SINK CUTOUTS 32" X 21" DOUBLE BOWL STAINLESS STEEL SINK WITH SINGLE LEVER FAUCET FURNISHED AND INSTALLED WITHIN 5 FEET OF EXISTING STACK 15" X 15" BAR SINK WITH BAR FAUCET (KITCHEN NO. 4) FURNISHED AND INSTALLED WITHIN 5 FEET OF EXISTING STACK
REPLACE SWITCHES AND LIGHTS IN EXISTING LOCATIONS AS NECESSARY	SWITCHES AND LIGHTS AS SHOWN DUPLEX OUTLETS TO CODE APPLIANCE OUTLET 220-V RANGE OUTLET
FURNISH AND INSTALL ELECTRIC COOKTOP WITH DOWN-DRAFT DUCTED EXHAUST AND ELECTRIC WALL OVEN IN SAME LOCATION 30" BUILT-IN MICROWAVE OVEN 1/2 HP DISPOSER ON EXISTING CIRCUIT AND PLUMBING DISHWASHER ON EXISTING CIRCUIT AND PLUMBING 16 CF REFRIGERATOR, DOUBLE DOOR, TOP FREEZER, FROST FREE, WHITE, BUILDER GRADE	FURNISH AND INSTALL ELECTRIC COOKTOP WITH DOWN-DRAFT DUCTED EXHAUST AND ELECTRIC WALL OVEN, INCLUDING DEDICATED CIRCUIT AS NECESSARY 30" BUILT-IN MICROWAVE OVEN 1/2 HP DISPOSER ON NEW CIRCUIT AND PLUMBING DISHWASHER ON NEW CIRCUIT AND PLUMBING 16 CF REFRIGERATOR, DOUBLE DOOR, TOP FREEZER, FROST FREE, WHITE, BUILDER GRADE
INSTALL 3/8" PLYWOOD UNDERLAYMENT AND SHEET VINYL LAID IN ADHESIVE, INCLUDING 4" VINYL BASE AND CORNERS	INSTALL 3/8" PLYWOOD UNDERLAYMENT AND SHEET VINYL LAID IN ADHESIVE, INCLUDING 4" VINYL BASE AND CORNERS
PRIME ROOM AS NECESSARY AND PAINT WITH 2 COATS TOP QUALITY SEMI-GLOSS ENAMEL	PRIME ENTIRE ROOM AND PAINT WITH 2 COATS TOP QUALITY SEMI-GLOSS ENAMEL
REMOVE OLD CABINETS, FIXTURES, APPLIANCES AND TRASH FROM PREMISES AND FINAL CLEAN-UP, DUMPING FEE **NOT** INCLUDED	REMOVE TRASH FROM PREMISES AND FINAL CLEAN-UP, DUMPING FEE **NOT** INCLUDED

SPECIFICATIONS			UNIT	JOB COST			PRICE	LOCAL AREA MODIFICATION			
				MATLS	LABOR	TOTAL		MATLS	LABOR	TOTAL	PRICE
REMODEL KITCHEN	REMODEL BUILDER QUALITY KITCHEN	NO. 1	EA	4912.00	2722.00	7,634.00	11,451.00				
		NO. 2	EA	8549.00	3703.00	12,252.00	18,378.00				
		NO. 3	EA	7053.00	3119.00	10,172.00	15,258.00				
		NO. 4	EA	7663.00	3357.00	11,020.00	16,530.00				
INSTALL NEW KITCHEN	INSTALL NEW BUILDER QUALITY KITCHEN	NO. 1	EA	4845.00	3209.00	8,054.00	12,081.00				
		NO. 2	EA	8547.00	4563.00	13,110.00	19,665.00				
		NO. 3	EA	7059.00	4027.00	11,086.00	16,629.00				
		NO. 4	EA	7845.00	4431.00	12,276.00	18,414.00				

KITCHEN -- EXTRAS AND ALLOWANCES

SPECIFICATIONS		UNIT	JOB COST			PRICE	LOCAL AREA MODIFICATION				
			MATLS	LABOR	TOTAL		MATLS	LABOR	TOTAL	PRICE	
ADDITIONAL AREA	ADDITIONAL FLOOR, WALL AND CEILING AREA FOR KITCHEN (ADDITIONAL CABINETS AND APPLIANCES MAY REQUIRE ADDITIONAL FLOOR SPACE)	SF	5.23	5.30	10.53	15.80					
	SF = FLOOR AREA OF ADDITIONAL SPACE										
SUBSTITUTE CABINET QUALITY	SUBSTITUTE ECONOMY QUALITY CABINETS										
	SMALL KITCHEN 1 **DEDUCT**	EA	402.00	--	402.00	603.00					
	2 **DEDUCT**	EA	876.00	--	876.00	1,314.00					
	3 **DEDUCT**	EA	884.00	--	884.00	1,326.00					
	4 **DEDUCT**	EA	634.00	--	634.00	951.00					
	LARGE KITCHEN 1 **DEDUCT**	EA	964.00	--	964.00	1,446.00					
	2 **DEDUCT**	EA	1448.00	--	1,448.00	2,172.00					
	3 **DEDUCT**	EA	1010.00	--	1,010.00	1,515.00					
	4 **DEDUCT**	EA	1294.00	--	1,294.00	1,941.00					
	ISLAND KITCHEN 1 **DEDUCT**	EA	596.00	--	596.00	894.00					
	2 **DEDUCT**	EA	1532.00	--	1,532.00	2,298.00					
	3 **DEDUCT**	EA	1038.00	--	1,038.00	1,557.00					
	4 **DEDUCT**	EA	1128.00	--	1,128.00	1,692.00					
	SUBSTITUTE PREMIUM QUALITY CABINETS										
	SMALL KITCHEN 1 **ADD**	EA	384.00	--	384.00	576.00					
	2 **ADD**	EA	856.00	--	856.00	1,284.00					
	3 **ADD**	EA	858.00	--	858.00	1,287.00					
	4 **ADD**	EA	612.00	--	612.00	918.00					
	LARGE KITCHEN 1 **ADD**	EA	956.00	--	956.00	1,434.00					
	2 **ADD**	EA	1146.00	--	1,146.00	1,719.00					
	3 **ADD**	EA	988.00	--	988.00	1,482.00					
	4 **ADD**	EA	1244.00	--	1,244.00	1,866.00					
	ISLAND KITCHEN 1 **ADD**	EA	604.00	--	604.00	906.00					
	2 **ADD**	EA	1598.00	--	1,598.00	2,397.00					
	3 **ADD**	EA	1056.00	--	1,056.00	1,584.00					
	4 **ADD**	EA	1498.00	--	1,498.00	2,247.00					
ADD OR OMIT CABINETS	ADD OR OMIT BASE CABINETS										
	ECONOMY **ADD** OR **DEDUCT**	LF	61.00	15.00	76.00	114.00					
	BUILDER **ADD** OR **DEDUCT**	LF	95.00	15.00	110.00	165.00					
	PREMIUM **ADD** OR **DEDUCT**	LF	132.00	16.00	148.00	222.00					
	LF = FRONT OF CABINETS										
	ADD OR OMIT WALL CABINETS										
	ECONOMY **ADD** OR **DEDUCT**	LF	36.00	16.00	52.00	78.00					
	BUILDER **ADD** OR **DEDUCT**	LF	60.00	16.00	76.00	114.00					
	PREMIUM **ADD** OR **DEDUCT**	LF	80.00	16.00	96.00	144.00					
	LF = FRONT OF CABINETS										
BROOM CLOSET	SUBSTITUTE 18" X 24" BROOM CLOSET FOR BASE AND WALL CABINETS										
	ECONOMY **ADD**	EA	30.00	--	30.00	45.00					
	BUILDER **ADD**	EA	46.00	--	46.00	69.00					
	PREMIUM **ADD**	EA	82.00	--	82.00	123.00					
	ADD 18" X 24" BROOM CLOSET **ADD**	EA	273.00	49.00	322.00	483.00					

SPECIFICATIONS		UNIT	JOB COST			PRICE	LOCAL AREA MODIFICATION				
			MATLS	LABOR	TOTAL		MATLS	LABOR	TOTAL	PRICE	
COUNTER-TOP	SUBSTITUTE COUNTERTOP WITH 4" BACKSPLASH										
	CORIAN **ADD**	LF	62.00	16.00	78.00	117.00					
	BUTCHER BLOCK **ADD**	LF	30.00	12.00	42.00	63.00					
	CERAMIC TILE **ADD**	LF	--	12.00	12.00	18.00					
BULKHEAD	DRYWALL BULKHEAD FROM CEILING TO WALL CABINETS, MOULDING AT JOINT **ADD**	LF	1.80	9.50	11.30	16.95					
PLUMBING	SUBSTITUTE 33" X 21" CAST IRON PORCELAIN FINISHED DOUBLE BOWL SINK FOR SPECIFIED STAINLESS STEEL SINGLE BOWL **ADD**	EA	10.00	--	10.00	15.00					
	ADD ICEMAKER TO FACTORY EQUIPPED REFRIGERATOR **ADD**	EA	25.00	75.00	100.00	150.00					
	INSTALL INSTANT HOT AT SINK, INCLUDING ELECTRICAL HOOKUP **ADD**	EA	96.00	22.00	118.00	177.00					
	EXTEND EXISTING PLUMBING AND ROUGH IN FOR KITCHEN SINK, OVER FIVE FEET FROM EXISTING SINK LF = TOTAL DISTANCE FROM EXISTING SINK	LF	16.00	12.00	28.00	42.00					
ELECTRICAL	ADDITIONAL LIGHT AND SWITCH (DOES **NOT** INCLUDE COST OF FIXTURE) **ADD**	EA	24.00	50.00	74.00	111.00					
	ADDITIONAL DUPLEX OUTLET **ADD**	EA	13.00	25.00	38.00	57.00					
	SUBSTITUTE DIMMER SWITCH FOR SPECIFIED LIGHT AND SWITCH **ADD**	EA	11.00	3.00	14.00	21.00					
FLOOR	SUBSTITUTE KITCHEN FLOOR COVERING										
	CERAMIC TILE **ADD**	SF	2.63	9.31	11.94	17.91					
	12" X 12" MARBLE **ADD**	SF	5.94	8.26	14.20	21.30					
	T&G SELECT OAK FLOORING, SANDED AND FINISHED **ADD**	SF	2.01	1.66	3.67	5.51					
	12" X 12" SLATE **ADD**	SF	3.36	1.66	5.02	7.53					

OPEN PORCH

SPECIFICATIONS

PLANS AND PERMIT	PREPARE PLANS AND OBTAIN PERMITS AS REQUIRED, PERMIT FEE COST **NOT** INCLUDED
CONCRETE AND MASONRY	EXCAVATE TO 36" BELOW GRADE, FORM AND POUR CONTINUOUS CONCRETE FOOTINGS WITH REBAR BLOCK FOUNDATION WALL ON FOOTINGS TO 18" ABOVE GRADE, BACKFILL AND ROUGH GRADE AS REQUIRED 4" THICK CONCRETE FLOOR SLAB OVER GRAVEL BASE
WALL FRAMING	4" X 4" PRESSURE TREATED FIR OR PINE POSTS, 6'-0" O.C. 4" X 10" HEADERS ON POSTS AT PORCH PERIMETER 2" X 4" STUDS ON GABLE END WITH 1/2" PLYWOOD SHEATHING
ROOF FRAMING	GABLE TYPE ROOF WITH 12" OVERHANG, RAFTERS AND CEILING JOISTS AS REQUIRED 1/2" CDX PLYWOOD ROOF SHEATHING
ROOF COVERING, GUTTERS, FLASHING	#215 FIBERGLASS SHINGLES OVER #15 FELT PAPER ALUMINUM GUTTERS AND DOWNSPOUTS ALUMINUM FLASHING WHERE ROOF JOINS EXISTING HOUSE
EXTERIOR TRIM	TRIM HEADERS THREE SIDES WITH TIGHT KNOT PINE OR FIR TIGHT KNOT PINE OR FIR FASCIA AND SOFFIT RAKE AND RAKE MOULDING ON GABLE END
SIDING	VINYL SIDING ON GABLE END
CEILING	FINISH PORCH CEILING WITH 5/8" X 4" FIR BEADED CEILING WITH 3/4" CEILING COVE MOULDING AT HEADER INTERSECTIONS
CLEAN-UP	REMOVE TRASH FROM PREMISES, DUMPING FEE **NOT** INCLUDED
	NO PAINTING

SPECIFICATIONS		UNIT	JOB COST			PRICE	LOCAL AREA MODIFICATION			
			MATLS	LABOR	TOTAL		MATLS	LABOR	TOTAL	PRICE
OPEN PORCH	AS SPECIFIED	EA PLUS	1200.00	1350.00	2,550.00	3,825.00				
	SF = PORCH FLOOR	SF	6.00	5.50	11.50	17.25				

SPECIFICATIONS		UNIT	JOB COST			PRICE	LOCAL AREA MODIFICATION				
			MATLS	LABOR	TOTAL		MATLS	LABOR	TOTAL	PRICE	
DIFFERENT FOOTING DEPTH	*Depth of Bottom of Footing Below Grade*										
	12" **DEDUCT**	LF	3.68	10.28	13.96	20.94					
	24" **DEDUCT**	LF	1.84	5.13	6.97	10.46					
	48" **DEDUCT**	LF	1.83	5.12	6.95	10.43					
	LF = FOOTINGS										
SUBSTITUTE FLOOR SYSTEM	SUBSTITUTE MUDSILL, FLOOR JOISTS AND 3/4" T & G FIR OR PINE PORCH FLOORING FOR SPECIFIED CONCRETE **ADD**	SF	3.21	.34	3.55	5.33					
	SF = PORCH FLOOR										
DOOR	REMOVE EXISTING WINDOW AND WALL UNDER AND INSTALL 2-8 X 6-8 THREE PANEL, 4 LIGHTS 1-3/4" DOOR, FULLY TRIMMED, FROM HOUSE TO PORCH	EA	417.00	409.00	826.00	1,239.00					
	BREAK THROUGH EXISTING HOUSE WALL AND INSTALL 2-8 X 6-8 THREE PANEL, 4 LIGHTS 1-3/4" DOOR, FULLY TRIMMED, FROM HOUSE TO PORCH										
	FRAME WALL **ADD**	EA	440.00	480.00	920.00	1,380.00					
	BRICK & FRAME **ADD**	EA	440.00	518.00	958.00	1,437.00					
	BRICK & BLOCK **ADD**	EA	440.00	540.00	980.00	1,470.00					
SCREEN	ALUMINUM SCREENING ON EXISTING POSTS WITH PANEL STRIPS **ADD**	SF	1.42	1.34	2.76	4.14					
SCREEN DOOR, WOOD	WOOD FRAME SCREEN DOOR WITH ALUMINUM SCREENING, INCLUDING ALL NECESSARY HARDWARE FOR INSTALLATION **ADD**	EA	76.00	36.00	112.00	168.00					

BASEMENT RECREATION ROOM

SPECIFICATIONS	
PLANS AND PERMIT	PREPARE PLANS AND OBTAIN PERMITS AS REQUIRED, PERMIT FEE COST **NOT** INCLUDED
WALL FRAMING	1" X 3" FURRING, 16" O.C., AND/OR 2" X 3" STUDDING, 16" O.C., ON FOUR WALLS
DOOR	ONE 2-6 X 6-8 X 1-3/8" HOLLOW CORE FLUSH DOOR, INCLUDING HARDWARE AND PREFINISHED TRIM BOTH SIDES
ELECTRI-CAL	DUPLEX OUTLETS TO NATIONAL CODE ONE RECESSED CEILING FIXTURE AND SWITCH
WALLS	1/4" PREFINISHED PANELING ON FOUR WALLS @ $20.00 PER SHEET
TRIM	ON PANELING: 3-1/2" PREFINISHED CLAM BASE AND 1-5/8" PREFINISHED CEILING MOULDING TRIM ONE EXISTING WINDOW WITH PLYWOOD OR DRYWALL AND WOOD CORNER GUARD
CEILING	1" X 3" FURRING ACROSS JOISTS, 12" O.C., AND 12" X 24" PLAIN WHITE CEILING TILE @ .40 PSF RETAIL
CLEAN-UP	REMOVE TRASH FROM PREMISES, DUMPING FEE **NOT** INCLUDED
	NO PAINTING

SPECIFICATIONS		UNIT	JOB COST			PRICE	LOCAL AREA MODIFICATION			
			MATLS	LABOR	TOTAL		MATLS	LABOR	TOTAL	PRICE
RECREA-TION ROOM	AS SPECIFIED	EA PLUS	610.00	966.00	1,576.00	2,364.00				
	SF = FLOOR AREA	SF	4.29	3.13	7.42	11.13				

SPECIFICATIONS		UNIT	JOB COST			PRICE	LOCAL AREA MODIFICATION				
			MATLS	LABOR	TOTAL		MATLS	LABOR	TOTAL	PRICE	
INTERIOR PARTITION	BUILD COMPLETE PARTITION WITH FOLLOWING SPECIFICATIONS: • WALL FRAMING, 2 X 4 STUDS AND PLATES, 8 FOOT HIGH WALL • DUPLEX OUTLETS TO CODE BOTH SIDES OF WALL • 1/4" PREFINISHED PLYWOOD PANELING @ $15 PER SHEET BOTH SIDES OF WALL • 3-1/2" BASE AND SHOE BOTH SIDES OF WALL LF = LENGTH OF PARTITION	LF	19.00	21.00	40.00	60.00					
TEAR OUT OLD FLOOR, INSTALL NEW FLOOR	BREAK UP EXISTING FLOOR SLAB IN BASEMENT AND HAUL RUBBLE FROM PREMISES, INSTALL NEW BASEMENT FLOOR, 4" CONCRETE WITH PLASTIC VAPOR BARRIER UNDER	SF	1.66	4.25	5.91	8.87					
HEAT	INSTALL REGISTER IN EXISTING HEAT DUCT	EA	30.00	90.00	120.00	180.00					
INSULATION	INSULATE EXTERIOR WALLS AND/OR CEILING										
	3/4" FOIL FACED FOAM	SF	.20	.19	.39	.59					
	3-1/2" BLANKET	SF	.28	.20	.48	.72					

CONVERSION -- ATTACHED GARAGE TO ROOM

SPECIFICATIONS

PLANS AND PERMIT	PREPARE PLANS AND OBTAIN PERMITS AS REQUIRED, PERMIT FEE COST **NOT** INCLUDED
TEAR-OUT	REMOVE EXISTING OVERHEAD GARAGE DOOR AND TRIM
FLOOR FRAMING	BUILD UP A LEVEL FLOOR ABOVE EXISTING CONCRETE WITH 2" X 4", 2" X 6" OR 2" X 8" SLEEPERS, 16" O.C. 3/4" T & G FIR PLYWOOD SUBFLOOR, GLUED AND NAILED
WALL FRAMING	FRAME IN FOR THREE WINDOWS IN EXISTING GARAGE DOOR OPENING WITH 2" X 4" STUDS AND 1/2" CDX PLYWOOD SHEATHING FURR EXISTING HOUSE WALL WITH 1" X 3" FURRING
SIDING	8" HORIZONTAL VINYL SIDING, INSULATED, BELOW WINDOW OPENING
WINDOWS	THREE WOOD DOUBLE HUNG, DOUBLE GLAZED WINDOWS AND FRAMES, FULLY TRIMMED TRIM TWO EXISTING WINDOWS WITH CASINGS, STOOLS AND STOOL CAPS
HEATING	EXTEND HEAT FROM MAIN SYSTEM WITH WARM AIR HEAT DUCTS OR ELECTRIC BASEBOARD
ELECTRICAL	DUPLEX WALL OUTLETS TO CODE ON EXISTING SERVICE
INSULATION	3-1/2" FIBERGLASS BLANKET INSULATION IN EXTERIOR WALLS 6" FIBERGLASS BLANKET INSULATION IN FLOOR AND CEILING
INTERIOR WALLS	1/2" DRYWALL ON FOUR WALLS, TAPED AND FINISHED READY FOR PAINTING
CEILING COVERING	1/2" DRYWALL, TAPED AND FINISHED READY FOR PAINTING
INTERIOR TRIM	3-1/2" BASE AND SHOE MOULDING ON ALL FOUR WALLS
FLOOR COVERING	2-1/4" SELECT OAK FLOORING, SANDED AND FINISHED
CLEAN-UP	REMOVE TRASH FROM PREMISES, DUMPING FEE **NOT** INCLUDED
	NO PAINTING

SPECIFICATIONS		UNIT	JOB COST			PRICE	LOCAL AREA MODIFICATION			
			MATLS	LABOR	TOTAL		MATLS	LABOR	TOTAL	PRICE
CONVERT GARAGE TO ROOM	AS SPECIFIED ABOVE	EA PLUS	932.00	1308.00	2,240.00	3,360.00				
	SF = SQUARE FOOT FLOOR AREA OF NEW ROOM	SF	7.88	6.59	14.47	21.71				
	ADD OR OMIT TRIMMING ONE WOOD DOUBLE HUNG WINDOW **ADD** OR **DEDUCT**	EA	21.00	29.00	50.00	75.00				
	INSTALL ADDITIONAL VINYL SIDING ON FRONT, INCLUDING REMOVING EXISTING SIDING **ADD** SF = ADDITIONAL SIDING	SF	1.39	1.31	2.70	4.05				
	UNDERPIN EXISTING SLAB AND INTERMEDIATE POINTS WITH 24" X 24" CONCRETE FOOTINGS 48" BELOW GRADE **ADD**	EA	103.00	323.00	426.00	639.00				

SPECIFICATIONS

PLANS AND PERMIT	PREPARE PLANS AND OBTAIN PERMITS AS REQUIRED, PERMIT FEE COST **NOT** INCLUDED
TEAR-OUT	REMOVE EXISTING CORNER, END AND INTERMEDIATE POSTS, PORCH SCREENING AND CEILING COVERING
	INSTALL TEMPORARY SUPPORT FOR ROOF STRUCTURE
FLOOR FRAMING	BUILD UP A LEVEL FLOOR ABOVE EXISTING CONCRETE OR WOOD DECK WITH 2" X 4", 2" X 6" OR 2" X 8" SLEEPERS, 16" O.C.
	3/4" T & G FIR PLYWOOD SUBFLOOR, GLUED AND NAILED
WALL FRAMING	2" X 4" STUDS, 16" O.C., AND 1/2" CDX PLYWOOD SHEATHING AT PORCH PERIMETER
	FURR EXISTING HOUSE WALL WITH 1" X 3" FURRING
SIDING	8" HORIZONTAL VINYL SIDING, INSULATED
WINDOWS	ONE 2-8 X 4-6 WOOD DOUBLE HUNG, DOUBLE GLAZED WINDOW FOR EACH 100 SF OF LIVING AREA, FULLY TRIMMED
HEATING	EXTEND HEAT FROM MAIN SYSTEM WITH WARM AIR HEAT DUCTS OR ELECTRIC BASEBOARD
ELECTRI-CAL	DUPLEX WALL OUTLETS TO CODE ON EXISTING SERVICE
INSULA-TION	3-1/2" FIBERGLASS BLANKET INSULATION IN EXTERIOR WALLS
	6" FIBERGLASS BLANKET INSULATION IN FLOOR AND CEILING
INTERIOR WALL COVERING	1/2" DRYWALL ON WALLS, TAPED AND FINISHED READY FOR PAINTING
CEILING COVERING	1/2" DRYWALL, TAPED AND FINISHED READY FOR PAINTING
INTERIOR TRIM	3-1/2" BASE AND SHOE MOULDING ON ALL FOUR WALLS
FLOOR COVERING	2-1/4" SELECT OAK FLOORING, SANDED AND FINISHED
CLEAN-UP	REMOVE TRASH FROM PREMISES, DUMPING FEE **NOT** INCLUDED
	NO PAINTING

SPECIFICATIONS		UNIT	JOB COST			PRICE	LOCAL AREA MODIFICATION			
			MATLS	LABOR	TOTAL		MATLS	LABOR	TOTAL	PRICE
CONVERT OPEN PORCH TO ROOM	AS SPECIFIED ABOVE	EA	490.00	1350.00	1,840.00	2,760.00				
	SF = SQUARE FOOT FLOOR AREA OF NEW ROOM	PLUS SF	11.21	8.39	19.60	29.40				
EXTRAS & ALLOW-ANCES	ADD OR OMIT ONE WOOD DOUBLE HUNG WINDOW **ADD** OR **DEDUCT**	EA	193.00	59.00	252.00	378.00				
	UNDERPIN EXISTING CON-CRETE SLAB AND INTER-MEDIATE POINTS WITH 24" X 24" CONCRETE FOOTINGS 48" BELOW GRADE **ADD**	EA	103.00	323.00	426.00	639.00				
	OMIT REMOVAL OF PORCH FLOORING **DEDUCT** SF = FLOORING	SF	–	.50	.50	.75				

CONVERSION -- SINGLE APARTMENT

SPECIFICATIONS	
	CONVERT SINGLE FLOOR (OR BASEMENT) OF EXISTING HOUSE TO SINGLE APARTMENT AS SPECIFIED BELOW:
PLANS & PERMIT	PREPARE PLANS AND OBTAIN PERMITS AS REQUIRED, PERMIT FEE COST **NOT** INCLUDED
TEAR-OUT	REMOVE EXISTING NON-BEARING WALLS
	REMOVE WALL COVERING FROM BEARING WALLS
FRAMING	FRAME INTERIOR PARTITIONS ACCORDING TO PLAN
	FURR EXTERIOR WALLS WITH 1" X 3" FURRING STRIPS
DOORS	NEW PAINT GRADE INTERIOR FLUSH DOORS, 1-3/8"
	TRIM DOORS WITH RANCH CASING
WINDOWS	RE-TRIM WINDOWS WITH RANCH CASING, STOP, STOOL AND APRON
PLUMBING	FULL BATH WITH LAVATORY, TUB AND W.C., PLUMBING FOR KITCHEN SINK AND DISPOSAL
HEATING & AIR CONDITIONING	SEPARATE FORCED AIR HEAT AND COOLING SYSTEM, ELECTRIC OR GAS, INCLUDING DUCTWORK AND REGISTERS
ELECTRICAL	150-AMP SERVICE WITH CIRCUIT BREAKERS, WIRE APARTMENT TO CODE, INCLUDING WALL OUTLETS, SWITCHES, FIXTURES, KITCHEN APPLIANCE OUTLETS
INSULATION	INSULATE EXTERIOR WALL WITH ALUMINUM FOIL ON FURRING STRIPS
INTERIOR WALLS AND CEILINGS	1/2" DRYWALL ON ALL WALLS, 5/8" DRYWALL ON ALL CEILINGS: TAPE AND FINISH 3 COATS AND SAND, READY FOR PAINTING
	CERAMIC TILE WITH MASTIC 6 FEET ABOVE FLOOR IN BATHTUB AREA ONLY
MILLWORK, TRIM	RANCH BASE AND OAK SHOE THROUGHOUT
	LINEN CLOSET 24" DEEP, 36" WIDE WITH FULL DEPTH SHELVES
	CLOSET TRIM IN BEDROOM CLOSET
CABINETS AND APPLIANCES	10 LF OF UPPER AND LOWER KITCHEN CABINETS (RANGE, SINK AND REFRIGERATOR AREAS INCLUDED IN MEASUREMENT)
	COUNTERTOP WITH STAINLESS STEEL SINK AND DISPOSAL
	NEW GAS OR ELECTRIC RANGE
	NEW 30-GALLON GAS OR ELECTRIC HOT WATER HEATER
	DUCT TYPE KITCHEN FAN, HOOD AND VENT TO OUTSIDE
	BATHROOM FAN, VENTED TO OUTSIDE
SPECIALTIES	HINGED DOOR MEDICINE CABINET WITH TWO FLUORESCENT LIGHTS
	PAPER HOLDER AND TOWEL BAR IN BATHROOM
FLOOR	VINYL COMPOSITION TILE IN KITCHEN, CERAMIC TILE IN BATHROOM, AND $12/YD CARPET WITH PAD IN ALL OTHER ROOMS
PAINTING	PAINT ALL WALLS AND CEILINGS, 2 COATS FLAT PAINT
	PAINT ALL TRIM, BATH AND KITCHEN, WITH 2 COATS SEMI-GLOSS
CLEAN-UP	CLEAN UP ALL RUBBISH DURING AND AT COMPLETION OF JOB AND REMOVE TO DUMPING GROUND WITHIN FIVE MILES
	DUMPSTER AND DUMPING FEE **NOT** INCLUDED

SPECIFICATIONS		UNIT	JOB COST			PRICE	LOCAL AREA MODIFICATION			
			MATLS	LABOR	TOTAL		MATLS	LABOR	TOTAL	PRICE
SINGLE APART-MENT, AS SPECIFIED	*NOTE:* SF = SQUARE FOOT LIVING AREA, INCLUDING UTILITY ROOM									
	EFFICIENCY, ONE ROOM, ONE BATHROOM	EA PLUS SF	4750.00 10.90	6560.00 6.60	11,310.00 17.50	16,965.00 26.25				
	ONE BEDROOM, KITCHEN AND LIVING ROOM, ONE BATHROOM	EA PLUS SF	4750.00 11.90	6560.00 8.23	11,310.00 20.13	16,965.00 30.20				
	TWO BEDROOMS, KITCHEN AND LIVING ROOM, ONE BATHROOM	EA PLUS SF	4750.00 12.27	6560.00 8.87	11,310.00 21.14	16,965.00 31.71				
ALLOW-ANCES	ELIMINATE FURRING AND FOIL INSULATION AND IN-STALL 3-1/2" BLANKET INSU-LATION ON EXISTING EXTE-RIOR STUD WALLS **DEDUCT** SF = SQUARE FOOT WALLS	SF	.11	.20	.31	.47				
	ELIMINATE FURRING, INSULA-TION AND DRYWALL ON EXISTING PLASTERED EX-TERIOR WALLS -- PATCH EXISTING PLASTER WHERE REQUIRED **DEDUCT** SF = SQUARE FOOT WALLS	SF	.48	.80	1.28	1.92				

CONVERSION -- FULL HOUSE TO APARTMENTS

SPECIFICATIONS	
	CONVERT EXISTING ONE-FAMILY HOUSE TO FLATS (BASEMENT UNFINISHED) ACCORDING TO THE FOLLOWING SPECIFICATIONS:
PLANS & PERMIT	PREPARE PLANS AND OBTAIN PERMITS AS REQUIRED, PERMIT FEE COST **NOT** INCLUDED
TEAR-OUT AND DEMOLITION	GUT INTERIOR, LEAVING ONLY STUDS, FLOORS, STAIRS AND CHIMNEY
	REMOVE ALL PLUMBING, HEATING SYSTEM, WIRING AND ALL PLASTER EXCEPT ON INSIDE OF EXTERIOR MASONRY WALLS
MASONRY	INSTALL NEW FLUE IN EXISTING CHIMNEY: TEAR OUT PORTION OF CHIMNEY FROM INSIDE HOUSE, INSTALL NEW TERRA COTTA FLUE LINING, RE-BRICK
FRAMING	RE-BUILD STUD PARTITIONS ACCORDING TO PLANS
ROOFING, GUTTERS	INSTALL NEW ALUMINUM GUTTERS AND DOWNSPOUTS
DOORS	NEW EXTERIOR DOOR, 3-0 X 6-8, SOLID CORE, FLUSH
	SOLID CORE 1-3/4" DOORS AT ENTRANCE AND REAR OF EACH APARTMENT
	NEW INTERIOR DOORS, BIRCH FLUSH, 1-3/8"
	SOLID CORE 1-3/4" DOOR ON FURNACE ROOM ENCLOSURE
	TRIM ALL DOORS WITH RANCH CASING AND STOPS
WINDOWS	RE-TRIM ALL WINDOWS WITH RANCH CASING, STOPS, STOOL AND APRON
PLUMBING	IN EACH APARTMENT: BATHROOM WITH LAVATORY, TUB AND W.C., AND KITCHEN PLUMBING WITH NEW WATER PIPING AND STACK
HEATING & AIR CONDITIONING	NEW FORCED AIR HEAT AND COOLING SYSTEM, ELECTRIC OR GAS, INCLUDING HUMIDIFIER, ALL DUCT-WORK, REGISTERS FOR EACH APARTMENT (IF HEATING BY GAS, VENT TO OUTSIDE)
ELECTRICAL	SEPARATE 100-AMP SERVICE FOR EACH APARTMENT WITH CIRCUIT BREAKERS
	RE-WIRE HOUSE TO CODE, INCLUDING WALL OUTLETS, SWITCHES, FIXTURES
INSULATION	6" BLANKET INSULATION IN TOP FLOOR CEILING
	3-1/2" BLANKET INSULATION BETWEEN FLOORS
	ALUMINUM FOIL INSULATION ON FURRED EXTERIOR WALL
INTERIOR WALLS AND CEILINGS	5/8" FIRECODE DRYWALL ON ENTIRE BASEMENT CEILING, WALLS OF FURNACE ENCLOSURES, BASE-MENT BEAMS AND COLUMNS, HALLWAYS
	1/2" DRYWALL ON ALL OTHER WALLS AND CEILINGS, ALL DRYWALL FINISHED, 3 COATS, READY FOR PAINTING
	CERAMIC TILE ON WALLS OF BATHS, 4 FEET ABOVE THE FLOOR (6 FEET ABOVE IN BATHTUB AREAS)
MILLWORK, TRIM, STAIRS	RANCH BASE AND OAK SHOE THROUGHOUT HOUSE
	LINEN CLOSET 24" DEEP, 36" WIDE IN EACH APARTMENT
	CLOSET TRIM IN ALL BEDROOM CLOSETS
	REPAIR MAIN STAIRS, REPLACE HANDRAILS, BALUSTERS, ETC.
	NEW STAIRS TO BASEMENT, OR ELIMINATE AND CLOSE IN OPENING

SPECIFICATIONS

CABINETS & APPLIANCES	SPECIFICATIONS FOR FULL HOUSE CONVERSION, CONTINUED: INSTALL IN EACH APARTMENT: 12 LF OF UPPER AND LOWER KITCHEN CABINETS (RANGE, SINK AND REFRIGERATOR AREAS INCLUDED IN MEASUREMENT) BROOM CABINET COUNTERTOP, STAINLESS STEEL SINK, DISPOSAL NEW GAS OR ELECTRIC RANGE NEW GAS OR ELECTRIC HOT WATER HEATER DUCT TYPE KITCHEN FAN AND HOOD, VENTED TO OUTSIDE
SPECIALTIES	HINGED DOOR MEDICINE CABINETS IN BATHROOM WITH FLUORESCENT LIGHTS EACH SIDE OF MIRROR
FLOOR COVERING	CERAMIC TILE FLOOR IN BATHROOMS INLAID ROLL FLOORING IN KITCHENS $12/YD CARPET WITH PAD IN OTHER ROOMS
PAINTING	PAINT EXTERIOR TRIM, CORNICE, WINDOWS, DOORS, TWO COATS PAINT INTERIOR WALLS AND CEILINGS, DOORS, TRIM, TWO COATS
CLEAN-UP AND HAULING	CHUTE DEMOLITION RUBBISH TO TRUCK FROM BUILDING CLEAN UP TRASH DURING AND AT COMPLETION OF JOB DUMPSTER AND DUMPING FEE **NOT** INCLUDED

SPECIFICATIONS		UNIT	JOB COST			PRICE	LOCAL AREA MODIFICATION			
			MATLS	LABOR	TOTAL		MATLS	LABOR	TOTAL	PRICE
FULL HOUSE CONVERSION	TWO STORY HOUSE (2 APARTMENTS)	EA PLUS SF	11,260.00 16.17	9,220.00 13.23	20,480.00 29.40	30,720.00 44.10				
	THREE STORY HOUSE (3 APARTMENTS)	EA PLUS SF	16,800.00 16.17	13,900.00 13.23	30,700.00 29.40	46,050.00 44.10				
	FOUR STORY HOUSE (4 APARTMENTS) SF = TOTAL LIVING AREA OF HOUSE	EA PLUS SF	22,500.00 16.17	18,500.00 13.23	41,000.00 29.40	61,500.00 44.10				
ALLOWANCES	ELIMINATE FURRING AND FOIL INSULATION AND INSTALL 4" BLANKET INSULATION ON EXISTING EXTERIOR STUD WALLS **DEDUCT** SF = WALLS	SF	.11	.20	.31	.47				
	ELIMINATE FURRING, INSULATION AND DRYWALL ON EXISTING PLASTERED EXTERIOR WALLS — PATCH EXISTING PLASTER WHERE REQUIRED **DEDUCT** SF = WALLS	SF	.48	.80	1.28	1.92				

RENOVATION -- FULL HOUSE

SPECIFICATIONS	
PLANS & PERMIT	RENOVATE COMPLETE HOUSE ACCORDING TO THE FOLLOWING SPECIFICATIONS:
	PREPARE PLANS AND OBTAIN PERMITS AS REQUIRED, PERMIT FEE COST **NOT** INCLUDED
TEAR-OUT AND DEMOLITION	GUT INTERIOR, LEAVING ONLY STUDS, FLOORS, STAIRS AND CHIMNEY
	REMOVE ALL PLUMBING, HEATING SYSTEM, WIRING AND ALL PLASTER EXCEPT ON INSIDE OF EXTERIOR MASONRY WALLS
MASONRY	INSTALL NEW FLUE IN EXISTING CHIMNEY: TEAR OUT PORTION OF CHIMNEY FROM INSIDE HOUSE, INSTALL NEW TERRA COTTA FLUE LINING, RE-BRICK
FRAMING	RE-BUILD STUD PARTITIONS ACCORDING TO PLANS
	FURR EXTERIOR WALLS
ROOFING, GUTTERS	INSTALL NEW ALUMINUM GUTTERS AND DOWNSPOUTS
DOORS	NEW EXTERIOR DOOR, 3-0 X 6-8, SOLID CORE, FLUSH, 1-3/4"
	NEW REAR DOOR, 2-8 X 6-8, TWO PANELS, FOUR LIGHTS, 1-3/4"
	NEW PRE-HUNG INTERIOR DOORS, FLUSH, 1-3/8"
	TRIM ALL DOORS WITH RANCH CASING AND STOP, INCLUDING ALL HARDWARE
WINDOWS	RE-TRIM ALL WINDOWS WITH RANCH CASING, STOPS, STOOL AND APRON
PLUMBING	POWDER ROOM WITH LAVATORY AND W.C.
	TWO BATHS WITH LAVATORY, TUB AND W.C.
	NEW KITCHEN SINK AND DISPOSAL
	ALL NEW PLUMBING PIPES THROUGHOUT HOUSE
	NEW GAS OR ELECTRIC HOT WATER HEATER
HEATING & AIR CONDITIONING	NEW FORCED AIR HEAT AND COOLING SYSTEM, ELECTRIC OR GAS, INCLUDING HUMIDIFIER, ALL DUCT-WORK, REGISTERS
ELECTRICAL	HEAVY UP SERVICE TO 200 AMPS WITH CIRCUIT BREAKERS
	RE-WIRE HOUSE TO CODE, INCLUDING OUTLETS, SWITCHES, FIXTURES
INSULATION	6" BLANKET INSULATION IN TOP FLOOR CEILING
	ALUMINUM FOIL INSULATION ON FURRED EXTERIOR WALL
INTERIOR WALLS AND CEILINGS	1/2" DRYWALL ON ALL WALLS AND CEILINGS, FINISHED 3 COATS, READY FOR PAINTING
	CERAMIC TILE ON WALLS OF TWO BATHS, 4 FEET ABOVE THE FLOOR AND 6 FEET ABOVE THE FLOOR IN BATHTUB AREA
MILLWORK, TRIM, STAIRS	RANCH BASE AND OAK SHOE THROUGHOUT HOUSE
	LINEN CLOSET 24" DEEP, 36" WIDE WITH SHELVES FULL DEPTH OF CLOSET
	CLOSET TRIM IN ALL BEDROOM CLOSETS
	REPAIR MAIN STAIRS, REPLACE HANDRAILS AND BALUSTERS

SPECIFICATIONS

CABINETS & APPLIANCES	SPECIFICATIONS FOR FULL HOUSE RENOVATION, CONTINUED: 15 LF OF UPPER AND LOWER KITCHEN CABINETS (RANGE, SINK AND REFRIGERATOR AREAS INCLUDED IN MEASUREMENT) BROOM CABINET COUNTERTOP, STAINLESS STEEL SINK, DISPOSAL NEW GAS OR ELECTRIC RANGE REFRIGERATOR DUCT TYPE KITCHEN FAN AND HOOD, VENTED TO OUTSIDE CLOTHES WASHER/DRYER
SPECIAL-TIES	HINGED DOOR MEDICINE CABINETS IN BATHROOMS AND POWDER ROOM WITH FLUORESCENT LIGHTS EACH SIDE OF MIRRORS
FLOOR COVERING	CERAMIC TILE FLOOR IN BATHROOMS VINYL COMPOSITION TILE IN POWDER ROOM AND KITCHEN $12/YD CARPET WITH PAD IN OTHER ROOMS
PAINTING	PAINT EXTERIOR TRIM, CORNICE, WINDOWS, DOORS, TWO COATS PAINT INTERIOR WALLS AND CEILINGS, DOORS, TRIM, TWO COATS
CLEAN-UP AND HAULING	CHUTE DEMOLITION RUBBISH TO TRUCK FROM BUILDING CLEAN UP TRASH DURING AND AT COMPLETION OF JOB DUMPSTER AND DUMPING FEE **NOT** INCLUDED

SPECIFICATIONS		UNIT	JOB COST			PRICE	LOCAL AREA MODIFICATION			
			MATLS	LABOR	TOTAL		MATLS	LABOR	TOTAL	PRICE
FULL HOUSE RENOVA-TION	ONE STORY HOUSE, 2 BATHS	EA PLUS	16,260.00	16,440.00	32,700.00	49,050.00				
		SF	6.20	10.22	16.42	24.63				
	TWO STORY HOUSE, POWDER ROOM AND 2 BATHS	EA PLUS	15,312.00	11,052.00	26,364.00	39,546.00				
		SF	7.00	14.30	21.30	31.95				
	THREE STORY HOUSE, POW-DER ROOM & 4 BATHS SF = TOTAL LIVING AREA OF HOUSE (**NOT** INCLUDING BASEMENT UNLESS IT INCLUDES FINISHED LIVING AREA)	EA PLUS SF	15,770.00 7.00	13,000.00 14.30	28,770.00 21.30	43,155.00 31.95				
ALLOW-ANCES	ELIMINATE FURRING AND FOIL INSULATION AND INSTALL 3-1/2" BLANKET INSULATION ON EXISTING EXTERIOR STUD WALLS **DEDUCT** SF = WALLS	SF	.11	.20	.31	.47				
	ELIMINATE FURRING, INSULA-TION AND DRYWALL ON EXIST-ING PLASTERED EXTERIOR WALLS — PATCH EXISTING PLASTER WHERE REQUIRED **DEDUCT** SF = WALLS	SF	.48	.80	1.28	1.92				

PORCH AND SLAB ALTERATIONS

SPECIFICATIONS		UNIT	JOB COST			PRICE	LOCAL AREA MODIFICATION			
			MATLS	LABOR	TOTAL		MATLS	LABOR	TOTAL	PRICE
REMOVE ROOF AND FLOOR	• REMOVE EXISTING WOOD PORCH, INCLUDING ALL ROOF AND FLOOR FRAMING AND STEPS, REMOVE DEBRIS FROM PREMISES • PATCH BRICKWORK ON FRONT WHERE WOOD BEAMS ATTACH TO HOUSE, TOOTHING IN TO MATCH EXISTING BRICKWORK AS CLOSELY AS POSSIBLE • DIG, FORM & POUR REINFORCED CONCRETE PLATFORM WITH 4-0 WIDE STEPS SF = TOP SURFACE NEW PLATFORM & STEPS	EA PLUS SF	270.00 6.00	530.00 11.00	800.00 17.00	1,200.00 25.50				
REMOVE ROOF	REMOVE EXISTING PORCH ROOF, INCLUDING SUPPORT POSTS AND SCREENING — PATCH BRICKWORK ON FRONT WHERE ROOF ATTACHES TO HOUSE, TOOTHING IN TO MATCH EXISTING AS CLOSELY AS POSSIBLE SF = PORCH ROOF	EA PLUS SF	50.00 .50	150.00 1.50	200.00 2.00	300.00 3.00				
REPLACE RAILING	REMOVE EXISTING WOOD RAILING, INSTALL ECONOMY GRADE ORNAMENTAL IRON RAILING LF = RAILING	LF	15.00	5.00	20.00	30.00				
REMOVE FLOOR AND REPLACE	REMOVE EXISTING PORCH FLOOR, LEAVING PILLARS TO SUPPORT ROOF, REMOVE DEBRIS FROM PREMISES — DIG, FORM AND POUR REINFORCED CONCRETE PLATFORM WITH 4-0 WIDE STEPS SF = TOP SURFACE NEW PLATFORM AND STEPS	EA PLUS SF	320.00 4.50	510.00 9.10	830.00 13.60	1,245.00 20.40				
REPAIR METAL PLATFORM	WELD BROKEN PARTS TO METAL PORCH, LEVEL PLATFORM, STEPS AND RAILING AND ANCHOR IN PLACE	EA	25.00	129.00	154.00	231.00				
MUDJACK	RAISE EXISTING SETTLED CONCRETE WALKWAY OR SLAB BY FILLING VOID UNDER WITH PUMPED CONCRETE GROUT: • DRILL HOLES IN CONCRETE SLAB • PUMP CONCRETE GROUT UNDER PRESSURE THRU HOLES, FILLING VOID UNDER SLAB AND COMPACTING EARTH BELOW • WHEN SLAB RAISED AS MUCH AS 8" TO ORIGINAL LEVEL, PATCH DRILLED HOLES WITH MORTAR MIX SF = SLAB	EA PLUS SF	160.00 .50	430.00 1.25	590.00 1.75	885.00 2.63				

SPECIFICATIONS		UNIT	JOB COST			PRICE	LOCAL AREA MODIFICATION				
			MATLS	LABOR	TOTAL		MATLS	LABOR	TOTAL	PRICE	
AREAWAY FOR BASEMENT APART- MENT	• DIG OUT AND REMOVE DIRT AS REQUIRED FOR AREAWAY 5 FEET DEEP AND UP TO 18 FEET WIDE • POUR CONCRETE FOOT-INGS FOR 8" MASONRY WALL • FORM AND POUR REIN-FORCED (6 X 6 WWM) CONCRETE SLAB 4" THICK WITH BASEMENT DRAIN CONNECTED WITH SEWER • INSTALL CONCRETE STEPS TO GRADE, UP TO 8 STEPS 4 FEET WIDE • INSTALL 4" BRICK AND 4" BLOCK WALL WITH TOP ROLOK COURSE AROUND PERIMETER OF AREAWAY WALL EXTENDING FROM FOOTINGS TO GRADE SF = AREAWAY FLOOR	EA PLUS SF	220.00 15.00	380.00 23.00	600.00 38.00	900.00 57.00				
CLEAR AREAWAY	REMOVE OBSTRUCTION FROM BASEMENT AREAWAY DRAIN	EA	--	100.00	100.00	150.00				
WEEP HOLES IN EXISTING RETAINING WALL	• CUT 1 SF HOLE IN EXIST-ING RETAINING WALL • DIG OUT TO HOLE BEHIND WALL AND INSTALL 1 CF CRUSHED STONE • GROUT IN 3" PLASTIC PIPE 8" BRICK & BLOCK 12" BRICK & BLOCK 12" SOLID BLOCK EA = EACH WEEPHOLE	EA EA EA	7.00 9.00 10.00	19.00 21.00 24.00	26.00 30.00 34.00	39.00 45.00 51.00				
FRENCH DRAIN	• BREAK CONCRETE BASE-MENT FLOOR ALONG INSIDE OF FOOTINGS AND EXCAVATE TO DEPTH OF FOOTINGS • INSTALL PLASTIC PIPE ALONG INSIDE OF FOOT-INGS WITH 1/8" SLOPE TO SUMP IN BASEMENT FLOOR • DISCHARGE WATER WITH 1/3 HP SUMP PUMP TO STORM SEWER OR OUT-SIDE TO DAYLIGHT • BACKFILL AND PATCH FLOOR LF = FOOTINGS	EA PLUS LF	236.00 8.00	290.00 12.00	526.00 20.00	789.00 30.00				

FIREWALL, SOUND BARRIER, SHOWER PAN

SPECIFICATIONS		UNIT	JOB COST			PRICE	LOCAL AREA MODIFICATION				
			MATLS	LABOR	TOTAL		MATLS	LABOR	TOTAL	PRICE	
FIREWALL AROUND BOILER & HOT WATER HEATER	BUILD WALL ENCLOSING 6' X 8' SPACE AT END OF EXISTING BASEMENT • 8" X 8" X 16" BLOCK **OR** 2 X 4 STUD WALL AND 5/8" FIRECODE DRYWALL BOTH SIDES OF WALL • CLASS B FIREDOOR, APPROX. 2-6 X 6-8, WITH AUTOMATIC CLOSER • 5/8" FIRECODE DRYWALL ON CEILING, TAPED AND SPACKLED, FILL IN AROUND PIPES AS THOROUGHLY AS POSSIBLE • 8" X 12" DUCT FROM FURNACE ROOF 10 FEET OR LESS TO OUTSIDE FOR VENTILATION • PULL CHAIN CEILING LIGHT	EA	550.00	1250.00	1,800.00	2,700.00					
SOUND BARRIER	INSTALL SOUND BARRIER BETWEEN BASEMENT CEILING AND FIRST FLOOR APARTMENT WITH FULL THICK BLANKET INSULATION, RESILIENT CHANNEL 24" O.C. AND 1/2" DRYWALL, TAPED AND SPACKLED, READY FOR PAINTING	SF	1.50	2.50	4.00	6.00					
REPLACE LEAD SHOWER PAN	REPLACE EXISTING LEAD PAN IN SHOWER STALL • TEAR OUT TILE FLOOR AND WALLS UP THREE ROWS FROM BASE • SUPPLY AND INSTALL NEW DRAIN ASSEMBLY AND COVER • REPLACE EXISTING LEAD PAN WITH VINYL OR RUBBER PAN • INSTALL NEW TILE ON FLOOR AND UP THREE ROWS IN MUD, TILE TO MATCH EXISTING AS CLOSELY AS POSSIBLE	EA	226.00	470.00	696.00	1,044.00					

SPECIFICATIONS		UNIT	JOB COST			PRICE	LOCAL AREA MODIFICATION				
			MATLS	LABOR	TOTAL		MATLS	LABOR	TOTAL	PRICE	
CLOSE IN EXTERIOR DOORWAY OR WINDOW	REMOVE DOOR OR WINDOW AND CLOSE IN WITH STUDS, SHEATHING, SIDING, INSULATION, FURRING AND/OR MASONRY AS REQUIRED — FINISH INSIDE WITH DRYWALL AND BASE AND SHOE IF REQUIRED — TOOTH IN BRICK TO EXISTING WHEN BRICK IS SPECIFIED										
	FRAMED WALL WITH SIDING	EA	48.00	260.00	308.00	462.00					
	FRAME WITH BRICK VENEER	EA	110.00	496.00	606.00	909.00					
	BRICK AND BLOCK	EA	132.00	588.00	720.00	1,080.00					
CLOSE IN INTERIOR DOORWAY	REMOVE EXISTING INTERIOR DOORWAY AND CLOSE IN WALL WITH STUDS AND DRYWALL AND BASE AND SHOE BOTH SIDES OF WALL	EA	45.00	235.00	280.00	420.00					
CLOSE IN FIREPLACE	CLOSE IN EXISTING FIREPLACE WITH STUDS, DRYWALL AND BASE	EA	28.00	92.00	120.00	180.00					
	SAME AS ABOVE, AND REMOVE BRICK FACING	EA	30.00	144.00	174.00	261.00					
REMOVE FRONT HEARTH	REMOVE FRONT HEARTH AND PATCH WITH HARDWOOD FLOORING, SANDED AND FINISHED	EA	45.00	69.00	114.00	171.00					
CLOSE IN TRANSOM	REMOVE EXISTING TRANSOM OVER DOOR AND CLOSE IN WITH DRYWALL	EA	18.00	36.00	54.00	81.00					
PLASTER PATCHING	IF ANY OF THE ABOVE CLOSE-INS REQUIRE PLASTER PATCHING **ADD**	EA	18.00	56.00	74.00	111.00					

BREAK THROUGH WALLS FOR NEW DOORWAYS

SPECIFICATIONS		UNIT	JOB COST			PRICE	LOCAL AREA MODIFICATION			
			MATLS	LABOR	TOTAL		MATLS	LABOR	TOTAL	PRICE
NEW DOOR IN EXTERIOR WALL	BREAK OPENING IN EXISTING FRAME WALL, INSTALL HEADER OR LINTEL AS REQUIRED AND INSTALL DOOR, INCLUDING FRAME, TRIM AND HARDWARE									
	SOLID CORE BIRCH, PRE-HUNG, 3-0 X 6-8	EA	316.00	480.00	796.00	1,194.00				
	COLONIAL SIX-PANEL, PRE-HUNG, 3-0 X 6-8	EA	442.00	480.00	922.00	1,383.00				
	4-LIGHT SIDE OR REAR DOOR, 2-8 X 6-8	EA	401.00	485.00	886.00	1,329.00				
	WOOD GLIDER, BUILDING QUALITY, 6-0 X 6-8	EA	807.00	639.00	1,446.00	2,169.00				
	ALUMINUM GLIDER, 6-0 X 6-8	EA	513.00	627.00	1,140.00	1,710.00				
	SAME AS ABOVE, REMOVE EXISTING WINDOW AND WALL UNDER, OMIT HEADER OR LINTEL AND INSTALL DOOR **DEDUCT**	EA	25.00	71.00	96.00	144.00				
	SAME AS ABOVE, IN BRICK VENEER WALL **ADD**	EA	--	38.00	38.00	57.00				
	SAME AS ABOVE, IN BRICK AND BLOCK WALL **ADD**	EA	--	60.00	60.00	90.00				
NEW DOOR IN INTERIOR WALL	BREAK THROUGH EXISTING NON-BEARING INTERIOR WALL AND HANG DOOR IN OPENING, INCLUDING DOOR, STOPS, TRIM, HINGES AND PASSAGE SET									
	BIRCH FLUSH, HOLLOW CORE, 2-8 X 6-8	EA	133.00	271.00	404.00	606.00				
	6 PANEL COLONIAL, 2-8 X 6-8	EA	217.00	271.00	488.00	732.00				
DOUBLE DOORS	DOUBLE DOORS: DOUBLE COST OF ONE DOOR AND **ADD**	EA	30.00	20.00	50.00	75.00				
	SAME AS ABOVE IN LOAD-BEARING WALL **ADD**	EA	20.00	120.00	140.00	210.00				
REMOVE WALL AND MAKE CASED OPENING	BREAK THROUGH EXISTING BEARING OR NON-BEARING WALL AND MAKE NEW CASED OPENING WITH JAMBS AND 2 SIDES OF DOOR TRIM *Non-Bearing Wall*									
	3-0 OPENING	EA	68.00	276.00	344.00	516.00				
	6-0 OPENING	EA	84.00	352.00	436.00	654.00				
	Bearing Wall									
	3-0 OPENING	EA	84.00	388.00	472.00	708.00				
	6-0 OPENING	EA	124.00	536.00	660.00	990.00				

SPECIFICATIONS		UNIT	JOB COST			PRICE	LOCAL AREA MODIFICATION			
			MATLS	LABOR	TOTAL		MATLS	LABOR	TOTAL	PRICE
REMOVE WALL AND MAKE OPENING	BREAK THROUGH EXISTING LOAD-BEARING INTERIOR WALL AND MAKE NEW OPENING WITH MINIMUM OF 6" WALL SHOWING ON EACH SIDE AND 6" HEADER SHOWING ABOVE, PATCH WALLS, CEILING, MOULDINGS AND FLOOR TO MATCH EXISTING, 10 TO 16 LF OPENING EA = EACH JOB	EA	290.00	430.00	720.00	1,080.00				
	SAME AS ABOVE, WITH FLUSH CEILING AND 6" MINIMUM WALLS SHOWING EACH SIDE **ADD**	EA	30.00	120.00	150.00	225.00				
	SAME AS ABOVE, WITH FLUSH CEILING AND FLUSH WALLS **ADD**	EA	36.00	230.00	266.00	399.00				
DOOR ADJUSTMENT	CUT OFF BOTTOM OF SOLID WOOD DOOR FOR CARPETING CLEARANCE INTERIOR	EA	--	12.00	12.00	18.00				
	EXTERIOR	EA	--	42.00	42.00	63.00				
	CUT OFF BOTTOM RAIL OF HOLLOW CORE DOOR AND RAISE BOTTOM RAIL INSIDE THE VENEER	EA	--	22.00	22.00	33.00				
	RE-SET SCREWS IN HINGES EA = UP TO 3 HINGES ON ONE DOOR	EA	--	12.00	12.00	18.00				
	SAND OR PLANE DOOR TO FIT EXISTING OPENING	EA	--	12.00	12.00	18.00				
	ADD HINGE IN MIDDLE OF DOOR TO CORRECT WARP	EA	2.50	9.50	12.00	18.00				
	ALTER SWING OF DOOR, RE-USING SAME HARDWARE	EA	--	30.00	30.00	45.00				
	REMOVE EXISTING LOCKSET AND REPLACE WITH NEW LOCKSET WITHOUT BORING NEW HOLES PASSAGE OR PRIVACY @ $6	EA	6.25	11.75	18.00	27.00				
ADJUST POCKET DOOR	OPEN WALL, UNDERCUT, RE-SET, ADJUST AND LUBRICATE POCKET DOOR, CLOSE WALL WITH DRYWALL AND INSTALL NEW ONE-PIECE BASE AND CASING	EA	24.00	74.00	98.00	147.00				
	FOR PLASTERING WALL ABOVE INSTEAD OF DRYWALL **ADD**	EA	2.25	4.25	6.50	9.75				
	SUBSTITUTE 1" X 4" BASE, OG AND SHOE IN ABOVE CLOSE-IN **ADD**	EA	2.50	3.50	6.00	9.00				

NEW WINDOWS IN EXISTING WALLS

SPECIFICATIONS		UNIT	JOB COST			PRICE	LOCAL AREA MODIFICATION				
			MATLS	LABOR	TOTAL		MATLS	LABOR	TOTAL	PRICE	
REPLACE EXISTING DOOR WITH NEW WINDOW	• REMOVE EXISTING DOOR AND FRAME FROM FRAME WALL • INSTALL NEW BUILDER QUALITY WINDOW AND FRAME NOT MORE THAN WIDTH OF OLD DOOR • CLOSE IN BELOW NEW WINDOW WITH STUDS, SHEATHING AND BEVELED WOOD OR ALUMINUM SIDING • 1/2" DRYWALL, PLASTER OR PLYWOOD PANELING ON INSIDE OF WALL • INSTALL BASE AND SHOE ON INSIDE OF WALL TO MATCH EXISTING AS CLOSELY AS POSSIBLE										
	WOOD DOUBLE HUNG, 3-0 X 4-6	EA	273.00	249.00	522.00	783.00					
	WOOD CASEMENT, 2-4 X 5-6	EA	262.00	238.00	500.00	750.00					
	WOOD SLIDER, 3-0 X 3-6	EA	256.00	232.00	488.00	732.00					
	ALUMINUM SLIDER, 3-0 X 4-0	EA	187.00	229.00	416.00	624.00					
	SAME AS ABOVE WITH DIFFERENT EXTERIOR WALLS										
	BRICK VENEER **ADD**	EA	--	76.00	76.00	114.00					
	BRICK & BLOCK **ADD**	EA	--	86.00	86.00	129.00					
NEW WINDOW IN EXISTING WALL	BREAK OPENING IN EXISTING EXTERIOR FRAME WALL, INSTALL HEADER, PATCH OPENING INCLUDING SIDING AND INTERIOR WALL COVERING AND INSTALL NEW WINDOW										
	WOOD DOUBLE HUNG, 3-0 X 4-6	EA	279.00	291.00	570.00	855.00					
	WOOD CASEMENT, 2-4 X 5-6	EA	279.00	279.00	558.00	837.00					
	WOOD SLIDER, 3-0 X 3-6	EA	189.00	273.00	462.00	693.00					
	ALUMINUM SLIDER, 3-0 X 4-0	EA	164.00	270.00	434.00	651.00					
	SAME AS ABOVE WITH DIFFERENT EXTERIOR WALLS										
	BRICK VENEER **ADD**	EA	--	10.00	10.00	15.00					
	BRICK & BLOCK **ADD**	EA	--	30.00	30.00	45.00					

SPECIFICATIONS		UNIT	JOB COST			PRICE	LOCAL AREA MODIFICATION				
			MATLS	LABOR	TOTAL		MATLS	LABOR	TOTAL	PRICE	
RE-SET WINDOW	REMOVE EXISTING WOOD WINDOW, RE-SET AND IN-STALL NEW INTERIOR TRIM	EA	22.00	34.00	56.00	84.00					
REPLACE GLASS	REMOVE OLD BROKEN GLASS AND PUTTY AND RE-GLAZE WOOD WINDOW PANE										
	UP TO 8" X 12"	EA	4.50	11.50	16.00	24.00					
	UP TO 36" X 24"	EA	10.00	16.00	26.00	39.00					
	REMOVE OLD BROKEN GLASS AND PUTTY FROM METAL WINDOW AND RE-GLAZE										
	WITH STANDARD GLASS	EA	11.00	13.00	24.00	36.00					
	WITH PLATE OR WIRED	EA	16.00	18.00	34.00	51.00					
	REPLACE BROKEN STORM WINDOW PANE	EA	14.00	10.00	24.00	36.00					
STUCK WINDOW	FREE STUCK WINDOW TO OPERATE FREELY AND MOVE STOPS AS NECESSARY	EA	--	18.00	18.00	27.00					
SASH REPAIRS	INSTALL SASH FASTENER ON WINDOW	EA	1.50	3.50	5.00	7.50					
	KEYLOCK TYPE FASTENER ON EXISTING SASH	EA	4.00	14.00	18.00	27.00					
	REPLACE CASEMENT WIN-DOW CRANK	EA	6.50	3.50	10.00	15.00					
	REPLACE ROTTED EXTERIOR WINDOW SILL	EA	8.50	23.50	32.00	48.00					
	REPLACE EXISTING SASH CORDS WITH SASH CHAINS EA = EACH SASH CORD	EA	4.50	15.50	20.00	30.00					
	REPLACE WINDOW STOPS AND PARTING BEADS ON EX-ISTING WINDOW EA = EACH WINDOW	EA	4.50	19.50	24.00	36.00					

SKYLIGHT IN EXISTING ROOF

SPECIFICATIONS		UNIT	JOB COST			PRICE	LOCAL AREA MODIFICATION				
			MATLS	LABOR	TOTAL		MATLS	LABOR	TOTAL	PRICE	
FIXED SKYLIGHT	• OPEN EXISTING ASPHALT OR FIBERGLASS SHINGLED ROOF • INSTALL HEADERS ON EXISTING RAFTERS • WOOD CURB WITH 2" X 6" • ALUMINUM FLASHING • PATCH ROOFING AS REQUIRED • FINISH INTERIOR WITH WOOD TRIM AND/OR DRYWALL TAPED AND FINISHED (**NOT** PAINTED) • WOOD FRAMED SKYLIGHT WITH DOUBLE GLAZED TEMPERED GLASS, EXTERIOR OF FRAME ALUMINUM CLAD WITH BAKED ON ENAMEL FINISH • CORD OPERATED ROLLER BLIND ***Outside Frame***										
	30" X 22"	EA	260.00	336.00	596.00	894.00					
	X 30"	EA	280.00	356.00	636.00	954.00					
	45-1/2" X 30"	EA	316.00	400.00	716.00	1,074.00					
	X 45-1/2"	EA	380.00	440.00	820.00	1,230.00					
VENTILATING SKYLIGHT	SAME AS ABOVE, VENTILATING SKYLIGHT WITH INSECT SCREEN **ADD**	EA	16.00	30.00	46.00	69.00					
SKYLIGHT ACCESSORIES	ROD CONTROL FOR BLINDS AND/OR WINDOWS	EA	30.00	12.00	42.00	63.00					
	ELECTRIC WINDOW MOTOR CONTROL	EA	180.00	40.00	220.00	330.00					
LIGHT SHAFT	BUILD LIGHT SHAFT FROM SLOPED RAFTERS TO FLAT CEILING BELOW • INSTALL HEADERS ON EXISTING CEILING JOISTS • FRAME OUT LIGHT SHAFT FROM EXISTING RAFTER HEADERS TO NEW JOIST HEADERS • INSULATE LIGHT SHAFT • FINISH INTERIOR OF LIGHT SHAFT WITH DRYWALL, TAPED AND FINISHED (**NOT** PAINTED) TO FIXED AND VENTILATING SKYLIGHT PRICES **ADD** AS SHOWN BELOW: DISTANCE FROM ROOM'S FINISHED CEILING TO CLOSEST POINT OF ROOF ABOVE										
	UP TO 2'-0" **ADD**	EA	60.00	176.00	236.00	354.00					
	2'-0" TO 4'-0" **ADD**	EA	85.00	235.00	320.00	480.00					
	4'-0" TO 6'-0" **ADD**	EA	90.00	310.00	400.00	600.00					
	6'-0" TO 8'-0" **ADD**	EA	115.00	375.00	490.00	735.00					
	8'-0" TO 10'-0" **ADD**	EA	145.00	425.00	570.00	855.00					

SPECIFICATIONS		UNIT	JOB COST			PRICE	LOCAL AREA MODIFICATION				
			MATLS	LABOR	TOTAL		MATLS	LABOR	TOTAL	PRICE	
ROOF WINDOW IN EXISTING ROOF	• OPEN EXISTING ASPHALT OR FIBERGLASS SHINGLED ROOF • INSTALL HEADERS ON EXISTING RAFTERS • INSTALL WOOD CURB WITH 2" X 6" • ALUMINUM FLASHING • PATCH ROOFING AS RE-QUIRED • FINISH INTERIOR WITH WOOD TRIM AND/OR DRY-WALL, TAPED AND FIN-ISHED (**NOT** PAINTED) • WOOD FRAMED ROOF WINDOW WITH DOUBLE GLAZED TEMPERED GLASS, EXTERIOR OF FRAME ALUMINUM CLAD WITH BAKED ON ENAMEL FINISH, INTERIOR OF FRAME UNFINISHED WOOD • INSECT SCREEN • CENTER PIVOT										
	Outside Frame 21-1/2" X 27-1/2"	EA	186.00	470.00	656.00	984.00					
	X 38-1/2"	EA	245.00	445.00	690.00	1,035.00					
	27-1/2" X 46-1/2"	EA	235.00	555.00	790.00	1,185.00					
	30-5/8" X 38-1/2"	EA	280.00	480.00	760.00	1,140.00					
	X 55"	EA	310.00	546.00	856.00	1,284.00					
	36-7/8" X 62-7/8"	EA	385.00	595.00	980.00	1,470.00					
	44-3/4" X 46-1/2"	EA	390.00	566.00	956.00	1,434.00					
	52-5/8" X 38-1/2"	EA	385.00	595.00	980.00	1,470.00					
	X 55"	EA	415.00	595.00	1,010.00	1,515.00					
	SAME AS ABOVE, TOP HUNG WITH STILL MOUNTED SCIS-SOR TYPE OPERATOR										
	Outside Frame 21-1/2" X 38-1/2"	EA	246.00	460.00	706.00	1,059.00					
	30-5/8" X 38-1/2"	EA	280.00	516.00	796.00	1,194.00					
	X 55"	EA	310.00	576.00	886.00	1,329.00					
	44-3/4" X 46-1/2"	EA	390.00	600.00	990.00	1,485.00					
INSTALL ROOF WINDOW OR SKY-LIGHT IN ROOF WITH OTHER ROOF COVERING	BUILT-UP **ADD**	EA	60.00	196.00	256.00	384.00					
	SLATE **ADD**	EA	76.00	150.00	226.00	339.00					
	CEDAR SHAKES **ADD** AND SHINGLES	EA	40.00	62.00	102.00	153.00					
	CLAY TILE **ADD**	EA	96.00	170.00	266.00	399.00					
	CONCRETE ROOF- **ADD** ING TILES	EA	80.00	130.00	210.00	315.00					

FRAMING ALTERATIONS

SPECIFICATIONS		UNIT	JOB COST			PRICE	LOCAL AREA MODIFICATION				
			MATLS	LABOR	TOTAL		MATLS	LABOR	TOTAL	PRICE	
REPLACE COLUMN	REPLACE EXISTING WOOD OR MASONRY COLUMN WITH NEW 3" HOLLOW STEEL COL-UMN WITH BASE AND CAP, UP TO 8'-0" LONG — INCLUDES TEAR-OUT OF OLD COLUMN AND SHORING	EA	45.00	73.00	118.00	177.00					
REPLACE WOOD BEAM	REPLACE EXISTING WOOD BEAM WITH NEW 8" #13 STEEL BEAM — INCLUDES TEAR-OUT OF OLD BEAM	LF	160.00	242.00	402.00	603.00					
REPLACE SILL PLATE	REMOVE DEFECTIVE SILL PLATE AND REPLACE WITH NEW 2" X 6" OR 4" X 6" SILL — INCLUDES JACKING UP WALL, BUT **NO** WALL PAT-CHING ABOVE										
	IN BASEMENT WITH GOOD HEADROOM AND EASY ACCESS	LF	3.80	40.00	43.80	65.70					
	IN CRAWL SPACE WITH AT LEAST 18" HEAD-ROOM AND EASY ACCESS	LF	3.80	65.00	68.80	103.20					
DOUBLE UP JOISTS (SISTERING)	DOUBLE UP EXISTING JOISTS WITH NEW JOISTS SPIKED OR BOLTED TO EXISTING JOISTS, EASY ACCESS										
	2" X 6"	LF	.58	1.80	2.38	3.57					
	2" X 8"	LF	.93	1.80	2.73	4.10					
	LF = NEW JOISTS										
DOUBLE UP RAFTERS (SISTERING)	DOUBLE UP EXISTING RAF-TERS WITH NEW 2 X 6 RAF-TERS SPIKED OR BOLTED TO EXISTING RAFTERS										
	GROUND FLOOR RAFTERS	LF	.58	1.50	2.08	3.12					
	SECOND FLOOR RAFTERS	LF	.58	1.80	2.38	3.57					
	THIRD FLOOR RAFTERS	LF	.58	2.10	2.68	4.02					
DOUBLE UP RIDGE-BOARD	DOUBLE UP RIDGEBOARD WITH NEW RIDGEBOARD SPIKED OR BOLTED TO EXISTING RIDGEBOARD										
	2" X 10"	LF	1.18	.87	2.05	3.08					
	2" X 12"	LF	1.70	.90	2.60	3.90					
STRUC-TURAL RIDGE BEAM	ADD TO EXISTING RIDGE-BOARD, SPIKED OR BOLTED										
	(2) 2" X 10"	LF	2.30	1.60	3.90	5.85					
	(2) 2" X 12"	LF	3.34	1.65	4.99	7.49					
	STRUCTURAL RIDGE BEAM INSTALLED BELOW EXISTING RIDGEBOARD WITH 2" X 4" BRACES FROM ALL RAFTERS TO RIDGE BEAM										
	(2) 2" X 10"	LF	2.34	3.11	5.45	8.18					
	(3) 2" X 10"	LF	3.46	3.76	7.22	10.83					
	(2) 2" X 12"	LF	3.30	3.15	6.45	9.68					
	(3) 2" X 12"	LF	4.90	3.89	8.79	13.19					

HomeTech
Remodeling and Renovation
Cost Estimator

Section II

1. PLANS AND PERMITS

SPECIFICATIONS		UNIT	JOB COST			PRICE	LOCAL AREA MODIFICATION				DATA BASE ITEM NO.
			MATLS	LABOR	TOTAL		MATLS	LABOR	TOTAL	PRICE	
ALL REMODEL-ING AND RENOVA-TION WORK	IN ADDITION TO BUILDING PERMIT FEE. COST AND PRICE ARE BASED ON TOTAL AMOUNT OF JOB FOUNDATION PLANS, FLOOR PLANS, ELEVATIONS, SEC-TIONS, AS REQUIRED INCLUDES ON-SITE SURVEY & MEASUREMENTS BY DESIGN-ER, BUT DOES **NOT** INCLUDE CONFERENCES WITH OR AP-PROVAL OF PLANS BY OWNER EA = TOTAL JOB PER $1,000 = PER $1,000 OF TOTAL JOB COST (CON-TRACTOR'S COST) OR JOB PRICE (PRICE TO CUSTOMER)	EA PLUS PER $1,000	-- --	-- --	200.00 10.00	300.00 10.00					.000 .001
	SAME AS ABOVE AND IN-CLUDING CONFERENCES WITH AND APPROVAL OF PLANS BY OWNER	EA PLUS PER $1,000	-- --	-- --	600.00 55.00	900.00 55.00					.002 .003
	SAME AS ABOVE PLUS SE-LECTION OF PRIME CON-TRACTOR **ADD**	PER $1,000	--	--	6.00	6.00					.004
	SAME AS ABOVE PLUS ARCHITECTURAL SUPERVI-SION TO COMPLETION **ADD**	PER $1,000	--	--	25.00	25.00					.005
	FULL FEE FOR ALL OF ABOVE SERVICES	EA PLUS PER $1,000	-- --	-- --	600.00 96.00	900.00 96.00					.006 .007
BUILDING PERMIT (AVERAGE AMOUNT FOR U.S.)	BUILDING PERMIT FEE, BASED ON TOTAL AMOUNT OF JOB PER $1,000 = PER $1,000 OF TOTAL JOB COST (CONTRAC-TOR'S COST) OR JOB PRICE (PRICE TO CUSTOMER)	EA PLUS PER $1,000	-- --	-- --	50.00 10.00	75.00 10.00					.008 .009

SPECIFICATIONS	UNIT	JOB COST			PRICE	LOCAL AREA MODIFICATION				DATA BASE ITEM NO.
		MATLS	LABOR	TOTAL		MATLS	LABOR	TOTAL	PRICE	
NOTE — ARCHITECTURAL SUPERVISION INCLUDES INSPECTION OF WORK AT LEAST EVERY OTHER DAY WHILE JOB IS PROGRESSING, APPROVAL OF PAYMENTS TO CONTRACTOR AND ON-THE-JOB CONFERENCES WITH CONTRACTOR AND CUSTOMER WHEN REQUIRED. NO DIRECT SUPERVISION OF WORKERS OR PREPARATION OF MATERIAL LISTS IS INCLUDED. A RECOMMENDED ARCHITECTURAL FEE FOR FULL SERVICE IS 13% OF TOTAL JOB PRICE (THE AMOUNT THAT THE CUSTOMER PAYS FOR THE WORK).										1

2. PAVING AND SIDEWALK DEMOLITION

SPECIFICATIONS		UNIT	JOB COST			PRICE	LOCAL AREA MODIFICATION				DATA BASE ITEM NO.
			MATLS	LABOR	TOTAL		MATLS	LABOR	TOTAL	PRICE	
	NOTE: TEAR-OUT AND DE-MOLITION COSTS INCLUDE PILING OF DEBRIS ON SITE BUT DO NOT INCLUDE RE-MOVAL FROM PREMISES. FOR REMOVAL OF DEBRIS FROM PREMISES, SEE "CLEAN-UP" ON PAGE 285.										
BITUMI-NOUS PAVING	REMOVE DRIVEWAY WITH SHOVEL OR LOADER	EA PLUS	—	309.00	309.00	463.50					.000
		SF	—	.32	.32	.48					.001
CONCRETE PAVING	6" THICK CONCRETE DRIVE-WAY REMOVED:										
	WITH PNEUMATIC TOOL	SF	—	1.14	1.14	1.71					.002
	WITH SHOVEL OR BULL-DOZER	EA PLUS	—	309.00	309.00	463.50					.003
		SF	—	.32	.32	.48					.004
CURB AND GUTTER	WITH COMPRESSOR AND LA-BORERS, BREAK OUT 12" HIGH, 6" THICK CONCRETE CURB	EA PLUS	—	128.00	128.00	192.00					.005
		SF	—	2.71	2.71	4.07					.006
SIDEWALK	BREAK UP BRICK OR CON-CRETE SIDEWALK WITH PNEUMATIC TOOL	SF	—	1.12	1.12	1.68					.007
	BREAK UP FLAGSTONE SET IN SAND OR CRUSHED STONE, BY HAND	SF	—	1.05	1.05	1.58					.049
	BREAK UP FLAGSTONE SET IN MORTAR, WITH PNEU-MATIC TOOL	SF	—	1.10	1.10	1.65					.050

SPECIFICATIONS		UNIT	JOB COST			PRICE	LOCAL AREA MODIFICATION				DATA BASE ITEM NO.
			MATLS	LABOR	TOTAL		MATLS	LABOR	TOTAL	PRICE	
CONCRETE FOOTING	REMOVE CONCRETE FOOTING WITH PNEUMATIC TOOL										
	8" X 16"	LF	--	2.58	2.58	3.87					.008
	12" X 24"	LF	--	4.86	4.86	7.29					.009
CONCRETE SLAB	BREAK UP EXISTING SLAB IN BASEMENT WITH PNEUMATIC TOOL AND HAUL RUBBLE OUTSIDE TO GRADE										
	2"	SF	--	1.34	1.34	2.01					.010
	4"	SF	--	1.56	1.56	2.34					.011
	6"	SF	--	1.83	1.83	2.75					.012
	BREAK UP EXISTING SLAB OUTSIDE ON GRADE WITH PNEUMATIC TOOL, 6" THICK										
	NO REINFORCEMENT	SF	--	1.22	1.22	1.83					.013
	REINFORCED	SF	--	1.74	1.74	2.61					.014
SUSPENDED SLAB AND STEPS	BREAK UP EXISTING SUSPENDED SLAB AND STEPS WITH PNEUMATIC TOOL, NOT OVER 10 FEET ABOVE GRADE, UP TO 4" THICK SF = TOTAL SLAB AND STEP AREA	SF	--	1.40	1.40	2.10					.015
CONCRETE WALL	REMOVE CONCRETE WALL WITH PNEUMATIC TOOL, 12" WALL										
	NO REINFORCEMENT	SF	--	4.11	4.11	6.17					.016
	REINFORCED	SF	--	6.11	6.11	9.17					.017
CONCRETE AND/OR BRICK PLATFORM & STEPS	BREAK UP EXISTING BRICK OR CONCRETE PLATFORM AND STEPS WITH PNEUMATIC TOOL (SOLID SIDES AND FILLED WITH RUBBLE)	SF	--	3.46	3.46	5.19					.018
CONCRETE SAWING	CUT CONCRETE WITH GAS CONCRETE SAW, PER INCH DEEP										
	CONCRETE SLAB W/MESH	LF	--	1.60	1.60	2.40					.051
	CONCRETE WALL	LF	--	4.40	4.40	6.60					.052
	BRICK	LF	--	3.98	3.98	5.97					.053
	BLOCK	LF	--	3.30	3.30	4.95					.054

2. MASONRY DEMOLITION

SPECIFICATIONS		UNIT	JOB COST			PRICE	LOCAL AREA MODIFICATION				DATA BASE ITEM NO.
			MATLS	LABOR	TOTAL		MATLS	LABOR	TOTAL	PRICE	
BLOCK WALL	REMOVE BLOCK WALL BY HAND, **NO** REINFORCEMENT										
	4"	SF	--	.61	.61	.92					.020
	8"	SF	--	.77	.77	1.16					.021
	12"	SF	--	.98	.98	1.47					.022
BRICK WALL	REMOVE BRICK WALL BY HAND										
	4" VENEER	SF	--	1.13	1.13	1.70					.023
	8" SOLID	SF	--	3.12	3.12	4.68					.024
	12" SOLID	SF	--	4.89	4.89	7.34					.025
	REMOVE BRICK WALL WITH PNEUMATIC TOOL OR LARGE ELECTRIC HAMMER										
	4" VENEER	SF	--	1.13	1.13	1.70					.026
	8" SOLID	SF	--	2.61	2.61	3.92					.027
	12" SOLID	SF	--	3.82	3.82	5.73					.028
BRICK & BLOCK WALL	REMOVE BRICK AND BLOCK WALL WITH PNEUMATIC TOOL										
	8"	SF	--	1.79	1.79	2.69					.029
	12"	SF	--	2.54	2.54	3.81					.030
STONE WALL	REMOVE STONE WALL WITH PNEUMATIC TOOL, MORTAR JOINTS, 12" - 16" WALL	SF	--	5.25	5.25	7.88					.031

SPECIFICATIONS		UNIT	JOB COST			PRICE	LOCAL AREA MODIFICATION				DATA BASE ITEM NO.
			MATLS	LABOR	TOTAL		MATLS	LABOR	TOTAL	PRICE	
REMOVE MASONRY WALL AND INSTALL STEEL OVER	REMOVE ENTIRE BRICK OR BRICK AND BLOCK WALL (ONE STORY) AND INSTALL STEEL ANGLES OVER										
	2 STEEL ANGLES	LF	12.00	46.00	58.00	87.00					.032
	3 STEEL ANGLES	LF	18.00	46.00	64.00	96.00					.033
	SAME AS ABOVE AND INSTALL STEEL BEAM OR FLITCH PLATE SUPPORT	LF	19.00	45.00	64.00	96.00					.034
	IF STEEL COLUMNS ARE REQUIRED FOR STEEL BEAM OR FLITCH PLATE SUPPORT **ADD** EA = EACH COLUMN	EA	62.00	14.00	76.00	114.00					.035
NEEDLE & SHORE	NEEDLE & SHORE MASONRY WALL OPENING, INCLUDING WALL BREAKTHROUGH AND REMOVAL										
	8" BLOCK	LF	23.00	60.00	83.00	124.50					.036
	4" BRICK & 8" BLOCK	LF	30.00	77.00	107.00	160.50					.037
	8" BRICK	LF	28.00	70.00	98.00	147.00					.038
	12" BRICK	LF	36.00	89.00	125.00	187.50					.039
	SAME AS ABOVE AND INSTALL 3 STEEL ANGLES **ADD**	LF	31.50	3.20	34.70	52.05					.040
	SAME AS ABOVE AND INSTALL STEEL BEAM **ADD** 8" WF 13#	LF	18.90	7.10	26.00	39.00					.041
	8" WF 17#	LF	26.20	7.80	34.00	51.00					.042
CHIMNEY	TEAR OUT EXISTING CHIMNEY FROM ROOF TO FOOTING, **NO** FLOOR OR WALL PATCHING										
	SMALL CHIMNEY, UP TO 20" X 20" EXTERIOR	LF Down	--	7.50	7.50	11.25					.043
	INTERIOR	LF Down	--	9.65	9.65	14.48					.044
	LARGE CHIMNEY, 20" X 20" TO 48" X 48" EXTERIOR	LF Down	--	11.80	11.80	17.70					.045
	INTERIOR	LF Down	--	18.00	18.00	27.00					.046
FIREPLACE	REMOVE BRICK FIREPLACE WITH 30" X 24" OPENING SOFT MORTAR	EA	--	120.00	120.00	180.00					.047
	HARD MORTAR	EA	--	162.00	162.00	243.00					.048

2. EXTERIOR TEAR-OUT

SPECIFICATIONS		UNIT	JOB COST			PRICE	LOCAL AREA MODIFICATION				DATA BASE ITEM NO.
			MATLS	LABOR	TOTAL		MATLS	LABOR	TOTAL	PRICE	
REMOVE FRAME WALL AND INSTALL WOOD HEADER OR STEEL OVER	REMOVE ENTIRE WALL (ONE STORY) AND INSTALL WOOD HEADER OVER AS REQUIRED	LF	5.60	27.00	32.60	48.90					.100
	EXTERIOR WALL MATERIALS WOOD, COMPOSITION, STUCCO OR METAL SIDING OVER STUDS AND SHEATHING										
	INTERIOR WALL MATERIALS DRYWALL, PLASTER AND LATH OR WOOD PANELING										
	SAME AS ABOVE AND INSTALL STEEL BEAM OR FLITCH PLATE SUPPORT	LF	3.00	32.00	62.00	93.00					.101
EXTERIOR WALL COVERING	REMOVE SIDING FROM EXISTING EXTERIOR WALL										
	DROP OR BEVELED WOOD	SF	--	.34	.34	.51					.102
	PLYWOOD	SF	--	.25	.25	.38					.103
	BOARD AND BATTEN	SF	--	.31	.31	.47					.104
	CEDAR SHINGLES	SF	--	.31	.31	.47					.105
	ALUMINUM OR VINYL	SF	--	.31	.31	.47					.106
	STUCCO & WOOD LATH	SF	--	.37	.37	.56					.107
	STUCCO & METAL LATH	SF	--	.59	.59	.89					.108
EXTERIOR TRIM	FASCIA BOARD OR SOFFIT 1" X 6"	LF	--	.39	.39	.59					.109
	1" X 12"	LF	--	.45	.45	.68					.110
	1" X 24"	LF	--	.52	.52	.78					.111
ROOF OVERHANG	REMOVE ROOF OVERHANG, INCLUDING SOFFIT, FASCIA, AND GUTTER										
	6"	LF	--	3.65	3.65	5.48					.112
	12"	LF	--	4.00	4.00	6.00					.113
	18"	LF	--	4.72	4.72	7.08					.114
	24"	LF	--	5.90	5.90	8.85					.115
GUTTERS	REMOVE EDGE HUNG GUTTERS FROM BUILDING										
	ALUMINUM, COPPER OR GALVANIZED	LF	--	.64	.64	.96					.116
	WOOD	LF	--	.89	.89	1.34					.117
DOWNSPOUTS	REMOVE DOWNSPOUTS FROM BUILDING	LF	--	.39	.39	.59					.118
SKYLIGHT	REMOVE GLAZED ROOF SKYLIGHT & CURB UP TO 24 SF	SF	--	4.00	4.00	6.00					.119
METAL PLATFORM AND STEPS	REMOVE EXISTING METAL PLATFORM AND STEPS FROM BUILDING (ONE STORY)	EA	--	108.00	108.00	162.00					.120
WOOD RETAINING WALL	REMOVE 6 X 6 OR 6 X 8 WOOD TIE WALL	SF	--	.62	.62	.93					.137

SPECIFICATIONS		UNIT	JOB COST			PRICE	LOCAL AREA MODIFICATION				DATA BASE ITEM NO.
			MATLS	LABOR	TOTAL		MATLS	LABOR	TOTAL	PRICE	
GARAGE	COMPLETE GARAGE, BUT **NOT** INCLUDING FLOOR SLAB AND DRIVEWAY										
	FRAME	SF	--	.98	.98	1.47					.200
	SOLID BRICK	SF	--	1.55	1.55	2.33					.201
	BRICK AND BLOCK	SF	--	1.33	1.33	2.00					.202
	SF = FLOOR AREA										
WOOD PORCH	REMOVE WOOD PORCH UP TO 240 SF. **NO** PATCHING OR REPAIRS INCLUDED										
	ONE STORY	EA	--	409.00	409.00	613.50					.203
	TWO STORY	EA	--	612.00	612.00	918.00					.204
PORCH SCREEN-ING	REMOVE PORCH SCREENING AND INTERMEDIATE POSTS, ETC. INSTALL TEMPORARY SUPPORTS FOR ROOF EA = EACH JOB	EA	--	116.00	116.00	174.00					.205
PORCH FLOORING	REMOVE TONGUE AND GROOVE PORCH FLOORING	SF	--	.50	.50	.75					.206
PORCH CEILING	REMOVE WOOD TONGUE & GROOVE PORCH CEILING	SF	--	.51	.51	.76					.233
	REMOVE 4x8 1/4" TO 1/2" PLY-WOOD PORCH CEILING WITH OR WITHOUT BATTENS	SF	--	.43	.43	.65					.234
PORCH RAILING	REMOVE WOOD RAILING AND SPINDLES	LF	--	.38	.38	.57					.235
	REMOVE METAL/WROUGHT IRON RAILING AND SPINDLES	LF	--	.47	.47	.71					.236
PORCH ROOF	TEAR OFF ENTIRE ROOF STRUCTURE OVER PORCH, UP TO 240 SF, LEAVING ONLY PLATFORM, **NO** POINTING OR PATCHING INCLUDED	EA	--	272.00	272.00	408.00					.207
DECK	REMOVE COMPLETE DECK INCLUDING SURFACE AND FRAMING	SF	--	1.07	1.07	1.61					.208
	REMOVE DECK SURFACE ONLY IN PREPARATION FOR REPLACING	SF	--	.64	.64	.96					.209
CHAIN LINK FENCING	REMOVE CHAIN LINK FENCE AND DISPOSE, INCLUDING POSTS SET IN CONCRETE	LF	--	2.38	2.38	3.57					.210
	REMOVE CHAIN LINK FENCE AND STORE FOR RE-USE	LF	--	1.47	1.47	2.21					.211
WOOD FENCING	6'-0" HIGH, 1" X 4" OR 1" X 6" PRIVACY FENCE OR STOCK-ADE FENCE, INCL. POSTS NOT SET IN CONCRETE	LF	--	1.56	1.56	2.34					.212
	SAME AS ABOVE WITH POSTS SET IN CONCRETE	LF	--	2.50	2.50	3.75					.213

2. FRAMING AND SHEATHING TEAR-OUT

SPECIFICATIONS		UNIT	JOB COST			PRICE	LOCAL AREA MODIFICATION				DATA BASE ITEM NO.
			MATLS	LABOR	TOTAL		MATLS	LABOR	TOTAL	PRICE	
BEAM	STEEL I-BEAM OR WIDE FLANGE	LF	--	5.14	5.14	7.71					.121
	WOOD BEAM										
	6 X 8	LF	--	3.89	3.89	5.84					.122
	10 X 12	LF	--	8.50	8.50	12.75					.123
JOISTS	REMOVE WOOD FLOOR JOISTS										
	2 X 6	SF	--	.43	.43	.65					.124
	2 X 8	SF	--	.44	.44	.66					.125
	2 X 10	SF	--	.46	.46	.69					.126
	2 X 12	SF	--	.48	.48	.72					.127
SUBFLOOR	REMOVE WOOD SUBFLOOR 1 X 6 OR 1 X 8	SF	--	.52	.52	.78					.128
	REMOVE PLYWOOD SUB-FLOOR, NAILED ONLY	SF	--	.45	.45	.68					.129
	REMOVE PLYWOOD SUB-FLOOR, GLUED AND NAILED	SF	--	.58	.58	.87					.130
ROOF SHEATHING	REMOVE SHEATHING										
	1" X 8"	SF	--	.31	.31	.47					.131
	1" X 12"	SF	--	.26	.26	.39					.132
	1/2" PLYWOOD	SF	--	.24	.24	.36					.133
	NOTE: FOR TEARING OUT ROOF COVERING, SEE PAGE 144.										
RAFTERS	REMOVE RAFTERS	SF	--	.44	.44	.66					.134
CEILING	REMOVE CEILING JOISTS	SF	--	.35	.35	.53					.135
TRUSSES	REMOVE TRUSSES	EA	--	11.00	11.00	16.50					.136

SPECIFICATIONS	UNIT	JOB COST			PRICE	LOCAL AREA MODIFICATION				DATA BASE ITEM NO.
		MATLS	LABOR	TOTAL		MATLS	LABOR	TOTAL	PRICE	
EXTERIOR DOOR REMOVE DOOR, FRAME AND TRIM FROM EXISTING EXTERIOR WALL, 3-0 X 6-8	EA	--	26.20	26.20	39.30					.214
REMOVE WOOD OR ALUMINUM SLIDING GLASS DOOR, FRAME AND TRIM	EA	--	40.00	40.00	60.00					.215
REMOVE OVERHEAD GARAGE DOOR 9 X 7	EA	--	58.00	58.00	87.00					.216
16 X 7	EA	--	96.00	96.00	144.00					.217
INTERIOR DOOR REMOVE INTERIOR DOOR, FRAME AND TRIM FROM INTERIOR WALL SINGLE DOOR	EA	--	13.00	13.00	19.50					.218
DOUBLE DOORS	EA	--	21.00	21.00	31.50					.219
WINDOW REMOVE WINDOW FROM WALL, INCLUDING TRIM *IN FRAME WALL* WOOD, TO 15 SF	EA	--	11.20	11.20	16.80					.220
OVER 15 SF	SF	--	.76	.76	1.14					.221
ALUMINUM, TO 15 SF	EA	--	7.20	7.20	10.80					.222
OVER 15 SF	SF	--	.49	.49	.74					.223
STEEL CASEMENT, TO 15 SF	EA	--	22.00	22.00	33.00					.224
OVER 15 SF	SF	--	1.42	1.42	2.13					.225
IN MASONRY WALL WOOD, TO 15 SF	EA	--	16.00	16.00	24.00					.226
OVER 15 SF	SF	--	1.07	1.07	1.61					.227
ALUMINUM, TO 15 SF	EA	--	11.50	11.50	17.25					.228
OVER 15 SF	SF	--	.79	.79	1.19					.229
STEEL CASEMENT, TO 15 SF	EA	--	33.00	33.00	49.50					.230
OVER 15 SF	SF	--	2.20	2.20	3.30					.231
IF WINDOWS HAVE BEEN CAULKED WITH SILICON **ADD**	EA	--	7.10	7.10	10.65					.232

2

2. FLOORING TEAR-OUT

SPECIFICATIONS		UNIT	JOB COST			PRICE	LOCAL AREA MODIFICATION				DATA BASE ITEM NO.
			MATLS	LABOR	TOTAL		MATLS	LABOR	TOTAL	PRICE	
HARD-WOOD FLOOR	REMOVE NAILED HARDWOOD FLOORING 2-1/4" X 25/32"	SF	--	.43	.43	.65					.300
	REMOVE HARDWOOD FLOOR-ING IN MASTIC	SF	--	.76	.76	1.14					.301
	REMOVE WOOD FLOORING AND SLEEPERS FROM CONCRETE SLAB	SF	--	.48	.48	.72					.302
BRICK FLOOR	REMOVE BRICK FLOORING FROM WOOD BASE	SF	--	.60	.60	.90					.303
	FROM CONCRETE BASE	SF	--	1.89	1.89	2.84					.304
CERAMIC TILE FLOOR	REMOVE CERAMIC TILE THIN SET	SF	--	.52	.52	.78					.305
	MUD SET	SF	--	.76	.76	1.14					.306
FLAG-STONE FLOOR	REMOVE SLATE OR FLAG-STONE FLOOR THIN SET	SF	--	.53	.53	.80					.348
	MUD SET	SF	--	.89	.89	1.34					.349
RESILIENT FLOORING	REMOVE RESILIENT TILE OR SHEET GOODS	SF	--	.19	.19	.29					.307
CARPET	REMOVE CARPET BONDED	SF	--	.29	.29	.44					.308
	TACKLESS	SF	--	.09	.09	.14					.309
UNDER-LAYMENT	REMOVE UNDERLAYMENT, NAILED OR STAPLED	SF	--	.29	.29	.44					.310
	NAILED OR STAPLED AND GLUED	SF	--	.40	.40	.60					.311

SPECIFICATIONS		UNIT	JOB COST			PRICE	LOCAL AREA MODIFICATION				DATA BASE ITEM NO.
			MATLS	LABOR	TOTAL		MATLS	LABOR	TOTAL	PRICE	
INTERIOR GUTTING	GUT INTERIOR OF BUILDING LEAVING ONLY LOAD BEARING STUDS, JOISTS, AND CHIMNEY: • REMOVE ALL PLUMBING, HEATING AND ELECTRICAL SYSTEMS • REMOVE INTERIOR WALL COVERINGS EXCEPT PLASTER APPLIED DIRECTLY ON INSIDE OF EXTERIOR MASONRY WALLS • LOAD AND HAUL RUBBISH TO DUMP WITHIN 5 MILES										
	INCLUDING CHUTING RUBBISH TO TRUCK	SF	--	1.47	1.47	2.21					.312
	HAND LOADING INTO TRUCK	SF	--	1.68	1.68	2.52					.313
	SF = TOTAL FLOOR (LIVING) AREA OF BUILDING, INCLUDING BASEMENT										
CEILING COVERING	REMOVE CEILING COVERING FROM CEILING JOISTS										
	PLASTER & GYPSUM LATH	SF	--	.26	.26	.39					.314
	PLASTER & WOOD LATH	SF	--	.36	.36	.54					.315
	PLASTER & METAL LATH	SF	--	.58	.58	.87					.316
	GYPSUM DRYWALL	SF	--	.19	.19	.29					.317
	REMOVE WOOD TONGUE & GROOVE CEILING	SF	--	.52	.52	.78					.318
CEILING TILE	12" X 12" COMPOSITION TILE										
	GLUED	SF	--	.40	.40	.60					.319
	STAPLED	SF	--	.20	.20	.30					.320
	24" X 48" TILE IN METAL GRID SYSTEM, INCLUDING GRIDS	SF	--	.45	.45	.68					.321
INSULATION	REMOVE INSULATION FROM OPEN WALL OR CEILING										
	LOOSE	SF	--	.08	.08	.12					.322
	BATT	SF	--	.12	.12	.18					.323

2. INTERIOR WALL TEAR-OUT

SPECIFICATIONS		UNIT	JOB COST			PRICE	LOCAL AREA MODIFICATION				DATA BASE ITEM NO.
			MATLS	LABOR	TOTAL		MATLS	LABOR	TOTAL	PRICE	
COMPLETE INTERIOR WALL	REMOVE NON-BEARING WALL AND COVERINGS, INCLUDING STUDS AND WALL COVERING BOTH SIDES										
	W/GYPSUM DRYWALL	LF	--	5.60	5.60	8.40					.324
	W/PLASTER & GYPSUM LATH	LF	--	6.64	6.64	9.96					.325
	W/PLASTER & WOOD LATH	LF	--	9.38	9.38	14.07					.326
	REMOVE BEARING WALL AND COVERINGS, INCLUDING STUDS AND WALL COVERING BOTH SIDES, INSTALL WOOD HEADER OR STEEL ANGLE SUPPORT AS REQUIRED										
	W/GYPSUM DRYWALL	LF	4.00	10.00	14.00	21.00					.327
	W/PLASTER & GYPSUM LATH	LF	4.00	11.00	15.00	22.50					.328
	W/PLASTER & WOOD LATH	LF	4.00	14.00	18.00	27.00					.329
	SAME AS ABOVE, WITH STEEL BEAM OR FLITCH PLATE SUPPORT										
	W/GYPSUM DRYWALL	LF	27.00	13.60	40.60	60.90					.330
	W/PLASTER & GYPSUM LATH	LF	27.00	14.30	41.30	61.95					.331
	W/PLASTER & WOOD LATH	LF	27.00	17.00	44.00	66.00					.332
HEADER OR STEEL ANGLE	SAME AS ABOVE, INSTALL HEADER OR STEEL ANGLE FLUSH WITH CEILING										
	W/GYPSUM DRYWALL	LF	4.00	18.75	22.75	34.13					.333
	W/PLASTER & GYPSUM LATH	LF	4.00	19.25	23.25	34.88					.334
	W/PLASTER & WOOD LATH	LF	4.00	20.25	24.25	36.38					.335
	SAME AS ABOVE, WITH STEEL BEAM OR FLITCH PLATE FLUSH WITH CEILING										
	W/GYPSUM DRYWALL	LF	27.00	27.00	54.00	81.00					.336
	W/PLASTER & GYPSUM LATH	LF	27.00	28.00	55.00	82.50					.337
	W/PLASTER & WOOD LATH	LF	27.00	29.00	56.00	84.00					.338
COMPLETE METAL STUDWALL	REMOVE METAL STUDWALL FINISHED BOTH SIDES WITH DRYWALL	LF	--	7.70	7.70	11.55					.339
INTERIOR WALL COVERING FROM STUDWALL	GYPSUM DRYWALL	SF	--	.14	.14	.21					.340
	GYPSUM LATH & PLASTER	SF	--	.27	.27	.41					.341
	METAL LATH & PLASTER	SF	--	.67	.67	1.01					.342
	WOOD LATH & PLASTER	SF	--	.49	.49	.74					.343
	PLYWOOD PANELING	SF	--	.14	.14	.21					.344
	SOLID WOOD PANELING	SF	--	.32	.32	.48					.345
	CERAMIC TILE IN MUD	SF	--	.79	.79	1.19					.346
	REMOVE PLASTER FROM MASONRY WALL	SF	--	1.12	1.12	1.68					.347

SPECIFICATIONS		UNIT	JOB COST			PRICE	LOCAL AREA MODIFICATION				DATA BASE ITEM NO.
			MATLS	LABOR	TOTAL		MATLS	LABOR	TOTAL	PRICE	
STAIRS	REMOVE SET OF STAIRS, **NO** PATCHING										
	BASEMENT	EA	--	26.20	26.20	39.30					.500
	ATTIC	EA	--	70.60	70.60	105.90					.501
	REMOVE SET OF MAIN STAIRS INCLUDING NEWELS, RAILS AND BALUSTERS, **NO** PATCHING										
	AVERAGE	EA	--	122.00	122.00	183.00					.502
	DIFFICULT	EA	--	146.00	146.00	219.00					.503
	VERY DIFFICULT	EA	--	182.00	182.00	273.00					.504
KITCHEN CABINETS	REMOVE WOOD OR STEEL KITCHEN CABINETS — BASE, WALL OR ISLAND EA = EACH CABINET	LF	--	7.50	7.50	11.25					.505
COUNTER-TOP	REMOVE COUNTERTOP FROM BASE CABINET (**NOT** INCLUDING DISCONNECT)	LF	--	1.28	1.28	1.92					.506
BASE-BOARD	REMOVE BASEBOARD AND SHOE MOULDING, UP TO 1" X 8" BASEBOARD	LF	--	.21	.21	.32					.507
	REMOVE BASEBOARD AND CLEAN AND STORE FOR RE-USE	LF	--	.30	.30	.45					.508
MOULDING	REMOVE DOOR OR WINDOW CASING										
	1-PIECE	LF	--	.33	.33	.50					.509
	2-PIECE	LF	--	.42	.42	.63					.510
	REMOVE BASE, CHAIR RAIL OR CEILING MOULD										
	1-PIECE	LF	--	.33	.33	.50					.511
	2-PIECE	LF	--	.42	.42	.63					.512
	3-PIECE	LF	--	.52	.52	.78					.513
	REMOVE BASE, CHAIR RAIL OR CEILING MOULD, CLEAN AND STORE FOR RE-USE										
	1-PIECE	LF	--	.37	.37	.56					.514
	2-PIECE	LF	--	.61	.61	.92					.515
	3-PIECE	LF	--	.86	.86	1.29					.516

2. PLUMBING TEAR-OUT

SPECIFICATIONS		UNIT	JOB COST			PRICE	LOCAL AREA MODIFICATION				DATA BASE ITEM NO.
			MATLS	LABOR	TOTAL		MATLS	LABOR	TOTAL	PRICE	
KITCHEN PLUMBING	DISCONNECT, REMOVE AND CAP										
	KITCHEN SINK	EA	--	68.00	68.00	102.00					.400
	DISHWASHER	EA	--	51.00	51.00	76.50					.401
	DISPOSAL	EA	--	22.00	22.00	33.00					.402
	BAR SINK	EA	--	57.00	57.00	85.50					.403
	20 GAL. LAUNDRY SINK	EA	--	45.00	45.00	67.50					.404
	DISCONNECT, REMOVE AND CAP ANY TWO OF ABOVE AT SAME TIME **DEDUCT**	EA	--	10%	10%	10%					.405
	ANY THREE OR MORE, **DEDUCT**	EA	--	15%	15%	15%					.406
BATHROOM PLUMBING	DISCONNECT, REMOVE AND CAP										
	WATER CLOSET	EA	--	51.00	51.00	76.50					.407
	BIDET	EA	--	68.00	68.00	102.00					.408
	STEEL TUB	EA	--	124.00	124.00	186.00					.409
	FIBERGLASS TUB	EA	--	114.00	114.00	171.00					.410
	FIBERGL. SHOWER/TUB	EA	--	137.00	137.00	205.50					.411
	WHIRLPOOL	EA	--	171.00	171.00	256.50					.412
	WALL-HUNG OR PEDESTAL LAVATORY	EA	--	51.00	51.00	76.50					.413
	19" VANITY AND TOP	EA	--	60.00	60.00	90.00					.414
	60" VANITY AND TOP	EA	--	79.00	79.00	118.50					.415
	DISCONNECT, REMOVE AND CAP ANY TWO OF ABOVE AT SAME TIME **DEDUCT**	EA	--	10%	10%	10%					.405
	ANY THREE OR MORE, **DEDUCT**	EA	--	15%	15%	15%					.406
	DISCONNECT, REMOVE AND CAP CAST IRON TUB	EA	--	171.00	171.00	256.50					.416
	DISCONNECT, REMOVE AND CAP HOT WATER HEATER 30 GALLON	EA	--	51.00	51.00	76.50					.417
	50 GALLON	EA	--	57.00	57.00	85.50					.418
	82 GALLON	EA	--	62.00	62.00	93.00					.419
	REMOVE SHOWER STALL LEAD PAN IN MUD SET TILE	EA	--	175.00	175.00	262.50					.420

SPECIFICATIONS		UNIT	JOB COST			PRICE	LOCAL AREA MODIFICATION				DATA BASE ITEM NO.
			MATLS	LABOR	TOTAL		MATLS	LABOR	TOTAL	PRICE	
HEATING TEAR-OUT	REMOVE BOILER (WITH **NO** ASBESTOS COMPLICATIONS)	EA	--	492.00	492.00	738.00					.517
	REMOVE GAS FURNACE (WITHOUT DUCTWORK)	EA	--	88.00	88.00	132.00					.518
	REMOVE OIL FURNACE (WITHOUT DUCTWORK)	EA	--	110.00	110.00	165.00					.519
	REMOVE BASEBOARD RADIATORS	LF	--	14.00	14.00	21.00					.520
	REMOVE FLUE	LF	--	1.69	1.69	2.54					.521
DRAIN SYSTEM	DRAIN HEATING SYSTEM AND CAP	EA	--	96.00	96.00	144.00					.522
DUCTWORK	TEAR OUT DUCTWORK ONLY 4" X 8"	LF	--	.96	.96	1.44					.523
	6" X 8"	LF	--	1.24	1.24	1.86					.524
	10" X 12"	LF	--	1.64	1.64	2.46					.525
	6" ROUND	LF	--	.87	.87	1.31					.535
	8" FLEX	LF	--	.72	.72	1.08					.536
REMOVE SYSTEM	REMOVE ENTIRE BOILER AND RADIATOR SYSTEM ONE STORY	EA	--	604.00	604.00	906.00					.526
	TWO STORY	EA	--	808.00	808.00	1,212.00					.527
	THREE STORY	EA	--	1016.00	1,016.00	1,524.00					.528
	TEAR OUT FURNACE AND ENTIRE DUCTWORK SYSTEM ONE STORY	EA	--	240.00	240.00	360.00					.529
	TWO STORY	EA	--	342.00	342.00	513.00					.530
	THREE STORY	EA	--	474.00	474.00	711.00					.531
WINDOW A/C	REMOVE WINDOW AIR CONDITIONERS SMALL	EA	--	20.00	20.00	30.00					.532
	MEDIUM	EA	--	34.00	34.00	51.00					.533
	LARGE	EA	--	50.00	50.00	75.00					.534

2

2. ELECTRICAL TEAR-OUT

SPECIFICATIONS		UNIT	JOB COST			PRICE	LOCAL AREA MODIFICATION				DATA BASE ITEM NO.
			MATLS	LABOR	TOTAL		MATLS	LABOR	TOTAL	PRICE	
LIGHT FIXTURES	REMOVE LIGHT FIXTURES:										
	SMALL CEILING FIXTURE	EA	--	10.00	10.00	15.00					.421
	LARGE CEILING FIXTURE	EA	--	16.00	16.00	24.00					.422
	LARGE CHANDELIER	EA	--	21.00	21.00	31.50					.423
	RECESSED FIXTURE	EA	--	15.00	15.00	22.50					.424
	48" FLUORESCENT FIXTURE	EA	--	12.00	12.00	18.00					.425
SWITCHES, OUTLETS	REMOVE SWITCHES AND DUPLEX OUTLETS	EA	--	6.00	6.00	9.00					.426
	REMOVE 220 VOLT OUTLET	EA	--	8.00	8.00	12.00					.427
	REMOVE ATTIC FAN	EA	--	49.00	49.00	73.50					.428
	REMOVE KITCHEN OR BATH-ROOM EXHAUST FAN	EA	--	14.00	14.00	21.00					.429
	REMOVE HARD WIRED SMOKE ALARM	EA	--	23.00	23.00	34.50					.430
	REMOVE 60 AMP SERVICE	EA	--	34.00	34.00	51.00					.431
	REMOVE 100 AMP SERVICE	EA	--	41.00	41.00	61.50					.432
	REMOVE 150 AMP SERVICE	EA	--	52.00	52.00	78.00					.433
	REMOVE BX CABLE	LF	--	.57	.57	.86					.434
	REMOVE ROMEX WIRING	LF	--	.57	.57	.86					.435
	REMOVE WHOLE HOUSE CEILING EXHAUST FAN	EA	--	54.00	54.00	81.00					.436
	REMOVE BASEBOARD HEAT-ING SYSTEM	LF	--	8.75	8.75	13.13					.437

2. CLEANING AND EXTERMINATION

SPECIFICATIONS		UNIT	JOB COST			PRICE	LOCAL AREA MODIFICATION				DATA BASE ITEM NO.
			MATLS	LABOR	TOTAL		MATLS	LABOR	TOTAL	PRICE	
CLEAN & STORE BRICKS	CLEAN BRICKS WITH PNEUMATIC TOOL AND STORE FOR RE-USE IN RESTORATION WORK	EA	--	.14	.14	.21					.600
	CLEAN BRICKS BY HAND AND STORE FOR RE-USE IN RESTORATION WORK	EA	--	.31	.31	.47					.601
CLEAN & STORE FLAGSTONE OR SLATE	CLEAN FLAGSTONE OR SLATE WITH PNEUMATIC TOOL AND STORE FOR RE-USE IN RESTORATION WORK	EA	--	.28	.28	.42					.625
	CLEAN FLAGSTONE OR SLATE BY HAND AND STORE FOR RE-USE IN RESTORATION WORK	EA	--	.45	.45	.68					.626
CLEAN WALL OR FLOOR	CLEAN BRICK OR CERAMIC TILE FLOOR OR WALL WITH MURIATIC ACID SOLUTION	SF	.01	.24	.25	.38					.602
EXTERMINATING	EXTERMINATE ROACHES AND VERMIN FROM PREMISES EA = 1 TO 3-STORY BUILDING	EA	--	134.00	134.00	201.00					.603
SANDBLAST CLEANING	SANDBLAST EXTERIOR WALL, **NO** REPOINTING OF MORTAR AND **NO** SCAFFOLDING INCLUDED SF = TOTAL AREA, INCLUDING DOOR & WINDOW OPENINGS	SF	--	--	.92	1.38					.604
CLEAN CHIMNEY	TWO STORY HOUSE										
	1 FLUE	EA	--	--	63.00	94.50					.605
	2 FLUES	EA	--	--	114.00	171.00					.606
	3 FLUES	EA	--	--	166.00	249.00					.607

2. SMOKE, FIRE AND FLOOD DAMAGE

SPECIFICATIONS		UNIT	JOB COST			PRICE	LOCAL AREA MODIFICATION				DATA BASE ITEM NO.
			MATLS	LABOR	TOTAL		MATLS	LABOR	TOTAL	PRICE	
FIRE DAMAGE TEAR-OUT	IF ANY TEAR-OUT MATERIALS SHOWN ON THE PRECEDING PAGES ARE FIRE DAMAGED **ADD**	EA	--	20%	20%	20%					.608
SCRAPE	SCRAPE FIRE DAMAGED MATERIAL										
	SIDING, BOARDS AND TRIM										
	1 X 6	LF	--	1.94	1.94	2.91					.609
	1 X 8	LF	--	2.70	2.70	4.05					.610
	1 X 10	LF	--	3.34	3.34	5.01					.611
	1 X 12	LF	--	3.88	3.88	5.82					.612
	FRAMING										
	2 X 4	LF	--	1.96	1.96	2.94					.613
	2 X 6	LF	--	2.39	2.39	3.59					.614
	2 X 8	LF	--	3.08	3.08	4.62					.615
	2 X 10	LF	--	3.57	3.57	5.36					.616
	2 X 12	LF	--	4.31	4.31	6.47					.617
	BEAMS										
	4 X 6	LF	--	2.96	2.96	4.44					.618
	6 X 10	LF	--	4.38	4.38	6.57					.619
SMOKE DAMAGE	REMOVE SMOKE FROM WALLS AND CEILINGS	SF	--	--	.35	.53					.620
	SEAL FIRE OR SMOKE DAMAGE										
	AVERAGE	SF	.10	.34	.44	.66					.621
	EXTENSIVE	SF	.10	.60	.70	1.05					.622
FLOOD DAMAGE CLEANING	CLEAN UNDAMAGED MATERIAL OF RESIDUE AND MILDEW AND DISINFECT										
	WALLS	SF	.25	.90	1.15	1.73					.623
	FLOOR	SF	.28	.95	1.23	1.85					.624

SPECIFICATIONS	UNIT	JOB COST			PRICE	LOCAL AREA MODIFICATION				DATA BASE ITEM NO.
		MATLS	LABOR	TOTAL		MATLS	LABOR	TOTAL	PRICE	

THE COST OF ASBESTOS ABATEMENT THROUGHOUT THE COUNTRY VARIES CONSIDERABLY. THE DIFFERENCE IN COSTS MAY NOT BE CAUSED AS MUCH BY DIFFERENCES IN LOCAL WAGE RATES AS BY VARYING STATE AND LOCAL GOVERNMENT RULES AND REQUIREMENTS.

SOME FACTORS AFFECTING COSTS ARE THE SPECIALIZED EQUIPMENT THAT MAY BE REQUIRED AND THE COST OF DISPOSAL OF THE ASBESTOS-CONTAINING MATERIALS. THE COST OF MOVING THE WASTE TO EPA-APPROVED DUMP SITES WILL VARY ACCORDING TO DISTANCES TRAVELED, DUMPING FEES AND REGULATIONS CONCERNING THE HANDLING OF HAZARDOUS WASTES.

BEFORE BIDDING ON AN ASBESTOS ABATEMENT PROJECT, CHECK EPA AND LOCAL REGULATIONS AND OBTAIN FIRM BIDS FOR THE WORK FROM TECHNICIANS WHO YOU ARE CERTAIN ARE PROPERLY QUALIFIED TO COMPLETE THE WORK.

THE COSTS SHOWN BELOW ARE INTENDED AS A GENERAL GUIDE, AND ARE **NOT** COSTS FOR ANY PARTICULAR AREA OF THE COUNTRY.

SPECIFICATIONS		UNIT	MATLS	LABOR	TOTAL	PRICE					DATA BASE ITEM NO.
REMOVAL	REMOVE ASBESTOS FROM CEILING										
	MINIMUM	EA	--	--	1250.00	1,875.00					.700
		PLUS SF	--	--	3.75	5.63					.701
	MAXIMUM	EA	--	--	2490.00	3,735.00					.702
		PLUS SF	--	--	16.25	24.38					.703
	REMOVE ASBESTOS FROM UP TO 6" PIPING										
	MINIMUM	EA	--	--	1250.00	1,875.00					.704
		PLUS LF	--	--	3.75	5.63					.705
	MAXIMUM	EA	--	--	2490.00	3,735.00					.706
		PLUS LF	--	--	28.00	42.00					.707
ENCAPSU-LATION	ENCAPSULATE ASBESTOS PIPE COVERING WITH SEAL-ANTS										
	MINIMUM	EA	--	--	620.00	930.00					.708
		PLUS LF	--	--	3.75	5.63					.709
	MAXIMUM	EA	--	--	3750.00	5,625.00					.710
		PLUS LF	--	--	11.20	16.80					.711

3. EXCAVATION AND CLEARING BY HAND

SPECIFICATIONS		UNIT	JOB COST			PRICE	LOCAL AREA MODIFICATION				DATA BASE ITEM NO.
			MATLS	LABOR	TOTAL		MATLS	LABOR	TOTAL	PRICE	
EXCAVA-TION	NOTE: ALL COSTS ARE BASED ON CUBIC **FEET** OF DIRT EXCAVATED										
	HAND-DIGGING AND PLACING DIRT IN WHEELBARROW OR ON GROUND NEXT TO EXCA-VATION. EXCAVATION NOT OVER 6 FEET DEEP. EXCESS DIRT WHEELED 50 FEET OR LESS.										
	DIRT, GRAVEL OR TOPSOIL	CF	--	1.13	1.13	1.70					.000
	HARD CLAY OR LOOSE ROCK	CF	--	2.03	2.03	3.05					.001
FROZEN SOIL	EXCAVATE FROZEN SOIL WITH PICK AND SHOVEL	CF	--	3.85	3.85	5.78					.002
DIFFICULT EXCAVA-TION	EXCAVATE BY HAND AROUND PIPES, SERVICES AND OTHER OBSTRUCTIONS	CF	--	3.26	3.26	4.89					.003
CRAWL SPACE	DIG OUT EXISTING CRAWL SPACE FOR FULL 18" CLEAR-ANCE UNDER FLOOR JOISTS										
	EASY ACCESS (OPEN FLOOR)	CF	--	1.55	1.55	2.33					.004
	DIFFICULT ACCESS (DIG THROUGH OPEN JOISTS)	CF	--	3.04	3.04	4.56					.005
LOWER EX-ISTING BASEMENT FLOOR	EXCAVATE BY HAND WITH PNEUMATIC TOOL, WHEEL DIRT TO OUTSIDE OR RE-MOVE IT BY BELT CONVEY-OR. DOES **NOT** INCLUDE BREAK-UP OF EXISTING CON-CRETE FLOOR OR UNDER-PINNING EXISTING FOUNDA-TION WALLS										
	EASY ACCESS	CF	--	1.55	1.55	2.33					.006
	DIFFICULT ACCESS	CF	--	2.58	2.58	3.87					.007
	CF = CUBIC FEET OF DIRT EXCAVATED										
CLEAR TREES	CLEAR TREES BY HAND WITH CHAIN SAW, CUT INTO SHORT LENGTHS AND LOAD INTO TRUCK OR STACK										
	8" TREE	EA	--	47.00	47.00	70.50					.100
	12" TREE	EA	--	56.00	56.00	84.00					.101
	18" TREE	EA	--	73.00	73.00	109.50					.102
	24" TREE	EA	--	77.00	77.00	115.50					.103
REMOVE STUMP	REMOVE STUMP FROM GROUND WITH BULLDOZER AND LABORER										
	ADD TO ABOVE	EA	--	120%	120%	120%					.104

SPECIFICATIONS		UNIT	JOB COST			PRICE	LOCAL AREA MODIFICATION				DATA BASE ITEM NO.
			MATLS	LABOR	TOTAL		MATLS	LABOR	TOTAL	PRICE	
BACKFILL BY HAND	BACKFILL BY HAND FROM PILES NEXT TO EXCAVATION	CF	--	.40	.40	.60					.008
COMPACT	COMPACT BACKFILLED DIRT										
	WITH HAND TAMPER	CF	--	.31	.31	.47					.009
	WITH PNEUMATIC TAMPER	CF	--	.13	.13	.20					.010
MOVE FILL DIRT	MOVE AND SPREAD FILL DIRT BY WHEELBARROW AFTER DIRT DUMPED ON PREMISES IN PILES	CF	--	.31	.31	.47					.011
FINISH GRADING	PLACING TOPSOIL DELIVERED BY TRUCK, TOPSOIL @ $10 PER CUBIC YARD DELIVERED										
	4"	SF	.15	.09	.24	.36					.012
	6"	SF	.20	.12	.32	.48					.013
	RAKING AND SEEDING BY HAND	SF	.09	.11	.20	.30					.014
	SODDING, LABOR AND MATERIALS BY NURSERY PERSONNEL, 1-1/2" DEEP										
	BLUEGRASS	SF	--	.68	.68	1.02					.015
	SUNSHADE	SF	--	.76	.76	1.14					.016
AREAWAY	EXCAVATE FOR AND INSTALL CORRUGATED METAL RETAINING WALL AROUND BASEMENT WINDOW, WITH GRAVEL OR LOOSE BRICK LAID OVER 6" SAND BED, GALVANIZED CORRUGATED WALL 24" X 37" NOTE: FOR AREAWAY WITH BRICK WALLS, SEE "MASONRY"	EA	26.00	46.00	72.00	108.00					.017
SOIL TREATMENT	TREAT SOIL FOR PERMANENT TERMITE PREVENTION, OPEN ACCESS	SF	.14	.08	.22	.33					.018

3

3. MACHINE EXCAVATION

SPECIFICATIONS		UNIT	JOB COST			PRICE	LOCAL AREA MODIFICATION				DATA BASE ITEM NO.
			MATLS	LABOR	TOTAL		MATLS	LABOR	TOTAL	PRICE	
	THE LABOR COSTS SHOWN BELOW INCLUDE A MACHINE OPERATOR'S OVERHEAD AND PROFIT										
	NOTE: MACHINE DIGGING COSTS HERE ARE BASED ON CUBIC **FEET** OF DIRT. TO CONVERT CUBIC FEET TO CUBIC YARDS, DIVIDE THE TOTAL CUBIC FEET BY 27, AND THE RESULT IS TOTAL CUBIC YARDS.										
EXCAVA-TION	WITH BACK HOE OR FRONT END LOADER	EA PLUS	--	350.00	350.00	525.00					.105
		CF	--	.15	.15	.23					.106
BACKFILL	BACKFILL WITH BULLDOZER OR FRONT END LOADER. **NO** ROUGH GRADING INCLUDED	EA PLUS	--	250.00	250.00	375.00					.107
		SF	--	.13	.13	.20					.108
EXCAVATE BY MACHINE & REMOVE EXCESS DIRT	WITH FRONT END LOADER OR BACKHOE AND 2 TRUCKS, INCLUDING 30-MIN. TRIPS PER LOAD FOR DUMPING, DUMPING FEES **NOT** INCLUD-ED	EA PLUS	--	470.00	470.00	705.00					.109
		CF	--	.30	.30	.45					.110
ROOM AD-DITION EX-CAVATION	EXCAVATE FOR ROOM ADDI-TION, BASEMENT OR CRAWL SPACE WITH FRONT END LOADER OR BACKHOE, IN-CLUDING MINIMUM OF 2 FEET BEYOND ADDITION LINE AT ALL WALLS										
	<u>DEPTH BELOW GRADE</u>	EA PLUS	--	340.00	340.00	510.00					.124
	3-0	SF	--	.22	.22	.33					.125
	4-0	SF	--	.28	.28	.42					.126
	5-0	SF	--	.36	.36	.54					.127
	6-0	SF	--	.43	.43	.65					.128
	7-0	SF	--	.49	.49	.74					.129
	8-0	SF	--	.52	.52	.78					.130
	9-6	SF	--	.65	.65	.98					.131
	SF = PLAN DIMENSION OF BASEMENT OR GROUND FLOOR										

SPECIFICATIONS		UNIT	JOB COST			PRICE	LOCAL AREA MODIFICATION				DATA BASE ITEM NO.
			MATLS	LABOR	TOTAL		MATLS	LABOR	TOTAL	PRICE	
FOOTING DRAIN PIPE	4" CORRUGATED PERFORATED PLASTIC PIPE SURROUNDED WITH 6" OF 3/4" AGGREGATE EXTENDING ALONG EXTERIOR PERIMETER OF FOOTING, COVER WITH VAPOR BARRIER, LEVEL GROUND FOR INSTALLATION OF TILE, BUT NO DIGGING OUT	LF	1.40	1.83	3.23	4.85					.111
EXTEND DRAIN PIPE TO OUTLET	DIG AND EXTEND PERIMETER DRAIN PIPE SYSTEM TO "DAYLIGHT" OR A MINIMUM OF 30 FEET TO DRYWELL, SLOPE PIPE 1/8" PER LINEAL FOOT. COVER WITH VAPOR BARRIER, BACKFILL										
	AVERAGE BELOW GRADE										
	12"	LF	1.00	3.04	4.04	6.06					.112
	24"	LF	1.00	4.08	5.08	7.62					.113
	36"	LF	1.00	5.07	6.07	9.11					.114
	48"	LF	1.00	6.51	7.51	11.27					.115
	60"	LF	1.00	8.00	9.00	13.50					.116
DRYWELL	CONSTRUCT DRYWELL TO RECEIVE DRAINAGE FROM PERIMETER DRAIN PIPE SYSTEM AS FOLLOWS: DIG AND BURY 55-GALLON CAPACITY STEEL OR PLASTIC DRUM, PERFORATED WITH 2" DIAMETER HOLES AND FILLED WITH 1-1/2" AGGREGATE, COVER WITH VAPOR BARRIER, BACKFILL	EA	142.00	194.00	336.00	504.00					.117

3. TREATED WOOD RETAINING WALL

SPECIFICATIONS		UNIT	JOB COST			PRICE	LOCAL AREA MODIFICATION				DATA BASE ITEM NO.
			MATLS	LABOR	TOTAL		MATLS	LABOR	TOTAL	PRICE	
WOOD RE-TAINING WALL (TIE WALL)	• 6" X 8" PRESSURE TREATED TIMBERS • 6" GRAVEL FILL UNDER • 12" GRAVEL BACKFILL BEHIND WALL • 1/2" ROD CONNECTORS, 4'-0" O.C. • TIEBACKS NAILED TO DEADMAN TIMBER (WALL ANCHOR) 6'-0" O.C. EVERY OTHER COURSE HEIGHT OF WALL ABOVE BOTTOM OF WALL										
	24"	LF	24.26	10.50	34.76	52.14					.118
	36"	LF	37.90	15.80	53.70	80.55					.119
	48"	LF	48.50	21.00	69.50	104.25					.120
	60"	LF	62.00	26.00	88.00	132.00					.121
	72"	LF	73.00	31.50	104.50	156.75					.122
DRAIN PIPE	INSTALL 4" PERFORATED PLASTIC PIPE SURROUNDED WITH 6" OF 3/4" AGGREGATE ALONG BACK OF WALL EXTENDING TO DAYLIGHT OR DRYWELL (SEE "DRYWELL", PAGE 97)	LF	1.40	2.10	3.50	5.25					.123

4. TYPES OF CONCRETE FOOTINGS, WALLS, PIERS, SLABS

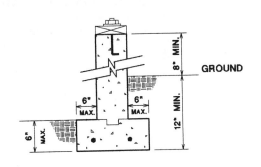

CONCRETE FOUNDATION WALL

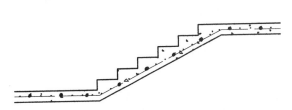

SUSPENDED CONCRETE SLAB & STAIR

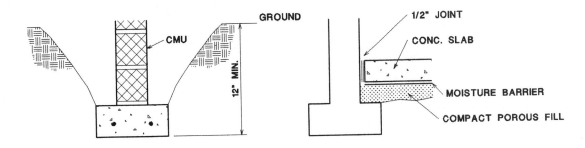

CONCRETE FOOTING FOR BLOCK WALL

FOOTING/FOUNDATION WITH SLAB

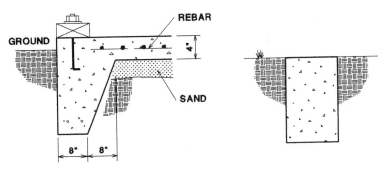

MONOLITHIC FOOTING & SLAB

FOOTING TO GRADE

CONCRETE WALL

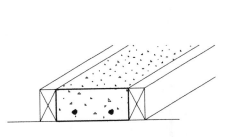

FOUNDATION FOOTING POURED IN FORMS

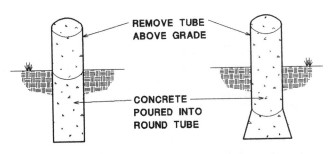

ROUND PIER

ROUND PIER WITH INTEGRAL FOOTING

4. FOOTINGS, MACHINE EXCAVATION

SPECIFICATIONS	UNIT	JOB COST			PRICE	LOCAL AREA MODIFICATION				DATA BASE ITEM NO.
		MATLS	LABOR	TOTAL		MATLS	LABOR	TOTAL	PRICE	

<table>
<tr><td colspan="11">THE LABOR COSTS SHOWN BELOW FOR MACHINE EXCAVATION INCLUDE A MACHINE OPERATOR'S OVERHEAD AND PROFIT</td></tr>
</table>

4 FOOTINGS FOR WALL

SPECIFICATIONS	UNIT	MATLS	LABOR	TOTAL	PRICE	MATLS	LABOR	TOTAL	PRICE	DATA BASE ITEM NO.
• DIG OUT BY MACHINE, WIDTH OF FOOTING • POUR • BACKFILL AFTER WALL BUILT ON FOOTING • DIG TO 12" BELOW GRADE (BOTTOM OF FOOTING)	EA PLUS	--	220.00	220.00	330.00					.100
SIZE OF FOOTING Height 8" Width 16"	LF	2.91	.86	3.77	5.66					.101
12" 12"	LF	3.26	.96	4.22	6.33					.103
10" 20"	LF	4.55	1.34	5.89	8.84					.104
12" 24"	LF	6.53	1.93	8.46	12.69					.105
12" 36"	LF	9.84	2.89	12.73	19.10					.106
SAME AS ABOVE, DIG OUT TO 24" BELOW GRADE	EA PLUS	--	220.00	220.00	330.00					.100
SIZE OF FOOTING Height 8" Width 16"	LF	2.91	1.71	4.62	6.93					.107
12" 12"	LF	3.26	1.93	5.19	7.79					.109
10" 20"	LF	4.55	2.68	7.23	10.85					.110
12" 24"	LF	6.53	3.85	10.38	15.57					.111
12" 36"	LF	9.84	5.78	15.62	23.43					.112
SAME AS ABOVE, DIG OUT TO 36" BELOW GRADE	EA PLUS	--	220.00	220.00	330.00					.100
SIZE OF FOOTING Height 8" Width 16"	LF	2.91	2.59	5.50	8.25					.113
12" 12"	LF	3.26	2.89	6.15	9.23					.114
10" 20"	LF	4.55	4.01	8.56	12.84					.115
12" 24"	LF	6.53	5.78	12.31	18.47					.116
12" 36"	LF	9.84	8.67	18.51	27.77					.117
SAME AS ABOVE, DIG OUT TO 48" BELOW GRADE	EA PLUS	--	220.00	220.00	330.00					.100
SIZE OF FOOTING Height 8" Width 16"	LF	2.91	3.45	6.36	9.54					.118
12" 12"	LF	3.26	3.85	7.11	10.67					.119
10" 20"	LF	4.55	5.35	9.90	14.85					.120
12" 24"	LF	6.53	7.70	14.23	21.35					.121
12" 36"	LF	9.84	11.56	21.40	32.10					.122
SAME AS ABOVE, DIG OUT TO 60" BELOW GRADE	EA PLUS	--	220.00	220.00	330.00					.100
SIZE OF FOOTING Height 8" Width 16"	LF	2.91	4.28	7.19	10.79					.123
12" 12"	LF	3.26	4.82	8.08	12.12					.124
10" 20"	LF	4.55	6.69	11.24	16.86					.125
12" 24"	LF	6.53	9.63	16.16	24.24					.126
12" 36"	LF	9.84	14.45	24.29	36.44					.127

SPECIFICATIONS		UNIT	JOB COST			PRICE	LOCAL AREA MODIFICATION				DATA BASE ITEM NO.
			MATLS	LABOR	TOTAL		MATLS	LABOR	TOTAL	PRICE	
FOOTINGS FOR WALL	• DIG OUT BY HAND, WIDTH OF FOOTING • POUR • BACKFILL AFTER WALL BUILT ON FOOTING • DIG TO 12" BELOW GRADE (BOTTOM OF FOOTING) ***SIZE OF FOOTING*** Height 8" Width 16"	LF	2.91	3.55	6.46	9.69					.200
	12" 12"	LF	3.26	3.00	6.26	9.39					.202
	10" 20"	LF	4.55	4.73	9.28	13.92					.203
	12" 24"	LF	6.53	6.00	12.53	18.80					.204
	12" 36"	LF	9.84	9.00	18.84	28.26					.205
	SAME AS ABOVE, DIG OUT TO 24" BELOW GRADE ***SIZE OF FOOTING*** Height 8" Width 16"	LF	2.91	6.23	9.14	13.71					.206
	12" 12"	LF	3.26	5.00	8.26	12.39					.208
	10" 20"	LF	4.55	8.07	12.62	18.93					.209
	12" 24"	LF	6.53	10.00	16.53	24.80					.210
	12" 36"	LF	9.84	15.00	24.84	37.26					.211
	SAME AS ABOVE, DIG OUT TO 36" BELOW GRADE ***SIZE OF FOOTING*** Height 8" Width 16"	LF	2.91	8.89	11.80	17.70					.212
	12" 12"	LF	3.26	7.00	10.26	15.39					.213
	10" 20"	LF	4.55	11.39	15.94	23.91					.214
	12" 24"	LF	6.53	14.00	20.53	30.80					.215
	12" 36"	LF	9.84	21.00	30.84	46.26					.216
	SAME AS ABOVE, DIG OUT TO 48" BELOW GRADE ***SIZE OF FOOTING*** Height 8" Width 16"	LF	2.91	11.54	14.45	21.68					.217
	12" 12"	LF	3.26	9.00	12.26	18.39					.218
	10" 20"	LF	4.55	14.73	19.28	28.92					.219
	12" 24"	LF	6.53	18.00	24.53	36.80					.220
	12" 36"	LF	9.84	27.00	36.84	55.26					.221
STEEL BARS IN FOOTING	CONTINUOUS 1/2" STEEL BAR REINFORCED 2 BARS	LF	.98	.22	1.20	1.80					.222
	3 BARS	LF	1.44	.32	1.76	2.64					.223
	DRILL 1/2" BAR 6" INTO EXISTING FOUNDATION BLOCK	EA	.98	3.75	4.73	7.10					.224
	BRICK	EA	.98	4.25	5.23	7.85					.225
	CONCRETE	EA	.98	5.15	6.13	9.20					.226

4. CONCRETE FOOTINGS TO GRADE

SPECIFICATIONS		UNIT	JOB COST			PRICE	LOCAL AREA MODIFICATION				DATA BASE ITEM NO.
			MATLS	LABOR	TOTAL		MATLS	LABOR	TOTAL	PRICE	
	THE LABOR COSTS SHOWN HERE FOR MACHINE TRENCHING INCLUDE A MACHINE OPERATOR'S OVERHEAD AND PROFIT										
MACHINE DUG FOOTINGS	DIG 8" WIDE TRENCH BY MACHINE AND FILL WITH CONCRETE TO GRADE, INCLUDING DIGGING AND POURING	EA PLUS	--	220.00	220.00	330.00					.300
	BOTTOM OF FOOTING UP TO GRADE										
	12"	LF	1.99	.46	2.45	3.68					.301
	24"	LF	3.96	.92	4.88	7.32					.302
	36"	LF	5.96	1.40	7.36	11.04					.303
	48"	LF	7.96	1.85	9.81	14.72					.304
	60"	LF	9.94	2.32	12.26	18.39					.305
	SAME AS ABOVE, 12" WIDE TRENCH	EA PLUS	--	220.00	220.00	330.00					.300
	BOTTOM OF FOOTING UP TO GRADE										
	12"	LF	3.25	.69	3.94	5.91					.306
	24"	LF	6.55	1.40	7.95	11.93					.307
	36"	LF	9.80	2.09	11.89	17.84					.308
	48"	LF	13.11	2.78	15.89	23.84					.309
	60"	LF	16.52	3.42	19.94	29.91					.310
HAND DUG FOOTINGS	DIG 8" WIDE TRENCH BY HAND AND FILL WITH CONCRETE TO GRADE, INCLUDING DIGGING AND POURING										
	BOTTOM OF FOOTING UP TO GRADE										
	12"	LF	1.99	2.15	4.14	6.21					.311
	24"	LF	3.96	4.28	8.24	12.36					.312
	36"	LF	5.96	6.43	12.39	18.59					.313
	48"	LF	7.96	8.56	16.52	24.78					.314
	SAME AS ABOVE, 12" WIDE TRENCH										
	BOTTOM OF FOOTING UP TO GRADE										
	12"	LF	3.25	2.58	5.83	8.75					.315
	24"	LF	6.55	5.15	11.70	17.55					.316
	36"	LF	9.80	7.73	17.53	26.30					.317
	48"	LF	13.11	10.27	23.38	35.07					.318
WALL ABOVE GRADE	• BUILD LEVEL FORMS ABOVE GRADE FOR 6" THICK WALL EXTENDING TO 12" ABOVE GRADE OVER ANY OF ABOVE FOOTINGS • POUR CONCRETE AND STRIP FORMS **ADD**	LF	3.25	3.90	7.15	10.73					.321
REBAR IN FOOTING	CONTINUOUS NO. 4 (1/2") STEEL BAR REINFORCEMENT										
	2 BARS	LF	.98	.21	1.19	1.79					.319
	3 BARS	LF	1.44	.30	1.74	2.61					.320

SPECIFICATIONS		UNIT	JOB COST			PRICE	LOCAL AREA MODIFICATION				DATA BASE ITEM NO.
			MATLS	LABOR	TOTAL		MATLS	LABOR	TOTAL	PRICE	
WALL FOOTINGS	CONTINUOUS CONCRETE FOOTINGS FOR BLOCK, BRICK OR CONCRETE WALL FOR FULL BASEMENT OR CRAWL SPACE • IN EXISTING EXCAVATION • BUILD LEVEL FORMS FOR FOOTINGS WITH 2 X LUMBER (ONE USE) • POUR CONCRETE • STRIP FORMS **SIZE OF FOOTING**										
	8" X 16"	LF	4.51	5.50	10.01	15.02					.400
	10" X 20"	LF	6.50	5.85	12.35	18.53					.401
	12" X 24"	LF	8.82	6.20	15.02	22.53					.402
	12" X 36"	LF	10.35	6.94	17.29	25.94					.403
KEYWAY	PLACE 2" X 4" KEYWAY FOR POURED CONCRETE WALL AND STRIP AFTER POURING	LF	.40	.75	1.15	1.73					.404
STEPPED FOOTING	CONCRETE WALL FOOTINGS BUILT ON INCLINE -- FOR EACH ONE FOOT STEP IN FOOTING, **ADD**										
	8" X 16"	EA	5.43	6.00	11.43	17.15					.405
	10" X 20"	EA	6.10	6.00	12.10	18.15					.406
	12" X 24"	EA	7.25	6.00	13.25	19.88					.407
	12" X 36"	EA	9.00	6.00	15.00	22.50					.408
	EA = EACH STEP										
REBAR	CONTINUOUS NO. 4 (1/2") STEEL BAR REINFORCEMENT										
	2 BARS	LF	.98	.21	1.19	1.79					.319
	3 BARS	LF	1.44	.30	1.74	2.61					.320
COLUMN FOOTING	IN EXISTING BASEMENT OR CRAWL SPACE EXCAVATION, FORM WITH 2 X AND POUR CONCRETE FOOTING FOR STEEL OR MASONRY COLUMN UNDER WOOD OR STEEL BEAM										
	8" THICK X 12" X 12"	EA	7.82	8.40	16.22	24.33					.411
	12" THICK X 12" X 12"	EA	9.42	8.40	17.82	26.73					.412
	12" THICK X 16" X 16"	EA	15.10	13.20	28.30	42.45					.413
	12" THICK X 24" X 24"	EA	23.44	15.20	38.64	57.96					.414

4. PIER FOOTINGS

SPECIFICATIONS			UNIT	JOB COST			PRICE	LOCAL AREA MODIFICATION				DATA BASE ITEM NO.
				MATLS	LABOR	TOTAL		MATLS	LABOR	TOTAL	PRICE	
PIER FOOTINGS	• DIG OUT BY HAND • POUR CONCRETE • BACKFILL AFTER PIER IS BUILT • BOTTOM OF PIER FOOTING 12" BELOW GRADE **SIZE OF PIER FOOTINGS** **Length Width Height**											
	16"	16"	12"	EA	14.00	16.00	30.00	45.00					.500
	24"	24"	12"	EA	25.00	21.00	46.00	69.00					.501
24" BELOW GRADE												
	16"	16"	12"	EA	14.00	30.00	44.00	66.00					.502
	24"	24"	12"	EA	25.00	30.00	55.00	82.50					.503
	30"	30"	18"	EA	42.00	34.00	76.00	114.00					.504
	36"	48"	18"	EA	75.00	54.00	129.00	193.50					.505
36" BELOW GRADE												
	16"	16"	12"	EA	14.00	33.00	47.00	70.50					.506
	24"	24"	12"	EA	25.00	33.00	58.00	87.00					.507
	30"	30"	18"	EA	42.00	41.00	83.00	124.50					.508
	36"	48"	18"	EA	75.00	81.00	156.00	234.00					.509
48" BELOW GRADE												
	16"	16"	12"	EA	14.00	42.00	56.00	84.00					.510
	24"	24"	12"	EA	25.00	42.00	67.00	100.50					.511
	30"	30"	18"	EA	42.00	47.00	89.00	133.50					.512
	36"	48"	18"	EA	75.00	108.00	183.00	274.50					.513
60" BELOW GRADE												
	16"	16"	12"	EA	14.00	48.00	62.00	93.00					.514
	24"	24"	12"	EA	25.00	48.00	73.00	109.50					.515
	30"	30"	18"	EA	42.00	71.00	113.00	169.50					.516
	36"	48"	18"	EA	75.00	135.00	210.00	315.00					.517

SPECIFICATIONS		UNIT	JOB COST			PRICE	LOCAL AREA MODIFICATION				DATA BASE ITEM NO.
			MATLS	LABOR	TOTAL		MATLS	LABOR	TOTAL	PRICE	
ROUND PIERS TO GRADE	• DIG OUT BY HAND • INSTALL ROUND SONO-TUBE ON GROUND OR EXISTING FOOTING • POUR CONCRETE • BACKFILL										
Diameter	Bottom of Footing Up To Grade										
8"	12"	EA	4.50	12.50	17.00	25.50					.518
	24"	EA	9.00	19.00	28.00	42.00					.519
	36"	EA	13.00	26.00	39.00	58.50					.520
	48"	EA	17.00	32.00	49.00	73.50					.521
	60"	EA	22.00	40.00	62.00	93.00					.522
12"	12"	EA	8.00	14.00	22.00	33.00					.523
	24"	EA	17.00	21.00	38.00	57.00					.524
	36"	EA	24.50	27.00	51.50	77.25					.525
	48"	EA	33.00	34.00	67.00	100.50					.526
	60"	EA	41.00	41.00	82.00	123.00					.527
	SAME AS ABOVE, INCLUDING CONCRETE INTEGRAL FOOTING UNDER, FLARED TO 16"										
Diameter	Bottom of Footing Up To Grade										
8"	12"	EA	8.00	17.00	25.00	37.50					.528
	24"	EA	13.00	24.00	37.00	55.50					.529
	36"	EA	17.00	30.00	47.00	70.50					.530
	48"	EA	21.00	37.00	58.00	87.00					.531
	60"	EA	26.00	44.00	70.00	105.00					.532
12"	12"	EA	9.00	18.00	27.00	40.50					.533
	24"	EA	17.00	25.00	42.00	63.00					.534
	36"	EA	26.00	32.00	58.00	87.00					.535
	48"	EA	34.00	38.00	72.00	108.00					.536
	60"	EA	42.00	45.00	87.00	130.50					.537
ROUND PIERS EXTENDING ABOVE GRADE	EXTEND ROUND PIERS ABOVE GRADE, CUTTING SONOTUBE OFF AT GRADE AFTER POURING										
	Diameter										
	8"	LF	4.37	2.84	7.21	10.82					.538
	12"	LF	8.23	3.48	11.71	17.57					.539
	LF = FROM GRADE TO TOP OF PIER										

4

4. CONCRETE FOOTING AND FOUNDATION WALL

| SPECIFICATIONS | | UNIT | JOB COST | | | PRICE | LOCAL AREA MODIFICATION | | | | DATA BASE ITEM NO. |
			MATLS	LABOR	TOTAL		MATLS	LABOR	TOTAL	PRICE	
	THE LABOR COSTS SHOWN BELOW FOR CONCRETE FOOTINGS AND FOUNDATION WALL INCLUDE A CONCRETE SUBCONTRACTOR'S OVERHEAD AND PROFIT										
CONCRETE FOOTINGS AND FOUNDATION WALL	CONCRETE FOOTINGS AND FOUNDATION WALL FOR FULL BASEMENT OR CRAWL SPACE										
	• FORM CONTINUOUS CONCRETE FOOTINGS TWICE THE THICKNESS OF WALL										
	• POUR FOOTINGS BY CHUTE FROM READY-MIX TRUCK, INCL. KEYWAY										
	• STRIP FORMS AND CLEAN										
	• SET UP PREBUILT SECTIONAL FORMS ON CONCRETE FOOTINGS										
	• SET WINDOWS & STOPS, CUTOUTS & CRAWL HOLES										
	• POUR 3000 PSI CONCRETE MIX INTO FORMS FROM READY-MIX TRUCK										
	• STRIP FORMS AND CLEAN										
	• HAND RUB WALLS WHERE REQUIRED										

COSTS	WALL THICKNESS	WALL HEIGHT ABOVE BOTTOM OF FOOTING	UNIT	MATLS	LABOR	TOTAL	PRICE					DATA BASE ITEM NO.
	8"	24"	LF	5.93	17.88	23.81	35.72					.000
		36"	LF	7.70	22.70	30.40	45.60					.001
		48"	LF	9.47	27.52	36.99	55.49					.002
		60"	LF	11.24	32.34	43.58	65.37					.003
		72"	LF	13.01	37.16	50.17	75.26					.004
		84"	LF	14.78	41.98	56.76	85.14					.005
		96"	LF	16.55	46.80	63.35	95.03					.006
	10"	24"	LF	8.10	18.94	27.04	40.56					.007
		36"	LF	10.31	23.82	34.13	51.20					.008
		48"	LF	12.52	28.70	41.22	61.83					.009
		60"	LF	14.73	33.58	48.31	72.47					.010
		72"	LF	16.94	38.46	55.40	83.10					.011
		84"	LF	19.15	43.34	62.49	93.74					.012
		96"	LF	21.36	48.22	69.58	104.37					.013
	12"	24"	LF	12.46	20.08	32.54	48.81					.014
		36"	LF	16.04	25.02	41.06	61.59					.015
		48"	LF	19.62	29.96	49.58	74.37					.016
		60"	LF	23.20	34.90	58.10	87.15					.017
		72"	LF	26.78	39.84	66.62	99.93					.018
		84"	LF	30.96	44.78	75.74	113.61					.019
		96"	LF	33.94	49.72	83.66	125.49					.020

SPECIFICATIONS		UNIT	JOB COST			PRICE	LOCAL AREA MODIFICATION				DATA BASE ITEM NO.
			MATLS	LABOR	TOTAL		MATLS	LABOR	TOTAL	PRICE	
STEEL BAR REIN-FORCE-MENT	CONTINUOUS 1/2" STEEL BAR REINFORCEMENT IN FOOT-INGS										
	2 BARS	LF	.98	.22	1.20	1.80					.021
	3 BARS	LF	1.44	.30	1.74	2.61					.022
	LF = FOOTINGS										
	3/8" STEEL BAR REINFORCE-MENT IN WALLS, 24" O.C., VERTICAL AND HORIZONTAL GRID SF = ONE FACE OF WALL	SF	.98	.22	1.20	1.80					.023
WHEEL CONCRETE	WHEEL CONCRETE FOR CON-CRETE FOUNDATION WALL FROM TRUCK UP TO 75 FEET AND POUR INTO FORMS, IN-CLUDING SETTING UP AND REMOVING RAMP IF RE-QUIRED AND TRANSIT-MIX TRUCK WAITING TIME **ADD**	LF	45%	12%	30%	30%					.024 .025
PUMP CON-CRETE	FOR CONCRETE FOUNDA-TION WALL, LESS THAN 300' FROM CONCRETE TRUCK, INCLUDING TRUCK AND OPERATOR AND TRANSIT-MIX TRUCK WAITING TIME **ADD**	EA PLUS	370.00	--	370.00	555.00					.026
		LF	25%	--	12%	12%					.027

4. MONOLITHIC FOOTING AND SLAB

SPECIFICATIONS		UNIT	JOB COST			PRICE	LOCAL AREA MODIFICATION				DATA BASE ITEM NO.
			MATLS	LABOR	TOTAL		MATLS	LABOR	TOTAL	PRICE	
MONO-LITHIC FOOTING AND SLAB (SINGLE POUR)	• DIG, FORM, PLACE REIN-FORCEMENT • 4" GRAVEL FILL • 6 X 6 #10 WOVEN WIRE MESH • VAPOR BARRIER • POUR AND FINISH CON-CRETE • BOTTOM OF SLAB FOOT-ING 8", TOP OF FOOTING BELOW SLAB 16" WIDE • SLAB THICKNESS 4" • FOOTING BELOW FROST 3 OR 4 SIDES • 1/2" STEEL BARS DRILLED INTO EXISTING BUILDING • TWO 1/2" CONTINUOUS STEEL BARS IN FOOTING										
	TOP SLAB	SF PLUS	1.70	1.47	3.17	4.76					.600
	FOOTING BELOW GRADE 12"	LF	5.61	2.84	8.45	12.68					.601
	24"	LF	9.21	5.36	14.57	21.86					.602
	36"	LF	12.86	7.88	20.74	31.11					.603
	48"	LF	16.35	10.50	26.85	40.28					.604
	LF = OUTSIDE PERIMETER MEASUREMENT OF FOOTING										
ADDI-TIONAL FOOTING	ADDITIONAL 12" WIDE FOOT-ING UNDER SLAB TO SUP-PORT PARTITION, BEAM OR SLAB ITSELF										
	BOTTOM OF FOOTING BELOW GRADE										
	12"	LF	5.61	2.84	8.45	12.68					.606
	24"	LF	9.21	5.36	14.57	21.86					.607
WHEEL CONCRETE	WHEEL CONCRETE FOR MONOLITHIC FOOTING AND SLAB FROM TRUCK UP TO 75 FEET AND POUR, INCLUDING SETTING UP AND REMOVING RAMP IF REQUIRED AND TRANSIT-MIX TRUCK WAITING TIME **ADD**	SF	45%	12%	33%	33%					.609 .610
	SF = TOP SURFACE OF SLAB										
PUMP CONCRETE	FOR MONOLITHIC FOOTING AND SLAB, LESS THAN 300' FROM CONCRETE TRUCK, IN-CLUDING TRUCK, OPERATOR AND READY-MIX TRUCK WAITING TIME **ADD**	EA PLUS	374.00	--	374.00	561.00					.611
	SF = TOP SURFACE OF SLAB	SF	25%	--	12%	12%					.612

SPECIFICATIONS		UNIT	JOB COST			PRICE	LOCAL AREA MODIFICATION				DATA BASE ITEM NO.
			MATLS	LABOR	TOTAL		MATLS	LABOR	TOTAL	PRICE	
SLAB AND STEP(S) ON GRADE	BUILD 6'-0" x 4'-0" CONCRETE SLAB AND STEP(S) ON GRADE • EXCAVATE TRENCH TO 36" BELOW GRADE • DRILL INTO EXISTING HOUSE WALL AND IN-STALL FOUR 1/2" STEEL BARS TO CONNECT FOOTING • POUR CONCRETE FOOT-ING TO GRADE • BUILD FORMS & POUR CONCR. FOUNDATION WALL ABOVE GRADE TO SUPPORT SLAB & STEP(S) • BUILD FORMS FOR SLAB AND STEP(S) • PLACE RUBBLE OR FORMS IN CENTER • PLACE STEEL • POUR CONCRETE SLAB AND STEP(S) • STRIP & CLEAN FORMS • FINISH CONCRETE • STEPS: 6"-7" RISERS, 12" TREADS	EA PLUS	131.00	127.00	258.00	387.00					.640
	Width of Tread										
	48"	Tread	61.00	79.00	140.00	210.00					.641
	60"	Tread	68.00	82.00	150.00	225.00					.642
	72"	Tread	72.00	84.00	156.00	234.00					.643
	EA = EACH JOB TREAD = EACH TREAD										

4. CONCRETE SLAB

SPECIFICATIONS		UNIT	JOB COST			PRICE	LOCAL AREA MODIFICATION				DATA BASE ITEM NO.
			MATLS	LABOR	TOTAL		MATLS	LABOR	TOTAL	PRICE	
BASEMENT FLOOR SLAB	• LEVEL GROUND WITHIN EXISTING FOUNDATION WALLS, BUT **NO** DIGGING OUT • FLOATED & TROWELED • 4" GRAVEL FILL • VAPOR BARRIER UNDER • 6 X 6 #10 WOVEN WIRE MESH										
	3"	SF	1.40	1.52	2.92	4.38					.613
	4"	SF	1.66	1.55	3.21	4.82					.614
	5"	SF	1.96	1.60	3.56	5.34					.615
	6"	SF	2.28	1.63	3.91	5.87					.616
SLAB, EXTERIOR	ON GRADE, SCREED COAT, FLOATED 2"	SF	1.23	1.04	2.27	3.41					.617
	ON GRADE • DIG OUT AND LEVEL GROUND • SPREAD 4" OF SAND, GRAVEL OR CINDERS • FORM AND POUR CONCRETE • FLOATED & TROWELED • VAPOR BARRIER UNDER • 6 X 6 #10 WWM • #4 BARS DRILLED INTO WALL OF EXISTING MASONRY BUILDING										
	4"	SF	1.53	1.45	2.98	4.47					.618
	5"	SF	1.74	1.47	3.21	4.82					.619
	6"	SF	2.05	1.51	3.56	5.34					.620
SLAB, SUSPENDED	SLAB TO BE PLACED ON EXISTING STEEL COLUMN, MASONRY PIERS OR FOUNDATION WALL, NOT OVER 8 FT. ABOVE GRADE • FORM • PLACE STEEL: 1/2" BARS, 12" O.C. • POUR 5"-6" CONCRETE SLAB • STRIP AND CLEAN FORMS FOR RE-USE • FINISH CONCRETE	EA PLUS	175.00	189.00	364.00	546.00					.628
		SF	3.05	2.84	5.89	8.84					.629
STEPS, SUSPENDED	SELF-SUPPORTING STEPS TO EXISTING PLATFORM OR AS PART OF SUSPENDED SLAB JOB, SUPPORTED BY EXISTING MASONRY OR UNDISTURBED EARTH • DIG AS REQUIRED • BUILD FORMS • PLACE STEEL AS REQUIRED • POUR CONCRETE STEPS • STRIP AND CLEAN FORMS • FINISH CONCRETE SF = TOTAL TOP SURFACE OF TREADS IN SQ FT	EA PLUS	175.00	77.00	252.00	378.00					.634
		SF	4.36	11.55	15.91	23.87					.635

4. DRIVEWAYS AND SIDEWALK

SPECIFICATIONS		UNIT	JOB COST			PRICE	LOCAL AREA MODIFICATION				DATA BASE ITEM NO.
			MATLS	LABOR	TOTAL		MATLS	LABOR	TOTAL	PRICE	
DRIVEWAY, CONCRETE	• LEVEL BY HAND • FILL WITH UP TO 4" SAND, CINDERS OR GRAVEL • FORM, PLACE #10 WWM AND EXPANSION JOINTS • POUR 4" SLAB, FINISH AND REMOVE FORMS • APRON **NOT** INCLUDED	SF	1.67	1.37	3.04	4.56					.700
SAND, CINDERS OR GRAVEL FILL UNDER	SAND, CINDER OR GRAVEL FILL UNDER DRIVEWAY, LEVELED BY HAND										
	4"	SF	.27	.21	.48	.72					.701
	6"	SF	.39	.26	.65	.98					.702
	8"	SF	.54	.33	.87	1.31					.703
SIDEWALK, CONCRETE	4" SAND, GRAVEL OR CINDERS BASE, 4" CONCRETE,	EA PLUS	57.00	21.00	78.00	117.00					.704
	LEVELING, FORMING, POURING AND FINISHING, REMOVE FORMS	SF	1.44	.95	2.39	3.59					.705
	THE COSTS SHOWN BELOW FOR ASPHALT DRIVEWAYS INCLUDE A SUBCONTRACTOR'S OVERHEAD AND PROFIT										
DRIVEWAY, ASPHALT	TOTAL JOB, INCLUDING FORMING, SPREADING AND ROLLING, OIL BASE, UNDER 250 SF										
	NEW, 4" OVER EXISTING STONE BASE	EA PLUS	--	--	416.00	624.00					.706
		SF	--	--	1.96	2.94					.707
	REPAVE, 2" ON GOOD BASE	EA PLUS	--	--	264.00	396.00					.708
		SF	--	--	1.20	1.80					.709
	SAME AS ABOVE, MORE THAN 250 SF 4"	EA PLUS	--	--	908.00	1,362.00					.710
		SF	--	--	1.80	2.70					.711
	2"	EA PLUS	--	--	568.00	852.00					.712
		SF	--	--	1.04	1.56					.713

5. BLOCK WALL

SPECIFICATIONS	UNIT	JOB COST			PRICE	LOCAL AREA MODIFICATION				DATA BASE ITEM NO.
		MATLS	LABOR	TOTAL		MATLS	LABOR	TOTAL	PRICE	
MORE THAN ONE STORY ABOVE GRADE — ALL MASONRY CONSTRUCTION ON THIS AND THE FOLLOWING PAGE ARE FOR WALLS NOT OVER ONE STORY ABOVE GRADE FOR MASONRY WALLS MORE THAN ONE STORY ABOVE GRADE 2 STORIES ABOVE GRADE **ADD**	SF	--	20%	20%	20%					.000
3 STORIES ABOVE GRADE **ADD**	SF	--	30%	30%	30%					.001
BLOCK WALL — • ON EXISTING FOOTINGS • TOOLED JOINTS AND SIDES • REINFORCING EVERY SECOND COURSE • STEEL ANGLES AS REQUIRED • TOP COURSE 4" SOLID CAP BLOCK 4 X 8 X 16	SF	1.33	2.03	3.36	5.04					.002
8 X 8 X 16	SF	1.84	2.47	4.31	6.47					.003
12 X 8 X 16	SF	2.58	3.22	5.80	8.70					.004
SF = ONE FACE OF WALL										
GLASS BLOCK — • 4" THICK GLASS BLOCK ON EXISTING SUPPORT • TOOLED JOINTS TWO SIDES • CLEAN BOTH SIDES AFTER INSTALLATION 6 X 6	SF	24.29	6.53	30.82	46.23					.005
8 X 8	SF	18.76	5.46	24.22	36.33					.006
12 X 12	SF	22.27	4.36	26.63	39.95					.007

SPECIFICATIONS		UNIT	JOB COST			PRICE	LOCAL AREA MODIFICATION				DATA BASE ITEM NO.
			MATLS	LABOR	TOTAL		MATLS	LABOR	TOTAL	PRICE	
WALL, BRICK, 4"	• BRICK VENEER WALL ON EXISTING FOOTINGS • OVER EXISTING FRAME WALL • STEEL ANGLES AS REQUIRED • WALL TIES AS REQUIRED • TOOLED JOINTS ONE SIDE • CLEAN ONE SIDE AFTER INSTALLATION	SF	2.40	4.52	6.92	10.38					.008
SOLID BRICK WALL, 8"	• ON EXISTING FOOTINGS • STEEL ANGLES AS REQUIRED • TOOLED JOINTS ONE SIDE • CLEAN TWO SIDES AFTER INSTALLATION	SF	4.79	6.85	11.64	17.46					.009
BRICK AND BLOCK WALL	• ON EXISTING FOOTINGS • STEEL ANGLES AS REQUIRED • TOOLED JOINTS ONE SIDE • CLEAN BRICK AFTER INSTALLATION										
	BRICK AND 4" BLOCK	SF	4.37	6.06	10.43	15.65					.010
	BRICK AND 8" BLOCK	SF	4.85	6.49	11.34	17.01					.011
REPOINT BRICK WALL	CUT JOINTS IN EXISTING BRICK WALL AND REPOINT										
	SOFT MORTAR	SF	.19	1.98	2.17	3.26					.012
	HARD MORTAR	SF	.19	4.29	4.48	6.72					.013
PATCH MASONRY	REMOVE DEFECTIVE MATERIALS AND PATCH EXISTING WALL, TOOTHING IN TO EXISTING WALL										
	BRICK	SF	2.75	8.72	11.47	17.21					.014
	BLOCK	SF	1.77	6.66	8.43	12.65					.015

5. PIERS

SPECIFICATIONS		UNIT	JOB COST			PRICE	LOCAL AREA MODIFICATION				DATA BASE ITEM NO.
			MATLS	LABOR	TOTAL		MATLS	LABOR	TOTAL	PRICE	
BLOCK PIER	• ON EXISTING FOOTING • TOOLED JOINTS • TOP 4" SOLID BLOCK OR CONCRETE FILLED										
	16" X 16"	LF UP	4.04	10.71	14.75	22.13					.100
	24" X 24"	LF UP	7.34	11.77	19.11	28.67					.101
	LF UP = TOP OF FOOTING TO TOP OF PIER										
BRICK PIER	• ON EXISTING FOOTING • TOOLED JOINTS • CLEAN BRICK AFTER IN-STALLATION										
	12" X 12"	LF UP	7.17	16.49	23.66	35.49					.102
	16" X 16"	LF UP	13.47	17.03	30.50	45.75					.103
	LF UP = TOP OF FOOTING TO TOP OF PIER										

5

SPECIFICATIONS		UNIT	JOB COST			PRICE	LOCAL AREA MODIFICATION				DATA BASE ITEM NO.
			MATLS	LABOR	TOTAL		MATLS	LABOR	TOTAL	PRICE	
SLAB AND STEP(S) ON GRADE	BUILD 6'-0" x 4'-0" CONCRETE SLAB AND STEP(S) ON GRADE • EXCAVATE TRENCH TO 36" BELOW GRADE • DRILL INTO EXISTING HOUSE WALL AND IN-STALL FOUR 1/2" STEEL BARS TO CONNECT FOOTING • POUR CONCRETE FOOT-ING TO GRADE • BUILD BRICK & BLOCK FOUNDATION WALL ABOVE GRADE TO SUP-PORT SLAB & STEP(S) • BUILD FORMS FOR SLAB AND STEP(S) • PLACE RUBBLE OR FORMS IN CENTER • PLACE STEEL • POUR CONCRETE SLAB AND STEP(S) • STRIP AND CLEAN FORMS • FINISH CONCRETE • STEPS: 6"-7" RISERS, 12" TREADS	EA PLUS	128.00	126.00	254.00	381.00					.104
	Width of Tread										
	48"	Tread	66.00	94.00	160.00	240.00					.105
	60"	Tread	72.00	98.00	170.00	255.00					.106
	72"	Tread	77.00	99.00	176.00	264.00					.107
	EA = EACH JOB TREAD = EACH TREAD										

5

5. CHIMNEY

SPECIFICATIONS		UNIT	JOB COST			PRICE	LOCAL AREA MODIFICATION				DATA BASE ITEM NO.
			MATLS	LABOR	TOTAL		MATLS	LABOR	TOTAL	PRICE	
CHIMNEY	• BRICK • CEMENT WASH AT TOP • ON EXISTING FOUNDA-TION • CLAY TILE FLUE LINERS										
	WITH ONE 8" X 8" FLUE	LF	18.86	34.65	53.51	80.27					.200
	WITH TWO 8" X 8" FLUES	LF	21.41	40.53	61.94	92.91					.201
	WITH ONE 8" X 12" FLUE	LF	23.76	46.20	69.96	104.94					.202
	WITH TWO 8" X 12" FLUES	LF	35.66	52.50	88.16	132.24					.203
	LF = LF UP FROM FOUN-DATION TO TOP OF CHIMNEY										
INSTALL FLUE(S) IN EXISTING CHIMNEY	TEAR OUT PORTION OF CHIMNEY FROM INSIDE HOUSE AND INSTALL CLAY TILE FLUE LINING(S), RE-PLACE BRICKWORK, **NO** PLASTER PATCHING										
	ONE FLUE	LF	9.29	30.61	39.90	59.85					.204
	TWO FLUES	LF	13.89	37.59	51.48	77.22					.205
	SAME AS ABOVE, WORKING FROM OUTSIDE ON SCAF-FOLDING										
	ONE FLUE	LF	9.29	28.19	37.48	56.22					.206
	TWO FLUES	LF	13.89	34.13	48.02	72.03					.207
POURED CONCRETE FLUE LINER IN EXISTING CHIMNEY	STRAIGHT CHIMNEY, PORT-LAND CEMENT AND INSULAT-ING AGGREGATE POURED AROUND TEMPORARY IN-FLATED RUBBER FORM OR STEEL LINER WITH 1/16" GLAZED FINISH ON INSIDE OF NEW FLUE LINER	EA	--	--	626.00	939.00					.208
		PLUS LF	--	--	38.00	57.00					.209
	FOR EACH ANGLE IN CHIM-NEY **ADD**	EA	--	--	216.00	324.00					.210
	CLEANOUT DOOR	EA	--	--	72.00	108.00					.211
	EA = EACH OPENING										
	TEAR OUT OLD FLUE LINER	LF	--	--	14.50	21.75					.212

SPECIFICATIONS		UNIT	JOB COST			PRICE	LOCAL AREA MODIFICATION				DATA BASE ITEM NO.
			MATLS	LABOR	TOTAL		MATLS	LABOR	TOTAL	PRICE	
FIREPLACE	• BRICK WITH FIREBRICK LINED FIRE HEARTH • DAMPER AND CLEANOUT • BRICK, SLATE OR TILE FRONT HEARTH • SIZE: WIDTH 36" 　　　　HEIGHT 29" 　　　　DEPTH 16"	EA	642.00	1786.00	2,428.00	3,642.00					.213
REPLACE LINTEL OVER EXISTING FIREPLACE	CUT AWAY BRICKWORK AND REMOVE EXISTING LINTEL, INSTALL NEW LINTEL AND REPLACE BRICKWORK										
	48" LINTEL	EA	48.00	276.00	324.00	486.00					.214
	72" LINTEL	EA	54.00	374.00	428.00	642.00					.215
RAISED HEARTH	HEARTH RAISED UP TO ONE FOOT ABOVE FLOOR **ADD**	EA	52.00	66.00	118.00	177.00					.216
DAMPER IN EXISTING FIREPLACE	REMOVE PORTION OF CHIMNEY ABOVE FIREPLACE AND INSTALL NEW DAMPER, REPLACE BRICKWORK	EA	125.00	295.00	420.00	630.00					.217
ALTER EXISTING FIREPLACE	BREAK THROUGH INTO EXISTING BRICKED UP FIREPLACE, CLEAN OUT AND REPAIR FOR USE WITH EXISTING DAMPER	EA	54.00	174.00	228.00	342.00					.218
	BREAK THROUGH BACK WALL OF EXISTING FIREPLACE, INSTALL NEW FIREPLACE OPENING FOR BACK-TO-BACK FIREPLACES WITH SAME FLUE AND OPEN FROM ROOM TO ROOM	EA	156.00	1038.00	1,194.00	1,791.00					.219
	GLASS DOORS INSTALLED ON ONE FIREPLACE OPENING ABOVE	EA	223.00	55.00	278.00	417.00					.220
REBUILD FIREBOX	REBUILD EXISTING FIREBOX WITH NEW FIREBOX APPROXIMATELY 42" X 21"	EA	144.00	180.00	324.00	486.00					.221
REBUILD FIREPLACE	REBUILD FIREBOX, INSTALL THROAT-TYPE DAMPER, REBUILD FIREPLACE UP TO 56" ABOVE FLOOR	EA	415.00	679.00	1,094.00	1,641.00					.222

5

5. UNDERPINNING

SPECIFICATIONS		UNIT	JOB COST			PRICE	LOCAL AREA MODIFICATION				DATA BASE ITEM NO.
			MATLS	LABOR	TOTAL		MATLS	LABOR	TOTAL	PRICE	
UNDERPIN EXISTING FOUNDA- TION WALL	EXCAVATE AS REQUIRED UNDER EXISTING FOOTING IN ALTERNATE 4-FOOT SEC- TIONS										
	POUR REINFORCED CON- CRETE UNDER AND BUILD NEW BLOCK WALL FROM TOP OF NEW FOOTING TO BOT- TOM OF OLD FOOTING										
	BREAK OFF EXISTING FOOT- ING PROJECTION										
	Below Existing Footing										
	24"	LF	19.36	120.00	139.36	209.04					.300
	48"	LF	19.84	132.00	151.84	227.76					.301
	72"	LF	22.89	162.00	184.89	277.34					.302
UNDERPIN CORNER OF EXISTING BRICK HOUSE (EXTERIOR)	EXCAVATE AS REQUIRED UNDER EXISTING CORNER AND POUR REINFORCED CONCRETE UNDER EXISTING FOOTING OR FOUNDATION WALL										
	Below Grade										
	48"	EA	103.00	323.00	426.00	639.00					.303
	72"	EA	103.00	443.00	546.00	819.00					.304
	96"	EA	103.00	763.00	866.00	1,299.00					.305

SPECIFICATIONS		UNIT	JOB COST			PRICE	LOCAL AREA MODIFICATION				DATA BASE ITEM NO.
			MATLS	LABOR	TOTAL		MATLS	LABOR	TOTAL	PRICE	
RETAINING WALL	• CONCRETE FOOTING • SOLID MASONRY WALL WITH BRICK FACING • 4" TILE WEEPHOLES 10 FT. O.C. WITH 1 CU.FT. CRUSHED STONE OR GRAVEL AT EACH WEEP-HOLE • PARGE BACK WALL WITH 1/2" CEMENT PLASTER OR 2 COATS HOT TAR • REINFORCING RODS AS REQUIRED **Height of Wall Above Bottom of Footing**										
	40"	LF	21.09	34.97	56.06	84.09					.306
	48"	LF	23.17	36.02	59.19	88.79					.307
	56"	LF	27.23	42.32	69.55	104.33					.308
	64"	LF	32.37	49.56	81.93	122.90					.309
	72"	LF	40.79	49.56	90.35	135.53					.310
	80"	LF	48.41	66.68	115.09	172.64					.311
DRAIN PIPE	INSTALL 4" PERFORATED PLASTIC PIPE SURROUNDED WITH 6" OF 3/4" AGGREGATE ALONG BACK OF FOOTING EXTENDING TO DAYLIGHT OR DRYWELL (SEE PAGE 97)	LF	1.03	2.00	3.03	4.55					.312

5. PATIO

SPECIFICATIONS		UNIT	JOB COST			PRICE	LOCAL AREA MODIFICATION				DATA BASE ITEM NO.
			MATLS	LABOR	TOTAL		MATLS	LABOR	TOTAL	PRICE	
PATIO, BRICK	ON GROUND IN SAND WITH SAND IN JOINTS, BRICKS LAID FLAT, INCLUDING EXCAVATION	SF	2.30	7.50	9.80	14.70					.400
	IN SAND OVER EXISTING CONCRETE SLAB, SAND IN JOINTS	SF	2.30	5.00	7.30	10.95					.401
	IN CONCRETE BED WITH CONCRETE JOINTS OVER EXISTING CONCRETE SLAB	SF	2.85	5.80	8.65	12.98					.402
	BRICK LAID ON EDGE IN CONCRETE BED WITH CONCRETE JOINTS OVER EXISTING CONCRETE SLAB	SF	5.10	11.30	16.40	24.60					.403
PATIO, FLAGSTONE	ON GROUND IN SAND WITH SAND IN JOINTS, INCLUDING EXCAVATION, 3/4" - 1"										
	IRREGULAR PA.	SF	1.25	5.80	7.05	10.58					.404
	REGULAR PA.	SF	2.60	5.45	8.05	12.08					.405
	IRREGULAR VT.	SF	1.60	5.90	7.50	11.25					.406
	REGULAR VT.	SF	8.15	5.45	13.60	20.40					.407
	REGULAR TENN.	SF	8.15	5.45	13.60	20.40					.408
	IN CONCRETE BED WITH CONCRETE JOINTS OVER EXISTING CONCRETE SLAB										
	IRREGULAR PA.	SF	2.30	4.70	7.00	10.50					.409
	REGULAR PA.	SF	3.45	4.15	7.60	11.40					.410
	IRREGULAR VT.	SF	2.60	4.70	7.30	10.95					.411
	REGULAR VT.	SF	9.20	4.15	13.35	20.03					.412
	REGULAR TENN.	SF	9.20	4.15	13.35	20.03					.413

5

SPECIFICATIONS	UNIT	JOB COST			PRICE	LOCAL AREA MODIFICATION				DATA BASE ITEM NO.
		MATLS	LABOR	TOTAL		MATLS	LABOR	TOTAL	PRICE	
	LABOR COSTS SHOWN BELOW FOR STUCCO WORK INCLUDE A SUBCONTRACTOR'S OVERHEAD AND PROFIT									
STUCCO — CEMENT SAND FINISH STUCCO ON CLEAN MASONRY WALL, TWO 1/2" COATS	SF	.58	3.65	4.23	6.35					.500
CEMENT SAND FINISH STUCCO OVER EXISTING SHEATHING • #15 FELT PAPER • SELF FURRING WIRE MESH • OUTSIDE CORNER BEADS AS NECESSARY • METAL LATH STRIPS AROUND WINDOWS AND DOORS • SCRATCH COAT, BROWN COAT AND FINISH COAT	SF	1.36	4.15	5.51	8.27					.501
E.I.F.S. (EXTERIOR INSULATION & FINISH SYSTEM) 1/8" FLEXIBLE STUCCO OVER 1" FOAM SHEATHING (INCLUDED)	SF	1.95	4.45	6.40	9.60					.502
PATCH — REMOVE DEFECTIVE MATERIALS AND CLEAN AND PREPARE AREA TO BE PATCHED, 2 OR 3 COATS STUCCO										
3 COATS ON FRAMING	SF	1.50	6.50	8.00	12.00					.503
2 COATS ON MASONRY	SF	.75	4.30	5.05	7.58					.504
WATER-PROOF PAINT — FILL CRACKS OVER 1/8" WIDE AND APPLY 2 COATS WATERPROOFING PAINT OVER MASONRY WALL	SF	.15	.29	.44	.66					.505
SILICON — CLEAR SILICON OVER EXISTING MASONRY WALL, 1 COAT SPRAYED	SF	.09	.15	.24	.36					.506
PARGING — CEMENT PARGING, 1/2" THICK, TWO COATS	SF	.29	.91	1.20	1.80					.507
ASPHALT COATING — ASPHALT COATING APPLIED TO MASONRY WALL										
1 COAT	SF	.16	.34	.50	.75					.508
2 COATS	SF	.21	.51	.72	1.08					.509
TWO LAYERS TARRED PAPER COVERED WITH HOT TAR OR ASPHALT ON MASONRY WALL	SF	.27	.73	1.00	1.50					.510

6. ALL WEATHER WOOD FOUNDATION

SPECIFICATIONS		UNIT	JOB COST			PRICE	LOCAL AREA MODIFICATION				DATA BASE ITEM NO.
			MATLS	LABOR	TOTAL		MATLS	LABOR	TOTAL	PRICE	
TREATED WOOD FOUNDA-TION	• 2" X 12" PRESSURE TREATED WOOD PLATE ON LAYER OF STONE OR GRAVEL • LAYER OF GRAVEL UP TO 3/4" OR CRUSHED STONE UP TO 1/2" EXTENDING MINIMUM 6" BEYOND PLATE AND 4" DEEP • CDX TREATED PANELS • CORROSION RESISTANT NAILS AND FASTENERS • BATT INSULATION BE-TWEEN STUDS • 15# FELT PAPER • 6-MIL POLYETHYLENE										
	Stud Size **CDX Panel Size**										
	2" X 4" 1/2"	SF	3.33	1.68	5.01	7.52					.010
	2" X 6" 1/2"	SF	4.79	1.82	6.61	9.92					.011
	2" X 8" 3/4"	SF	5.61	2.00	7.61	11.42					.012
	SF = ONE FACE OF WALL										

SPECIFICATIONS		UNIT	JOB COST			PRICE	LOCAL AREA MODIFICATION				DATA BASE ITEM NO.
			MATLS	LABOR	TOTAL		MATLS	LABOR	TOTAL	PRICE	
STEEL BEAM	STEEL BEAM BEARING ON EXISTING MASONRY OR STEEL COLUMN SUPPORTS, INCLUDING BEARING PLATES										
	8" #13	LF	12.50	9.45	21.95	32.93					.000
	8" #17	LF	16.50	10.76	27.26	40.89					.001
WOOD BEAM	WOOD BEAM BEARING ON EXISTING SUPPORTS, BEAM EITHER SOLID OR BUILT-UP, PRESSURE TREATED LUMBER										
	4 X 8	LF	2.59	1.55	4.14	6.21					.002
	4 X 10	LF	5.18	1.93	7.11	10.67					.003
	6 X 8	LF	5.67	2.08	7.75	11.63					.004
	6 X 10	LF	9.33	2.63	11.96	17.94					.005
LAMINATED WOOD BEAM	LAMINATED VENEER LUMBER ON EXISTING SUPPORTS										
	4" X 9-1/4"	LF	11.50	7.50	19.00	28.50					.013
	4" X 11-1/4"	LF	13.25	8.75	22.00	33.00					.014
FLITCH PLATE, STEEL	3/8" X 9" STEEL PLATE BETWEEN TWO 2x10s ON EXISTING SUPPORTS, INCLUDING PLATE, LUMBER, NUTS, BOLTS	LF	13.24	6.36	19.60	29.40					.006
FLITCH PLATE, WOOD	1/2" X 9" PLYWOOD GLUED AND BOLTED BETWEEN TWO 2 X 10s ON EXISTING SUPPORTS, INCL. PLYWOOD, LUMBER, GLUE, NUTS, BOLTS	LF	4.72	4.86	9.58	14.37					.007
STEEL COLUMN	3" HOLLOW STEEL COLUMN WITH BASE AND CAP UP TO 8'-0"	EA	37.50	14.50	52.00	78.00					.008
	3" HOLLOW ADJUSTABLE STEEL COLUMN WITH BASE AND CAP UP TO 8'-0"	EA	33.00	12.50	45.50	68.25					.009

6

6. FLOOR FRAMING

SPECIFICATIONS		UNIT	JOB COST			PRICE	LOCAL AREA MODIFICATION				DATA BASE ITEM NO.
			MATLS	LABOR	TOTAL		MATLS	LABOR	TOTAL	PRICE	
SLEEPERS ON EXISTING DECK	ON EXISTING WOOD, CONCRETE OR MASONRY DECK • SOLID BRIDGING • JOIST HEADERS AS REQ'D • VAPOR BARRIER • NAILS • SLEEPER JOISTS 16" O.C.										
	2" X 3"	SF	.51	.67	1.18	1.77					.100
	2" X 4"	SF	.62	.68	1.30	1.95					.101
	2" X 6"	SF	.80	.69	1.49	2.24					.102
	SAME AS ABOVE, USING PRESSURE-TREATED LUMBER										
	2" X 3"	SF	.57	.67	1.24	1.86					.103
	2" X 4"	SF	.68	.68	1.36	2.04					.104
	2" X 6"	SF	.91	.69	1.60	2.40					.105
	IF SLEEPERS ARE RIPPED LENGTHWISE TO FIT **ADD**	SF	--	.27	.27	.41					.106
	NOTE: FLOOR JOISTS COSTS BELOW ARE BASED ON USING JOISTS TO FULL PERMISSIBLE LENGTHS										
MUDSILL	• PRESSURE-TREATED DIMENSION FIR OR PINE • SILL PLATE • ANCHOR BOLTS, 4' OC										
	2" X 4"	LF	1.03	.60	1.63	2.45					.123
	2" X 6"	LF	1.25	.65	1.90	2.85					.124
	2" X 8"	LF	1.48	.69	2.17	3.26					.125
JOISTS, FIRST FLOOR	• FRAMING LUMBER • BRIDGING • JOIST HANGERS • LAG BOLTS • NAILS • JOIST HEADERS AS REQ. • JOISTS 16" O.C.										
	2" X 6"	SF	.83	.80	1.63	2.45					.107
	2" X 8"	SF	1.32	.80	2.12	3.18					.108
	2" X 10"	SF	1.71	.80	2.51	3.77					.109
	2" X 12"	SF	2.04	.80	2.84	4.26					.110
JOISTS, SECOND FLOOR	• FRAMING LUMBER • BRIDGING • JOIST HEADERS AS REQUIRED • JOIST HANGERS • LAG BOLTS • NAILS • JOISTS 16" O.C.										
	2" X 6"	SF	.83	.86	1.69	2.54					.111
	2" X 8"	SF	1.32	.86	2.18	3.27					.112
	2" X 10"	SF	1.71	.86	2.57	3.86					.113
	2" X 12"	SF	2.04	.86	2.90	4.35					.114

SPECIFICATIONS		UNIT	JOB COST			PRICE	LOCAL AREA MODIFICATION				DATA BASE ITEM NO.
			MATLS	LABOR	TOTAL		MATLS	LABOR	TOTAL	PRICE	
WOOD I-JOISTS	PREFABRICATED I-SHAPED JOISTS, 16" O.C. • 3/8" WEB • 1-5/8" TOP AND BOTTOM CHORDS • NAILS • JOIST HEADERS • JOIST HANGERS • **NO** BRIDGING										
	9-1/4"	SF	1.58	.69	2.27	3.41					.129
	12"	SF	1.98	.74	2.72	4.08					.130
SUBFLOOR	PLYWOOD CDX, GLUED AND NAILED TO JOISTS, INCLUDING ALL MATERIALS AS REQUIRED										
	3/8"	SF	.38	.42	.80	1.20					.115
	1/2"	SF	.45	.42	.87	1.31					.116
	5/8"	SF	.49	.43	.92	1.38					.117
	PLYWOOD, GLUED AND NAILED, INCLUDING ALL MATERIALS AS REQUIRED										
	3/4"	SF	.55	.44	.99	1.49					.118
	T & G 5/8"	SF	.57	.61	1.18	1.77					.119
	T & G 3/4"	SF	.67	.62	1.29	1.94					.120
	3/4" PINE, FIR OR HEMLOCK, NAILED										
	HORIZONTAL	SF	.68	.46	1.14	1.71					.121
	DIAGONAL	SF	.85	.59	1.44	2.16					.122
	ORIENTED STRAND BOARD										
	1/2"	SF	.49	.43	.92	1.38					.126
	5/8"	SF	.54	.44	.98	1.47					.127
	3/4"	SF	.62	.45	1.07	1.61					.128

6

7. WALL FRAMING AND FURRING

SPECIFICATIONS		UNIT	JOB COST			PRICE	LOCAL AREA MODIFICATION				DATA BASE ITEM NO.
			MATLS	LABOR	TOTAL		MATLS	LABOR	TOTAL	PRICE	
WOOD STUDS	BEARING WALL, EXTERIOR OR INTERIOR • SOLE PLATE • TWO 2" X 4" OR 2" X 6" CAP • AVERAGE NUMBER OF HEADERS FOR DOORS AND WINDOWS • 16" O.C.										
	2" X 4"	SF	.76	.62	1.38	2.07					.000
	2" X 6"	SF	1.09	.67	1.76	2.64					.001
	SF = TOTAL WALL AREA										
	NON-BEARING WALL • SOLE PLATE • ONE CAP • AVERAGE AMOUNT OF FRAMING FOR DOORS, CLOSETS AND CORNERS • 16" O.C.										
	2" X 3"	SF	.54	.55	1.09	1.64					.002
	2" X 4"	SF	.73	.65	1.38	2.07					.003
	2" X 6"	SF	1.05	.68	1.73	2.60					.004
STEEL STUDS	20 GA. STEEL STUDS AND TRACK										
	1-1/2"	SF	.29	.49	.78	1.17					.005
	2-1/2"	SF	.34	.51	.85	1.28					.006
	3-1/2"	SF	.39	.53	.92	1.38					.007
	5-1/2"	SF	.52	.55	1.07	1.61					.008
DIFFICULT WALL FRAMING	FOR DIFFICULT WALL FRAMING, CUT UP WITH BAY AND BOW WINDOWS, MANY CORNERS, ETC. **ADD**	SF	10%	40%	30%	30%					.009 .010
KNEEWALL	2 X 4 KNEEWALL SUPPORTING RAFTERS • SOLE PLATE • ONE CAP • STUDS CUT AT ANGLE TO FIT RAFTERS *Kneewall Height*										
	4'-0"	LF	2.81	3.80	6.61	9.92					.011
	6'-0"	LF	3.69	4.60	8.29	12.44					.012
	8'-0"	LF	4.83	5.30	10.13	15.20					.013
FURRING	OVER FRAMING, 16" O.C.										
	1" X 2"	SF	.11	.29	.40	.60					.014
	1" X 3"	SF	.17	.29	.46	.69					.015
	OVER STRAIGHT AND PLUMB MASONRY WALLS, 16" O.C.										
	1" X 2"	SF	.11	.44	.55	.83					.016
	1" X 3"	SF	.17	.44	.61	.92					.017
	OVER CROOKED MASONRY WALLS, INCLUDING SHIMS, 16" O.C.										
	1" X 2"	SF	.11	.60	.71	1.07					.018
	1" X 3"	SF	.17	.60	.77	1.16					.019

SPECIFICATIONS		UNIT	JOB COST			PRICE	LOCAL AREA MODIFICATION				DATA BASE ITEM NO.
			MATLS	LABOR	TOTAL		MATLS	LABOR	TOTAL	PRICE	
SHEATHING	PLYWOOD CDX										
	3/8"	SF	.38	.41	.79	1.19					.100
	1/2"	SF	.45	.42	.87	1.31					.101
	COMPOSITION SHEATHING, INCLUDING 1/2" PLYWOOD OR 1 X 6 DIAGONAL BRACING AT EACH CORNER, 1/2"x4x8 SHEETS										
	ASPHALT IMPREGNATED	SF	.30	.39	.69	1.04					.102
	SAME AS ABOVE, USING FOIL-FACED FOAM SHEATH-ING										
	1/4"	SF	.31	.39	.70	1.05					.103
	1/2"	SF	.37	.39	.76	1.14					.104
	5/8"	SF	.40	.39	.79	1.19					.105
	3/4"	SF	.47	.39	.86	1.29					.106
	SHEATHING PINE, 3/4"										
	HORIZONTAL	SF	.68	.50	1.18	1.77					.107
	DIAGONAL	SF	.85	.67	1.52	2.28					.108
	ORIENTED STRAND BOARD										
	3/8"	SF	.44	.42	.86	1.29					.109
	1/2"	SF	.49	.43	.92	1.38					.110

7

8. ROOF FRAMING

SPECIFICATIONS	UNIT	JOB COST			PRICE	LOCAL AREA MODIFICATION				DATA BASE ITEM NO.
		MATLS	LABOR	TOTAL		MATLS	LABOR	TOTAL	PRICE	
MEASURING SQUARE FOOTAGE OF NEW RAFTERS AND SHEATHING										
TO OBTAIN SQUARE FOOTAGE OF NEW ROOF RAFTERS AND SHEATHING, MEASURE THE ACTUAL LENGTH OF RAFTERS FROM RIDGEBOARD TO EAVE AND MULTIPLY BY THE OPPOSITE DIMENSION (WIDTH OF THE ROOF). BE SURE TO INCLUDE OVERHANGS AND ANY PORTION OF THE NEW ROOF WHICH WILL OVERLAY THE EXISTING ROOF.										
FLAT ROOF OR SHED ROOF — • RAFTERS 16" O.C. • BRIDGING • NAILS • FRAMING FOR OVERHANG AND EAVES										
2" X 6"	SF	.72	.64	1.36	2.04					.000
2" X 8"	SF	1.15	.71	1.86	2.79					.001
2" X 10"	SF	1.48	.97	2.45	3.68					.002
2" X 12"	SF	1.78	1.27	3.05	4.58					.003
CEILING JOISTS — ADD CEILING JOISTS TO SHED ROOF FRAMING										
2" X 6" **ADD**	SF	.63	.53	1.16	1.74					.004
2" X 8" **ADD**	SF	1.00	.58	1.58	2.37					.005
2" X 10" **ADD**	SF	1.26	.60	1.86	2.79					.006
SF = AREA COVERED BY CEILING JOISTS, **NOT** ROOF AREA										
GABLE ROOF — • RIDGE BOARD • RAFTERS 16" O.C. • CEILING JOISTS 16" O.C. • COLLAR TIES, 1" X 6", 48" O.C. • NAILS • FRAMING FOR OVERHANG AND EAVES										
Rafters Ceiling Joists										
2" X 6" 2" X 6"	SF	1.36	1.19	2.55	3.83					.007
2" X 8" 2" X 6"	SF	1.72	1.30	3.02	4.53					.008
2" X 10" 2" X 8"	SF	2.36	1.60	3.96	5.94					.009
2" X 12" 2" X 10"	SF	2.88	1.92	4.80	7.20					.010
OMIT CEILING JOISTS — OMIT CEILING JOISTS										
2" X 6" **DEDUCT**	SF	.63	.53	1.16	1.74					.011
2" X 8" **DEDUCT**	SF	1.00	.58	1.58	2.37					.012
2" X 10" **DEDUCT**	SF	1.26	.60	1.86	2.79					.013
SF = AREA COVERED BY CEILING JOISTS, **NOT** ROOF AREA										
OMIT COLLAR TIES — OMIT COLLAR TIES, 1" X 6", 48" O.C. **DEDUCT**	SF	.03	.03	.06	.09					.014
SF = ROOF AREA										
DIFFICULT ROOF FRAMING — DIFFICULT ROOF FRAMING, CUT UP WITH VALLEYS AND DORMERS										
SOME DIFFICULTY **ADD**	SF	--	.30	.30	.45					.015
DIFFICULT **ADD**	SF	.15	.48	.63	.95					.016
VERY DIFFICULT **ADD**	SF	.20	.80	1.00	1.50					.017

8

SPECIFICATIONS		UNIT	JOB COST			PRICE	LOCAL AREA MODIFICATION				DATA BASE ITEM NO.
			MATLS	LABOR	TOTAL		MATLS	LABOR	TOTAL	PRICE	
GAMBREL ROOF	• RIDGEBOARD • HEADERS • RAFTERS 16" O.C. • 2" X 6" CEILING JOISTS 16" O.C. • 2" X 4" STUDWALL 16" O.C. • NAILS • FRAMING FOR CHIMNEY, EAVES • 12" OVERHANG • SLOPE OF UPPER RAFTERS 7/12										
	2" X 6"	SF	2.14	1.60	3.74	5.61					.018
	2" X 8"	SF	2.60	1.68	4.28	6.42					.019
HIP ROOF	• RIDGEBOARD • HEADERS • HIP AND JACK RAFTERS 16" O.C. • CEILING JOISTS 16" O.C. • FRAMING FOR CHIMNEY, EAVES • 12" OVERHANG										
	Rafters / **Ceiling Joists**										
	2" X 6" / 2" X 6"	SF	1.60	1.40	3.00	4.50					.020
	2" X 8" / 2" X 6"	SF	2.15	1.53	3.68	5.52					.021
	2" X 10" / 2" X 8"	SF	2.89	1.87	4.76	7.14					.022
DIFFICULT ROOF FRAMING	DIFFICULT ROOF FRAMING, CUT UP WITH VALLEYS AND DORMERS										
	SOME DIFFICULTY **ADD**	SF	--	.30	.30	.45					.015
	DIFFICULT **ADD**	SF	.15	.48	.63	.95					.016
	VERY DIFFICULT **ADD**	SF	.15	.80	.95	1.43					.017
TRUSS ROOF	• SHOP BUILT BY OTHERS AND DELIVERED TO JOB • PLACED BY CRANE OR BY HAND • METAL PLATE CONNECTORS • 2" X 4" TOP AND BOTTOM CHORDS • INCLUDES 2 GABLE ENDS 24" O.C.										
	16" O.C.	SF	3.48	.80	4.28	6.42					.023
	24" O.C.	SF	2.42	.56	2.98	4.47					.024

8. ROOF FRAMING, SHEATHING

SPECIFICATIONS		UNIT	JOB COST			PRICE	LOCAL AREA MODIFICATION				DATA BASE ITEM NO.
			MATLS	LABOR	TOTAL		MATLS	LABOR	TOTAL	PRICE	
PLYWOOD	CDX PLYWOOD										
	3/8"	SF	.40	.36	.76	1.14					.100
	1/2"	SF	.45	.37	.82	1.23					.101
	5/8"	SF	.50	.38	.88	1.32					.102
	3/4"	SF	.56	.39	.95	1.43					.103
	ORIENTED STRAND BOARD										
	3/8"	SF	.49	.36	.85	1.28					.104
	1/2"	SF	.54	.37	.91	1.37					.105
	5/8"	SF	.62	.38	1.00	1.50					.106
STRIP SHEATHING	STRIP (SPACED) SHEATHING (UNDER WOOD SHAKES OR SHINGLES)										
	1" X 3", 5-1/2" O.C.	SF	.39	.36	.75	1.13					.107
	1" X 3", 7-1/2" O.C.	SF	.38	.28	.66	.99					.108
	1" X 4", 7-1/2" O.C.	SF	.50	.28	.78	1.17					.109
CATHE-DRAL CEILING, EXPOSED BEAMS	• GABLE ROOF • 4" X 8" RAFTERS, 48" O.C. • RIDGEBOARD • NAILS AND FASTENERS • FRAMING FOR OVERHANG AND EAVES • COVERED WITH 2" X 6" T&G SHEATHING • #2 CONSTRUCTION GRADE LUMBER SF = ROOF SURFACE AREA	SF	8.20	3.54	11.74	17.61					.110
	2" RIGID FOAM INSULATION AND 1/2" CDX PLYWOOD IN-STALLED OVER ABOVE **ADD**	SF	2.46	.84	3.30	4.95					.111
GABLE (DOG-HOUSE) DORMER	GABLE (DOGHOUSE) DOR-MER BUILT AT TIME OF CON-STRUCTING ANY OF ABOVE ROOFS, INCLUDING RAF-TERS, RIDGEBOARD, CEILING JOISTS, STUDS, PLATES, SHEATHING SF = DORMER FLOOR AREA	SF	4.00	3.70	7.70	11.55					.112
SOFFIT NAILER	2" X 4" FRAMING TO SUP-PORT SOFFIT ***Soffit Width***										
	12"	LF	.74	.46	1.20	1.80					.113
	18"	LF	.86	.53	1.39	2.09					.114
	24"	LF	.97	.60	1.57	2.36					.115

Measuring Square Footage of Roof

New roofs and re-roofing are measured the same way. To obtain the square footage of roof, measure from ridge to eave and multiply by the opposite dimension (the width of the roof). Be sure to include overhangs and any portion of the roof which overlays the existing roof.

In measuring for a re-roofing job, it is not always possible to measure the exact dimensions of the roof. In this case, you may estimate the number of inches exposure in each course of shingles, count the courses, and multiply that total by the number of inches per course. Convert that figure to feet and multiply by the width of the building (which will usually be approximately the same as the roof).

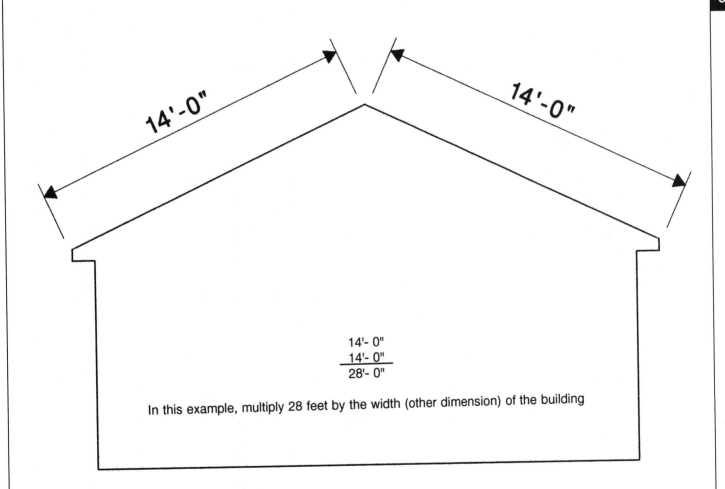

$$\begin{array}{r} 14'\text{-}0" \\ \underline{14'\text{-}0"} \\ 28'\text{-}0" \end{array}$$

In this example, multiply 28 feet by the width (other dimension) of the building

9. ASPHALT OR FIBERGLASS SHINGLES

SPECIFICATIONS		UNIT	JOB COST			PRICE	LOCAL AREA MODIFICATION				DATA BASE ITEM NO.
			MATLS	LABOR	TOTAL		MATLS	LABOR	TOTAL	PRICE	
	ALL ROOFING INSTALLATIONS ON THIS AND THE FOLLOWING PAGES ARE ON ONE STORY BUILDINGS										
MORE THAN 1 STORY ABOVE GRADE	FOR ROOFING MORE THAN ONE STORY ABOVE GRADE										
	2 STORIES ABOVE GRADE **ADD**	EA	--	60.00	60.00	90.00					.000
	3 STORIES ABOVE GRADE **ADD**	EA	--	90.00	90.00	135.00					.001
SMALL JOBS	SMALL JOB ROOFING LESS THAN 300 SF **ADD**	EA	--	70.00	70.00	105.00					.002
	300 TO 500 SF **ADD**	EA	--	40.00	40.00	60.00					.003
	500 TO 600 SF **ADD**	EA	--	20.00	20.00	30.00					.004
DIFFICULT ROOF COVERING	DIFFICULT ROOF COVERING, CUT UP WITH VALLEYS AND DORMERS										
	SOME DIFFICULTY **ADD**	SF	--	.20	.20	.30					.005
	DIFFICULT **ADD**	SF	--	.40	.40	.60					.006
	VERY DIFFICULT **ADD**	SF	--	.60	.60	.90					.007
ASPHALT OR FIBER-GLASS SHINGLES, 215-225 LB (20 YR)	ON PLAIN SHED OR GABLE WOOD ROOF DECK • LOAD ONTO ROOF • BUILD STAGING IF REQ'D • 3 TAB SQUARE BUTT SELF SEAL 215-225 LB. • 15 LB. FELT PAPER • GALVANIZED SHINGLE NAILS OR STAPLES • METAL DRIP EDGE 5"										
	Slope 4 TO 6 IN 12	SF	.38	.51	.89	1.34					.008
	7 TO 12 IN 12	SF	.38	.64	1.02	1.53					.009
250 LB. (25 YR.)	*Slope* 4 TO 6 IN 12	SF	.40	.51	.91	1.37					.010
	7 TO 12 IN 12	SF	.40	.64	1.04	1.56					.011
260 LB. (25 YR.)	*Slope* 4 TO 6 IN 12	SF	.59	.51	1.10	1.65					.012
	7 TO 12 IN 12	SF	.59	.64	1.23	1.85					.013
300 LB. (30 YR.)	*Slope* 4 TO 6 IN 12	SF	.85	.54	1.39	2.09					.014
	7 TO 12 IN 12	SF	.85	.67	1.52	2.28					.015
375 LB. (40 YR.)	*Slope* 4 TO 6 IN 12	SF	.93	.54	1.47	2.21					.016
	7 TO 12 IN 12	SF	.93	.67	1.60	2.40					.017

SPECIFICATIONS		UNIT	JOB COST			PRICE	LOCAL AREA MODIFICATION				DATA BASE ITEM NO.
			MATLS	LABOR	TOTAL		MATLS	LABOR	TOTAL	PRICE	
ROLL ROOFING	ON WOOD ROOF DECK WITH AVERAGE AMOUNT OF FITTING • ONE STORY ABOVE GRADE • MINERAL SURFACED 90 LB. • **NO** UNDERLAY • EXPOSED NAILS • TARRED SEAMS • 32" EXPOSURE (4" LAP) ***Slope***										
	4 TO 6 IN 12	SF	.20	.21	.41	.62					.018
	7 TO 12 IN 12	SF	.20	.23	.43	.65					.019
SELVAGE	ON WOOD ROOF DECK WITH AVERAGE AMOUNT OF FITTING • ONE STORY ABOVE GRADE • 19" SELVAGE, 110 TO 120 LB. • **NO** UNDERLAY • 17" EXPOSURE, 19" TO LAP • TARRED SEAMS ***Slope***										
	4 TO 6 IN 12	SF	.33	.30	.63	.95					.020
	7 TO 12 IN 12	SF	.33	.33	.66	.99					.021
OTHER TYPE ROOFS	INSTALL ASPHALT OR FIBERGLASS ROOFING ON OTHER TYPE ROOFS IN PLACE OF PLAIN SHED OR GABLE ROOFS										
	SM. GABLE DORMER **ADD**	SF	–	100%	50%	50%					.022
	GAMBREL **ADD**	SF	–	50%	25%	25%					.023
	HIP **ADD**	SF	15%	30%	22%	22%					.024
	NO FELT PAPER **DEDUCT**	SF	.04	.04	.08	.12					.026
	NO METAL DRIP EDGE **DEDUCT**	SF	.03	.05	.08	.12					.027
ICE DAM BARRIER	ICE AND WATER SHIELD 36" WIDE, RUBBERIZED MATERIAL PLACED UNDER STARTER COURSE AND IN VALLEYS	LF	1.19	.79	1.98	2.97					.028

9

9. SINGLE PLY & BUILT-UP ROOFS ON WOOD DECK

SPECIFICATIONS		UNIT	JOB COST			PRICE	LOCAL AREA MODIFICATION				DATA BASE ITEM NO.
			MATLS	LABOR	TOTAL		MATLS	LABOR	TOTAL	PRICE	
	THE ROOFING LABOR COSTS SHOWN ON THIS PAGE INCLUDE A ROOFING SUBCONTRACTOR'S OVERHEAD AND PROFIT										
SINGLE PLY	"RUBBERIZED" OVER WOOD ROOF DECK • SINGLE PLY FLEXIBLE SHEET MEMBRANE • MINIMUM SLOPE 1/8" • MAXIMUM SLOPE 4"	EA PLUS SF	-- .80	200.00 1.60	200.00 2.40	300.00 3.60					.029 .030
BITUMEN	MODIFIED BITUMEN REINFORCED WITH FABRIC MESH, HEAT APPLICATION (TORCH APPLIED) • MINIMUM SLOPE 1/8" • MAXIMUM SLOPE 4"	EA PLUS SF	-- .70	200.00 1.40	200.00 2.10	300.00 3.15					.031 .032
2 PLY	• #30 FELT NAILED TO DECK • 2 PLY #15 FELT • 3 COATS HOT ASPHALT • GRAVEL, SLAG OR MARBLE CHIPS • MINIMUM SLOPE 1/8" • MAXIMUM SLOPE 4"	EA PLUS SF	-- .65	315.00 1.30	315.00 1.95	472.50 2.93					.033 .034
3 PLY	• #30 FELT NAILED TO DECK • 3 PLY #15 FELT • 4 MOP COATS HOT ASPHALT • GRAVEL, SLAG OR MARBLE CHIPS • MINIMUM SLOPE 1/8" • MAXIMUM SLOPE 4"	EA PLUS SF	-- .70	325.00 1.40	325.00 2.10	487.50 3.15					.035 .036
4 PLY	• #30 FELT NAILED TO DECK • 4 PLY #15 FELT • 5 MOP COATS HOT ASPHALT • GRAVEL, SLAG OR MARBLE CHIPS • MINIMUM SLOPE 1/8" • MAXIMUM SLOPE 4"	EA PLUS SF	-- .80	325.00 1.60	325.00 2.40	487.50 3.60					.037 .038

9

SPECIFICATIONS	UNIT	JOB COST			PRICE	LOCAL AREA MODIFICATION				DATA BASE ITEM NO.
		MATLS	LABOR	TOTAL		MATLS	LABOR	TOTAL	PRICE	
SLATE ROOF COSTS BELOW ARE FOR ONE STORY BUILDINGS										
MORE THAN 1 STORY ABOVE GRADE — FOR ROOFING MORE THAN ONE STORY ABOVE GRADE										
2 STORIES ABOVE GRADE **ADD**	EA	--	60.00	60.00	90.00					.000
3 STORIES ABOVE GRADE **ADD**	EA	--	90.00	90.00	135.00					.001
SMALL JOBS — SMALL JOB ROOFING LESS THAN 300 SF **ADD**	EA	--	70.00	70.00	105.00					.002 .003
300 TO 500 SF **ADD**	EA	--	40.00	40.00	60.00					
500 TO 600 SF **ADD**	EA	--	20.00	20.00	30.00					.004
DIFFICULT ROOF COVERING, CUT UP WITH VALLEYS AND DORMERS										.106
SOME DIFFICULTY **ADD**	SF	--	.30	.30	.45					.107
DIFFICULT **ADD**	SF	--	.50	.50	.75					
VERY DIFFICULT **ADD**	SF	--	.80	.80	1.20					.108
BANGOR SLATE — ON PLAIN SHED OR GABLE WOOD ROOF DECK, 10" X 18" X 3/16" SLATE, 3" LAP, 7-1/2" EXPOSURE • LOAD ONTO ROOF • 1-1/4" COPPER SLATING NAILS • BUILD STAGING IF REQUIRED • 30 LB. ASPHALT FELT PAPER • RIDGE SLATES										
Slope 4 TO 6 IN 12	SF	3.55	1.45	5.00	7.50					.100
7 TO 12 IN 12	SF	3.55	1.76	5.31	7.97					.101
PENNSYLVANIA SLATE — SAME AS ABOVE, PENNSYLVANIA SLATE Slope 4 TO 6 IN 12	SF	4.50	1.45	5.95	8.93					.102
7 TO 12 IN 12	SF	4.50	1.76	6.26	9.39					.103
VERMONT SLATE — SAME AS ABOVE, VERMONT SLATE Slope 4 TO 6 IN 12	SF	4.50	1.45	5.95	8.93					.104
7 TO 12 IN 12	SF	4.50	1.76	6.26	9.39					.105
OTHER TYPE ROOFS — INSTALL SLATE ROOFING ON OTHER TYPE ROOFS IN PLACE OF PLAIN SHED OR GABLE ROOFS										
SM. GABLE DORMER **ADD**	SF	--	100%	50%	50%					.022
GAMBREL **ADD**	SF	--	50%	25%	25%					.023
HIP **ADD**	SF	15%	30%	22%	22%					.024

9. CEDAR SHAKE ROOFING

SPECIFICATIONS		UNIT	JOB COST			PRICE	LOCAL AREA MODIFICATION				DATA BASE ITEM NO.
			MATLS	LABOR	TOTAL		MATLS	LABOR	TOTAL	PRICE	
	CEDAR SHINGLE AND HAND-SPLIT SHAKE COSTS ARE FOR ONE STORY BUILDINGS										
MORE THAN 1 STORY ABOVE GRADE	FOR ROOFING MORE THAN ONE STORY ABOVE GRADE										
	2 STORIES ABOVE GRADE **ADD**	EA	--	60.00	60.00	90.00					.000
	3 STORIES ABOVE GRADE **ADD**	EA	--	90.00	90.00	135.00					.001
SMALL JOBS	SMALL JOB ROOFING LESS THAN 300 SF **ADD**	EA	--	70.00	70.00	105.00					.208
	300 TO 500 SF **ADD**	EA	--	40.00	40.00	60.00					.209
	500 TO 600 SF **ADD**	EA	--	20.00	20.00	30.00					.210
DIFFICULT ROOF COVERING	DIFFICULT ROOF COVERING, CUT UP WITH VALLEYS AND DORMERS										
	SOME DIFFICULTY **ADD**	SF	--	.30	.30	.45					.211
	DIFFICULT **ADD**	SF	--	.50	.50	.75					.212
	VERY DIFFICULT **ADD**	SF	--	.80	.80	1.20					.213
HANDSPLIT AND RESAWN, 24"	ON PLAIN SHED OR GABLE TYPE WOOD ROOF DECK HANDSPLIT AND RESAWN CEDAR SHAKES, 24" LONG, 10" EXPOSURE, 1/2" TO 3/4" THICKNESS • 15 LB. FELT PAPER BETWEEN COURSES • 6d RUST RESISTANT NAILS • STAGING AS REQUIRED										
	Slope 4 TO 6 IN 12	SF	1.93	1.18	3.11	4.67					.200
	7 TO 12 IN 12	SF	1.93	1.50	3.43	5.15					.201
	SAME AS ABOVE, 3/4" TO 1-1/4" THICKNESS										
	Slope 4 TO 6 IN 12	SF	2.14	1.28	3.42	5.13					.202
	7 TO 12 IN 12	SF	2.14	1.61	3.75	5.63					.203

SPECIFICATIONS		UNIT	JOB COST			PRICE	LOCAL AREA MODIFICATION				DATA BASE ITEM NO.
			MATLS	LABOR	TOTAL		MATLS	LABOR	TOTAL	PRICE	
16" CEDAR SHINGLES	ON PLAIN SHED OR GABLE TYPE WOOD ROOF DECK CEDAR SHINGLES 16" LONG, 5" EXPOSURE • 15 LB. FELT PAPER OVER SOLID WOOD ROOF SHEATHING OR **NO** FELT PAPER IF OVER SPACED SHEATHING • 3d RUST RESISTANT NAILS • STAGING AS REQUIRED										
	Slope 4 TO 6 IN 12	SF	2.39	1.06	3.45	5.18					.204
	7 TO 12 IN 12	SF	2.39	1.27	3.66	5.49					.205
18" CEDAR SHINGLES	SAME AS ABOVE, No. 1, 18" PERFECTIONS, 5-1/2" EXPOSURE										
	Slope 4 TO 6 IN 12	SF	2.45	1.06	3.51	5.27					.206
	7 TO 12 IN 12	SF	2.45	1.27	3.72	5.58					.207
OTHER TYPE ROOFS	INSTALL CEDAR SHINGLES AND HANDSPLIT SHAKES ON OTHER TYPE ROOFS IN PLACE OF PLAIN SHED OR GABLE ROOFS										
	SM. GABLE DORMER **ADD**	SF	--	100%	50%	50%					.022
	GAMBREL **ADD**	SF	--	50%	25%	25%					.023
	HIP **ADD**	SF	15%	30%	22%	22%					.024

9. WOOD FIBER ROOFING SHINGLE PANELS

SPECIFICATIONS		UNIT	JOB COST			PRICE	LOCAL AREA MODIFICATION				DATA BASE ITEM NO.
			MATLS	LABOR	TOTAL		MATLS	LABOR	TOTAL	PRICE	
	ROOFING SHINGLE AND SHAKE PANEL COSTS ARE FOR ONE STORY BUILDINGS										
MORE THAN 1 STORY ABOVE GRADE	FOR ROOFING MORE THAN ONE STORY ABOVE GRADE										
	2 STORIES ABOVE GRADE **ADD**	EA	—	60.00	60.00	90.00					.000
	3 STORIES ABOVE GRADE **ADD**	EA	—	90.00	90.00	135.00					.001
SMALL JOBS	SMALL JOB ROOFING LESS THAN 300 SF **ADD**	EA	—	70.00	70.00	105.00					.306
	300 TO 500 SF **ADD**	EA	—	40.00	40.00	60.00					.307
	500 TO 600 SF **ADD**	EA	—	20.00	20.00	30.00					.308
DIFFICULT ROOF COVERING	DIFFICULT ROOF COVERING, CUT UP WITH VALLEYS AND DORMERS										
	SOME DIFFICULTY **ADD**	SF	—	.20	.20	.30					.309
	DIFFICULT **ADD**	SF	—	.40	.40	.60					.310
	VERY DIFFICULT **ADD**	SF	—	.60	.60	.90					.311
MASONITE PANEL	• ON EXISTING PLAIN SHED OR GABLE WOOD ROOF DECK OR OVER ONE LAYER OF OLD ROOF • GALVANIZED ROOFING NAILS OR STAPLES • #30 FELT PAPER **Traditional L** 48" X 11-25/32" 230 LBS/SQ 9-1/4" EXPOSURE										
	Slope 4 TO 6 IN 12	SF	.92	.53	1.45	2.18					.300
	7 TO 12 IN 12	SF	.92	.79	1.71	2.57					.301
	Traditional C 48" X 11-25/32" 265 LBS/SQ 9-1/4" EXPOSURE										
	Slope 4 TO 6 IN 12	SF	1.21	.53	1.74	2.61					.302
	7 TO 12 IN 12	SF	1.21	.79	2.00	3.00					.303
OTHER TYPE ROOFS	WOOD FIBER SHINGLE PANELS ON OTHER TYPE ROOFS IN PLACE OF PLAIN SHED OR GABLE ROOFS										
	SM. GABLE DORMER **ADD**	SF	—	100%	50%	50%					.022
	GAMBREL **ADD**	SF	—	50%	25%	25%					.023
	HIP **ADD**	SF	15%	30%	22%	22%					.024

SPECIFICATIONS		UNIT	JOB COST			PRICE	LOCAL AREA MODIFICATION				DATA BASE ITEM NO.
			MATLS	LABOR	TOTAL		MATLS	LABOR	TOTAL	PRICE	
HANDSPLIT ROOF PANELS	• ON EXISTING RAFTERS OR EXISTING PLAIN SHED OR GABLE ROOF DECK • 24" X 96" HANDSPLIT ROOF PANEL • No. 1 GRADE CEDAR SHAKES BONDED TO 5" X 1/2" PLYWOOD STRIPS • HIP AND RIDGE UNITS AS REQUIRED • #30 LB. FELT PAPER BETWEEN COURSES • 10" EXPOSURE • RUST RESISTANT NAILS										
	Slope 4 TO 6 IN 12	SF	1.93	.53	2.46	3.69					.304
	7 TO 12 IN 12	SF	1.93	.79	2.72	4.08					.305
OTHER TYPE ROOFS	INSTALL HANDSPLIT ROOF PANELS ON OTHER TYPE ROOFS IN PLACE OF PLAIN SHED OR GABLE ROOFS										
	SM. GABLE DORMER **ADD**	SF	--	100%	50%	50%					.022
	GAMBREL **ADD**	SF	--	50%	25%	25%					.023
	HIP **ADD**	SF	15%	30%	22%	22%					.024

9

9. CONCRETE ROOFING TILES

SPECIFICATIONS		UNIT	JOB COST			PRICE	LOCAL AREA MODIFICATION				DATA BASE ITEM NO.
			MATLS	LABOR	TOTAL		MATLS	LABOR	TOTAL	PRICE	
	CONCRETE ROOFING TILE COSTS ARE FOR ONE STORY BUILDINGS										
MORE THAN 1 STORY ABOVE GRADE	FOR ROOFING MORE THAN ONE STORY ABOVE GRADE										
	2 STORIES ABOVE GRADE **ADD**	EA	--	60.00	60.00	90.00					.000
	3 STORIES ABOVE GRADE **ADD**	EA	--	90.00	90.00	135.00					.001
SMALL JOBS	SMALL JOB ROOFING LESS THAN 300 SF **ADD**	EA	--	70.00	70.00	105.00					.002
	300 TO 500 SF **ADD**	EA	--	40.00	40.00	60.00					.003
	500 TO 600 SF **ADD**	EA	--	20.00	20.00	30.00					.004
DIFFICULT ROOF COVERING	DIFFICULT ROOF COVERING, CUT UP WITH VALLEYS AND DORMERS										
	SOME DIFFICULTY **ADD**	SF	--	.20	.20	.30					.005
	DIFFICULT **ADD**	SF	--	.40	.40	.60					.006
	VERY DIFFICULT **ADD**	SF	--	.60	.60	.90					.007
CONCRETE ROOFING TILES (FLAT)	ON PLAIN SHED OR GABLE SOLID SHEATHING WOOD ROOF DECK • 11" X 17" X 1-1/4" THICK ROOFING TILES WITH LUGS AT HEAD • 2" WOODEN LATH STRIPS NAILED OVER EACH RAFTER ITS FULL LENGTH • 1" X 2" REDWOOD BATTENS NAILED ACROSS LATH STRIPS SPACED TO HANG TILES • NAIL EVERY 3rd COURSE OF TILE										
	Slope UNDER 3 IN 12	SF	1.47	1.29	2.76	4.14					.400
	OVER 3 IN 12 (OMIT HOT MOPPING)	SF	1.25	1.13	2.38	3.57					.401
	ON EXISTING 1" X 6" SPACED SHEATHING (SPACED TO HANG TILES) ONE STORY ABOVE GRADE • INTERLACE 16" STRIPS OF 30 LB. ROOFING FELT BETWEEN EACH COURSE										
	Slope OVER 4 IN 12	SF	.90	.92	1.82	2.73					.402
OTHER TYPE ROOFS	CONCRETE ROOFING TILES ON OTHER TYPE ROOFS IN PLACE OF PLAIN SHED OR GABLE ROOFS										
	SM. GABLE DORMER **ADD**	SF	--	100%	50%	50%					.022
	GAMBREL **ADD**	SF	--	50%	25%	25%					.023
	HIP **ADD**	SF	15%	30%	22%	22%					.024

SPECIFICATIONS		UNIT	JOB COST			PRICE	LOCAL AREA MODIFICATION				DATA BASE ITEM NO.
			MATLS	LABOR	TOTAL		MATLS	LABOR	TOTAL	PRICE	
	MINERAL FIBER TYPE SHINGLE COSTS ARE FOR ONE STORY BUILDINGS										
MORE THAN 1 STORY ABOVE GRADE	FOR ROOFING MORE THAN ONE STORY ABOVE GRADE										
	2 STORIES ABOVE GRADE **ADD**	EA	--	60.00	60.00	90.00					.000
	3 STORIES ABOVE GRADE **ADD**	EA	--	90.00	90.00	135.00					.001
SMALL JOBS	SMALL JOB ROOFING LESS THAN 300 SF **ADD**	EA	--	70.00	70.00	105.00					.002
	300 TO 500 SF **ADD**	EA	--	40.00	40.00	60.00					.003
	500 TO 600 SF **ADD**	EA	--	20.00	20.00	30.00					.004
DIFFICULT ROOF COVERING	DIFFICULT ROOF COVERING, CUT UP WITH VALLEYS AND DORMERS										
	SOME DIFFICULTY **ADD**	SF	--	.20	.20	.30					.005
	DIFFICULT **ADD**	SF	--	.40	.40	.60					.006
	VERY DIFFICULT **ADD**	SF	--	.60	.60	.90					.007
SUPRADUR	ON PLAIN SHED OR GABLE SOLID SHEATHING WOOD ROOF DECK • 9-7/16" X 16" X 1/4" THICK MINERAL FIBER ROOF SHINGLES • 500 LBS/SQUARE • 7" EXPOSURE • #15 FELT PAPER • RUST RESISTANT NAILS										
	Slope 3 TO 6 IN 12	SF	2.21	2.41	4.62	6.93					.403
	7 TO 12 IN 12	SF	2.21	2.67	4.88	7.32					.404
	• 14" X 30" X 3/16" THICK MINERAL FIBER ROOF SHINGLES • 325 LBS/SQUARE • 6" EXPOSURE • #15 FELT PAPER • RUST RESISTANT NAILS										
	Slope 4 TO 6 IN 12	SF	2.21	2.31	4.52	6.78					.405
	7 TO 12 IN 12	SF	2.21	2.57	4.78	7.17					.406
OTHER TYPE ROOFS	MINERAL FIBER SHINGLES ON OTHER TYPE ROOFS IN PLACE OF PLAIN SHED OR GABLE ROOFS SM. GABLE DORMER **ADD**	SF	--	100%	50%	50%					.022
	GAMBREL **ADD**	SF	--	50%	25%	25%					.023
	HIP **ADD**	SF	15%	30%	22%	22%					.024

9

9. CLAY ROOFING TILES

SPECIFICATIONS		UNIT	JOB COST			PRICE	LOCAL AREA MODIFICATION				DATA BASE ITEM NO.
			MATLS	LABOR	TOTAL		MATLS	LABOR	TOTAL	PRICE	
SPANISH TILE	ON PLAIN SHED OR GABLE WOOD ROOF DECK • 11" O.C., 15" EXPOSURE, 3" LAP, RED CLAY TILE • 30 LB. FELT PAPER • HOT MOP SURFACE • RUST RESISTANT NAILS • INCLUDING RIDGE AND ACCESSORY TILES ***Slope***										
	3 TO 6 IN 12	SF	3.81	2.03	5.84	8.76					.407
	7 TO 12 IN 12	SF	3.82	2.41	6.23	9.35					.408
MISSION (BARREL) TILE	ON WOOD ROOF DECK, 18" TILE, 10-3/4" O.C., 14" EXPOSURE, 4" LAP • 30 LB. FELT PAPER • 1" X 4" WOOD NAILING STRIPS • HOT MOP SURFACE • RUST RESISTANT NAILS • INCLUDING RIDGE AND ACCESSORY TILES ***Slope***										
	3 TO 6 IN 12	SF	5.26	2.34	7.60	11.40					.409
	7 TO 12 IN 12	SF	5.26	2.73	7.99	11.99					.410
	WOOD BATTEN STRIP UNDER	SF	.14	.20	.34	.51					.411
OTHER TYPE ROOFS	CLAY ROOFING TILES ON OTHER TYPE ROOFS IN PLACE OF PLAIN SHED OR GABLE ROOFS										
	SM. GABLE DORMER **ADD**	SF	--	100%	50%	50%					.022
	GAMBREL **ADD**	SF	--	50%	25%	25%					.023
	HIP **ADD**	SF	15%	30%	22%	22%					.024

SPECIFICATIONS	UNIT	JOB COST			PRICE	LOCAL AREA MODIFICATION				DATA BASE ITEM NO.
		MATLS	LABOR	TOTAL		MATLS	LABOR	TOTAL	PRICE	
ROOFING COSTS SHOWN ON THIS PAGE INCLUDE A ROOFING SUBCONTRACTOR'S OVERHEAD AND PROFIT										
COPPER OVER WOOD DECK • 16 OZ. (.020 INCH) COPPER • #15 FELT PAPER										
STANDING SEAM	EA	100.00	310.00	410.00	615.00					.412
	PLUS SF	4.86	5.18	10.04	15.06					.413
FLAT SEAM	EA	100.00	310.00	410.00	615.00					.412
	PLUS SF	4.12	5.18	9.30	13.95					.414
EA = EACH JOB LOCATION										
METAL OVER WOOD DECK • 30 GA. SHEET METAL WITH BAKED ENAMEL FINISH • #15 FELT PAPER										
STANDING SEAM	EA	50.00	190.00	240.00	360.00					.415
	PLUS SF	1.20	1.90	3.10	4.65					.416
FLAT SEAM	EA	50.00	190.00	240.00	360.00					.415
	PLUS SF	1.10	1.90	3.00	4.50					.417
EA = EACH JOB LOCATION										
NOTE: FOR ALUMINIZATION OF METAL ROOFS, SEE PAGE 146										

9

9. ROOFING TEAR-OFF

SPECIFICATIONS		UNIT	JOB COST			PRICE	LOCAL AREA MODIFICATION				DATA BASE ITEM NO.
			MATLS	LABOR	TOTAL		MATLS	LABOR	TOTAL	PRICE	
	FOR RE-ROOFING, USE THE SAME COSTS SHOWN FOR NEW ROOFS ON PAGES 132-143. COSTS OF TEARING OFF THE OLD ROOF, DEPOSITING DEBRIS DIRECTLY INTO TRUCK FROM ROOF AND HAULING TO DUMP WITHIN 5 MILES (**DUMPING FEE NOT INCLUDED**) ARE SHOWN BELOW FOR VARIOUS KINDS OF EXISTING ROOFS.										
ASPHALT OR FIBER-GLASS	ASPHALT OR FIBERGLASS SHINGLES — 1 LAYER UP TO 8 IN 12 SLOPE	SF	--	.27	.27	.41					.500
	STEEP SLOPE	SF	--	.35	.35	.53					.501
	ASPHALT OR FIBERGLASS SHINGLES — 2 LAYERS UP TO 8 IN 12 SLOPE	SF	--	.39	.39	.59					.502
	STEEP SLOPE	SF	--	.46	.46	.69					.503
	ASPHALT OR FIBERGLASS SHINGLES — 3 LAYERS UP TO 8 IN 12 SLOPE	SF	--	.50	.50	.75					.504
	STEEP SLOPE	SF	--	.58	.58	.87					.505
ROLL ROOF OR SELVAGE	ROLL ROOFING OR SELVAGE UP TO 8 IN 12 SLOPE	SF	--	.22	.22	.33					.506
	STEEP SLOPE	SF	--	.27	.27	.41					.507
BUILT-UP	3, 4 OR 5 PLY ROOF WITH GRAVEL	SF	--	.89	.89	1.34					.508
	WITHOUT GRAVEL	SF	--	.46	.46	.69					.509
SLATE	UP TO 8 IN 12 SLOPE	SF	--	.55	.55	.83					.510
	STEEP SLOPE	SF	--	.61	.61	.92					.511
CLAY TILE	1-PIECE OR 2-PIECE INTER-LOCKING UP TO 8 IN 12 SLOPE	SF	--	.46	.46	.69					.512
	STEEP SLOPE	SF	--	.56	.56	.84					.513
METAL ROOFING	COPPER, GALVANIZED OR ALUMINUM ROOFING	SF	--	.31	.31	.47					.514
CEDAR SHINGLES	16" OR 18" SHINGLES WITH 5" EXPOSURE UP TO 8 IN 12 SLOPE	SF	--	.37	.37	.56					.515
	STEEP SLOPE	SF	--	.46	.46	.69					.516
CEDAR SHAKES	24" CEDAR SHAKES WITH 10" EXPOSURE, 3/4" TO 5/4" UP TO 8 IN 12 SLOPE	SF	--	.25	.25	.38					.517
	STEEP SLOPE	SF	--	.35	.35	.53					.518

SPECIFICATIONS		UNIT	JOB COST			PRICE	LOCAL AREA MODIFICATION				DATA BASE ITEM NO.
			MATLS	LABOR	TOTAL		MATLS	LABOR	TOTAL	PRICE	
PREPARA-TION FOR RE-ROOFING	INSTALL 1/2" PLYWOOD OVER EXISTING ROOF COVERING BEFORE INSTALLING NEW ROOF	SF	.45	.39	.84	1.26					.519
	CUT BACK EXISTING ROOF-ING FROM EAVES AND/OR GABLE, AND FILL SPACE WITH 1" X 6" FIR OR PINE LF = EAVES AND/OR GABLE	LF	.25	.57	.82	1.23					.520
	NAIL BATTENS OVER EXIST-ING ROOF COVERING TO RE-CEIVE NEW ROOF COVERING										
	1" X 3"	LF	.08	.22	.30	.45					.521
	1" X 4"	LF	.13	.22	.35	.53					.522
	1" X 6"	LF	.20	.22	.42	.63					.523
	LF = BATTENS										

9

9. ROOF REPAIR

SPECIFICATIONS		UNIT	JOB COST			PRICE	LOCAL AREA MODIFICATION				DATA BASE ITEM NO.
			MATLS	LABOR	TOTAL		MATLS	LABOR	TOTAL	PRICE	
PATCH BUILT-UP ROOF	SCRAPE MATERIALS FROM DEFECTIVE AREA AND IN-STALL NEW PAPER, HOT AS-PHALT AND SLAG TO AREA	EA PLUS SF	-- --	90.00 1.11	90.00 1.11	135.00 1.67					.600 .601
METAL ROOF REPAIR	TAR OVER EXISTING METAL ROOF FOR TEMPORARY RE-PAIR OF LEAKS	SF	--	1.12	1.12	1.68					.602
ALUMINIZA-TION	• SCRAPE MATERIALS FROM DEFECTIVE AREA • PATCH HOLES WITH FI-BER TAPE AND BLACK ASPHALT ROOF CEMENT • APPLY 2 COATS OF ALUM-INUM ROOF COATING WITH FIBER (2nd COAT WHILE 1st COAT IS STILL TACKY)	SF	.41	4.22	4.63	6.95					.603
	SAME AS ABOVE, DIFFICULT CONDITIONS — STEEP ROOF SLOPE, ETC.	SF	.41	6.00	6.41	9.62					.604
	ALUMINUM BASED ROOF PAINT APPLIED ON NEW ROOF OR ROOF IN GOOD CONDITION, ONE COAT EACH 2 YEARS SF = ONE COAT	SF	.19	1.11	1.30	1.95					.605
	SAME AS ABOVE, DIFFICULT CONDITIONS — STEEP ROOF SLOPE, ETC.	SF	.19	1.38	1.57	2.36					.606
SLATE ROOF REPAIR	REPAIR SLATE ROOF LEAK: REMOVE INDIVIDUAL SLATE AND REPLACE WITH NEW EA = UP TO 10 SLATES	EA	--	194.00	194.00	291.00					.607
	FOR EACH SLATE MORE THAN 10 ADD	EA	--	10.00	10.00	15.00					.608
SLATE	REPLACE VENT PIPE COLLAR WITH NEW LEAD COLLAR	EA	--	162.00	162.00	243.00					.609
SLATE VALLEY	REPAIR SLATE VALLEY ON DORMER EA = BOTH VALLEYS	EA PLUS LF	-- --	322.00 54.60	322.00 54.60	483.00 81.90					.610 .611
	REPAIR JOB: CEMENT RIDGE, REPLACE 5 SLATES, SEAL AROUND CHIMNEY AND VENTS EA = ENTIRE JOB	EA	--	274.00	274.00	411.00					.612
ASBESTOS ROOFING	REMOVE DEFECTIVE ASBESTOS SHINGLES AND REPLACE WITH NEW (IF AVAILABLE) EA = UP TO 10 SHINGLES	EA	--	158.00	158.00	237.00					.613
CLAY TILE ROOF REPAIR	REMOVE DEFECTIVE FLAT OR CURVED TILES AND RE-PLACE WITH NEW TILES EA = UP TO 10 TILES	EA	--	158.00	158.00	237.00					.614

9

SPECIFICATIONS		UNIT	JOB COST			PRICE	LOCAL AREA MODIFICATION				DATA BASE ITEM NO.
			MATLS	LABOR	TOTAL		MATLS	LABOR	TOTAL	PRICE	
ASPHALT OR FIBER-GLASS SHINGLE REPAIR	REMOVE DEFECTIVE ASPHALT OR FIBERGLASS SHINGLES AND REPLACE WITH NEW EA = UP TO 10 SHINGLES	EA	3.00	99.00	102.00	153.00					.615
	FOR EACH SHINGLE MORE THAN 10 **ADD**	EA	.30	5.40	5.70	8.55					.616
REPAIR WOOD SHINGLES	REPAIR WOOD SHINGLE OR SHAKE LEAK BY NAILING DOWN, INSTALLING SHEET METAL PATCH UNDER OR CEMENTING CRACK EA = UP TO 10 SHINGLES OR SHAKES	EA	–	66.00	66.00	99.00					.617
	FOR EACH SHINGLE OR SHAKE MORE THAN 10 **ADD**	EA	–	4.30	4.30	6.45					.618
	REPAIR SHINGLE OR SHAKE LEAK BY REPLACING DEFECTIVE PIECES WITH NEW ONES EA = UP TO 10 SHINGLES OR SHAKES	EA	–	104.00	104.00	156.00					.619
	FOR EACH SHINGLE OR SHAKE MORE THAN 10 **ADD**	EA	–	5.40	5.40	8.10					.620

9. FLASHING AND ROOF VENTS

SPECIFICATIONS		UNIT	JOB COST			PRICE	LOCAL AREA MODIFICATION				DATA BASE ITEM NO.
			MATLS	LABOR	TOTAL		MATLS	LABOR	TOTAL	PRICE	
FLASH CHIMNEY	FLASH AVERAGE SIZE EXTERIOR OR INTERIOR CHIMNEY WITH STEPFLASHING AND COUNTERFLASHING TURNED 1/2" INTO OPEN JOINTS, **NOT** INCLUDING CUTTING JOINTS										
	ALUMINUM, .032 GA.	EA	13.00	37.00	50.00	75.00					.700
	COPPER, 16-20 OZ.	EA	42.00	50.00	92.00	138.00					.701
FLASH ROOF	FLASH GABLE ROOF TO WALL WITH STEPFLASHING										
	In Brick Open Joints										
	ALUMINUM	LF	1.40	3.75	5.15	7.73					.702
	COPPER	LF	6.00	4.60	10.60	15.90					.703
	On Wood Sheathing										
	ALUMINUM	LF	2.00	4.00	6.00	9.00					.704
	COPPER	LF	6.00	5.00	11.00	16.50					.705
	FLASH TOP OF FLAT OR SHED ROOF TO WALL IN OPEN JOINTS OR ON WOOD SHEATHING										
	ALUMINUM	LF	1.20	2.20	3.40	5.10					.706
	COPPER	LF	3.80	2.75	6.55	9.83					.707
	VALLEY FLASHING, 18" WIDE										
	ALUMINUM	LF	3.30	3.50	6.80	10.20					.708
	COPPER	LF	7.20	3.00	10.20	15.30					.709
CUT JOINTS	CUT JOINTS IN MASONRY FOR FLASHING										
	SOFT MORTAR	LF	--	1.50	1.50	2.25					.710
	HARD MORTAR	LF	--	2.60	2.60	3.90					.711
PARAPET	RE-FLASH PARAPET 12" HIGH AND 12" WIDE										
	ALUMINUM	LF	3.40	2.40	5.80	8.70					.712
	COPPER	LF	11.00	5.00	16.00	24.00					.713
PLUMBING FLASHING	REPLACE EXISTING FLASHING ON PLUMBING VENT	EA	8.70	14.50	23.20	34.80					.714
ROOF VENTILATOR	TURBINE VENTILATOR WITH WIND-DIRECTED ROTATING TOP, 12"	EA	40.00	23.00	63.00	94.50					.715
ROOF VENT	UP TO 9" DIAMETER SCREENED PLASTIC ROOF VENT IN FIBERGLASS, ROLL ROOFING & WOOD SHINGLE ROOFS	EA	18.00	15.00	33.00	49.50					.716
RIDGE VENT	CONTINUOUS ROOF RIDGE VENT WITH LOUVERED SIDE OPENINGS -- INCLUDES CUTTING EXISTING SHINGLES AND SHEATHING										
	ALUMINUM PAINTED	LF	3.30	2.75	6.05	9.08					.717
	VINYL COLORED	LF	3.40	2.75	6.15	9.23					.718
	SAME AS ABOVE, INSTALLATION ONLY										
	ALUMINUM PAINTED	LF	3.30	1.50	4.80	7.20					.719
	VINYL COLORED	LF	3.40	1.50	4.90	7.35					.720

SPECIFICATIONS		UNIT	JOB COST			PRICE	LOCAL AREA MODIFICATION				DATA BASE ITEM NO.
			MATLS	LABOR	TOTAL		MATLS	LABOR	TOTAL	PRICE	
ALUMINUM GUTTERS AND DOWN-SPOUTS	ALUMINUM GUTTER AND DOWNSPOUT SYSTEM WITH BAKED-ON WHITE ENAMEL FINISH, INCLUDING ACCES-SORIES LF = TOTAL COMBINED GUTTER AND DOWN-SPOUT LENGTH	LF	1.72	1.40	3.12	4.68					.721
	.032 SEAMLESS GUTTER AND DOWNSPOUT, WHITE ENAM-EL FINISH	LF	--	3.97	3.97	5.96					.722
GALV-ANIZED GUTTERS & DOWN-SPOUTS	GALVANIZED STEEL OG GUT-TER AND DOWNSPOUT, IN-CLUDING ACCESSORIES										
	5"	LF	1.46	1.30	2.76	4.14					.723
	6"	LF	2.00	1.47	3.47	5.21					.724
	LF = TOTAL COMBINED GUTTER AND DOWN-SPOUT LENGTH										
COPPER	HALF ROUND COPPER GUT-TER AND DOWNSPOUTS, IN-CLUDING ACCESSORIES										
	5"	LF	7.25	1.89	9.14	13.71					.725
	6"	LF	9.00	2.10	11.10	16.65					.726
	LF = TOTAL COMBINED GUTTER AND DOWN-SPOUT LENGTH										
WOOD	4" X 5" TREATED FIR OR HEMLOCK	LF	10.00	2.36	12.36	18.54					.727
VINYL	VINYL 5" COLORED OR WHITE GUTTER AND DOWNSPOUT, INCLUDING ACCESSORIES LF = TOTAL COMBINED GUTTER AND DOWN-SPOUT LENGTH	LF	1.42	1.29	2.71	4.07					.728
GUTTER GUARD	ALUMINUM MESH GUTTER GUARD, 6" WIDE	LF	.67	.47	1.14	1.71					.729
CANALES	CANALES, FOR FLAT ROOF APPLICATION ONLY	EA	19.00	23.00	42.00	63.00					.730

9

10. EXTERIOR TRIM

SPECIFICATIONS		UNIT	JOB COST			PRICE	LOCAL AREA MODIFICATION				DATA BASE ITEM NO.
			MATLS	LABOR	TOTAL		MATLS	LABOR	TOTAL	PRICE	
FASCIA OR FRIEZE	#2 PINE, SPRUCE, HEMLOCK, FIR										
	1 X 6	LF	.79	1.01	1.80	2.70					.000
	1 X 8	LF	1.04	1.05	2.09	3.14					.001
	1 X 10	LF	1.24	1.17	2.41	3.62					.002
	1 X 12	LF	1.70	1.17	2.87	4.31					.003
SOFFIT	#2 PINE, SPRUCE, HEMLOCK, FIR										
	6"	LF	.77	1.32	2.09	3.14					.006
	8"	LF	1.04	1.37	2.41	3.62					.007
	10"	LF	1.24	1.43	2.67	4.01					.008
	12"	LF	1.70	1.47	3.17	4.76					.009
	3/8" FIR PLYWOOD, AC EXTERIOR GRADE										
	12"	LF	.61	1.76	2.37	3.56					.010
	16"	LF	.82	1.97	2.79	4.19					.011
	18"	LF	1.25	2.10	3.35	5.03					.012
	24"	LF	1.24	2.43	3.67	5.51					.013
	30"	LF	2.47	2.76	5.23	7.85					.014
	36"	LF	2.47	3.08	5.55	8.33					.015
	48"	LF	2.47	2.66	5.13	7.70					.016
BED MOULD	PINE, FIR, HEMLOCK, 3/4" X 2-5/8"	LF	.82	.58	1.40	2.10					.017
RAKE	#2 PINE OR FIR RAKE 1 X 6 AND RAKE MOULD (SHINGLE MOULD)	LF	.99	1.21	2.20	3.30					.018
ATTIC LOUVER	REDWOOD, INCLUDING INSECT SCREEN										
	CIRCULAR 24" DIAMETER	EA	88.00	24.00	112.00	168.00					.019
	HALF CIRCLE 30" X 18"	EA	92.00	24.00	116.00	174.00					.020
	RECTANGULAR 16" X 24"	EA	54.00	24.00	78.00	117.00					.021
LATTICE WALL	CUSTOM MADE WITH PRESSURE TREATED LUMBER, KILN DRIED										
	1/4" X 1-1/8", 2-1/4" O.C.	SF	3.68	7.31	10.99	16.49					.022
	1/4" X 1-5/8", 3-1/4" O.C.	SF	3.90	5.87	9.77	14.66					.023
LATTICE PANEL	LATTICE PANEL, 4' X 8', PRE-ASSEMBLED, WITH SLATS 1/4" X 1-1/2", PRESSURE TREATED	SF	.66	1.23	1.89	2.84					.024
	PLASTIC PRE-FINISHED 4' X 8' LATTICE PANEL	SF	.90	1.25	2.15	3.23					.025

10

SPECIFICATIONS		UNIT	JOB COST			PRICE	LOCAL AREA MODIFICATION				DATA BASE ITEM NO.
			MATLS	LABOR	TOTAL		MATLS	LABOR	TOTAL	PRICE	
ALUMINUM SOFFIT	RIBBED SOFFIT, INCLUDING "F" CHANNEL										
	8"	LF	2.25	1.15	3.40	5.10					.100
	12"	LF	2.75	1.15	3.90	5.85					.101
	18"	LF	3.55	1.25	4.80	7.20					.102
	24"	LF	4.25	1.40	5.65	8.48					.103
ALUMINUM FASCIA OR RAKE	FASCIA OR RAKE TRIM										
	4"	LF	1.50	.90	2.40	3.60					.104
	6"	LF	1.65	.90	2.55	3.83					.105
	8"	LF	1.84	.95	2.79	4.19					.106
	10"	LF	2.05	.95	3.00	4.50					.107
SOFFIT VENT	PLUG TYPE, ROUND, ALUMINUM, INSTALLED IN EXISTING SOFFIT										
	1"	EA	2.20	3.66	5.86	8.79					.108
	3"	EA	3.65	3.66	7.31	10.97					.109
	CONTINUOUS SOFFIT VENT, INSTALLED IN NEW SOFFIT, 3"	LF	.75	1.90	2.65	3.98					.110
GRILLE	ALUMINUM GRILLE INSTALLED IN EXISTING SOFFIT, 8" X 12"	EA	8.50	15.50	24.00	36.00					.111
GABLE VENTS	GABLE LOUVER VENTS CENTER HEIGHT: 12"	EA	44.00	22.00	66.00	99.00					.112
	16"	EA	52.00	22.00	74.00	111.00					.113
	20"	EA	62.00	22.00	84.00	126.00					.114
ATTIC LOUVER VENTS	INCLUDING INSECT SCREEN, INSTALLED IN EXISTING HOLE										
	16" X 4"	EA	12.50	5.50	18.00	27.00					.115
	16" X 8"	EA	18.50	5.50	24.00	36.00					.116
	15" X 21"	EA	22.00	12.00	34.00	51.00					.117
	24" X 30"	EA	32.00	12.00	44.00	66.00					.118
	33" X 27"	EA	40.00	12.00	52.00	78.00					.119
VINYL SOFFIT	SOLID OR PERFORATED, INCLUDING "F" OR "J" CHANNEL										
	12"	LF	2.00	1.15	3.15	4.73					.120
	18"	LF	2.65	1.30	3.95	5.93					.121
	24"	LF	3.25	1.40	4.65	6.98					.122
WRAP TRIM	WRAP EXTERIOR WINDOW OR DOOR TRIM WITH ALUMINUM EA = PER OPENING	EA	7.30	23.25	30.55	45.83					.123

10

10. WOOD DECK

SPECIFICATIONS		UNIT	JOB COST			PRICE	LOCAL AREA MODIFICATION				DATA BASE ITEM NO.
			MATLS	LABOR	TOTAL		MATLS	LABOR	TOTAL	PRICE	
	NOTE: RUST-RESISTANT NAILS ARE INCLUDED IN ALL DECK INSTALLATIONS										
WOOD POSTS	SOLID WOOD POSTS ON EXISTING FOOTING UP TO 8 FT.										
	4 X 4 PRESSURE TREATED FIR OR PINE	EA	10.64	10.30	20.94	31.41					.200
	4 X 4 CONSTR. GRADE REDWOOD OR TIGHT KNOT CEDAR	EA	16.94	10.30	27.24	40.86					.201
PIPE COLUMN	PIPE COLUMN, INCL. PLATES, 3" GALVANIZED, UP TO 8 FT.	EA	83.00	27.00	110.00	165.00					.202
WOOD BEAM OR HEADER	SOLID OR BUILT-UP ON EXISTING SUPPORTS, WITH PRESS. TREATED LUMBER										
	4" X 8"	LF	2.09	1.61	3.70	5.55					.203
	6" X 8"	LF	3.14	2.04	5.18	7.77					.204
	4" X 10"	LF	2.70	1.90	4.60	6.90					.205
	6" X 10"	LF	4.12	2.76	6.88	10.32					.206
	4" X 12"	LF	3.40	2.20	5.60	8.40					.207
JOISTS	INCLUD. LEDGER BOLTED TO BUILDING, JOISTS DOUBLED AT ALL SIDES, JOIST HANGERS, WITH PRESSURE TREATED LUMBER, 16" O.C.										
	2" X 8"	SF	1.40	.88	2.28	3.42					.208
	2" X 10"	SF	1.77	.88	2.65	3.98					.209
	2" X 12"	SF	2.11	.88	2.99	4.49					.210
BAND	CEDAR OR REDWOOD, 1" X 12"	LF	3.00	1.00	4.00	6.00					.211
DECK SURFACE	DECK APPLIED FLAT W/ 1/4" SPACE BETWEEN EACH PIECE										
	2" X 6" PRESSURE TREATED FIR OR PINE	SF	1.50	.82	2.32	3.48					.212
	2" X 6" CONSTR. GRADE R'WOOD OR TIGHT KNOT CEDAR	SF	2.80	.82	3.62	5.43					.213
	5/4" X 6" CONSTR. GRADE R'WOOD OR TIGHT KNOT CEDAR	SF	1.85	.82	2.67	4.01					.214
	SAME AS ABOVE, DECK LAID DIAGONALLY										
	2" X 6" PRESSURE TREATED FIR OR PINE	SF	1.91	1.04	2.95	4.43					.215
	2" X 6" CONSTR. GRADE R'WOOD OR TIGHT KNOT CEDAR	SF	3.56	1.04	4.60	6.90					.216
	5/4" X 6" CONSTR. GRADE R'WOOD OR TIGHT KNOT CEDAR	SF	2.35	1.04	3.39	5.09					.217

10

SPECIFICATIONS		UNIT	JOB COST			PRICE	LOCAL AREA MODIFICATION				DATA BASE ITEM NO.
			MATLS	LABOR	TOTAL		MATLS	LABOR	TOTAL	PRICE	
DECK RAILING OR STEP RAILING	4" X 4" END AND INTER-MEDIATE POSTS EXTENDING TO 42" ABOVE SURFACE OF DECK, 48" O.C. 2" X 6" CAP LAID FLAT ON POSTS, 2" X 6" TOP RAIL UNDER CAP 2" X 6" MIDDLE AND BOTTOM RAILS										
	PRESSURE TREATED FIR OR PINE	LF	3.75	4.45	8.20	12.30					.300
	CONSTR. GRADE REDWOOD OR CEDAR	LF	6.72	4.45	11.17	16.76					.301
	4" X 4" END AND INTER-MEDIATE POSTS EXTENDING TO 42" ABOVE SURFACE OF DECK, 48" O.C. 2" X 6" CAP LAID FLAT ON POSTS, 2" X 6" TOP RAIL UNDER CAP 2" X 2" PICKETS 7-1/2" O.C. NAILED TO TOP RAIL & BAND										
	PRESSURE TREATED FIR OR PINE	LF	3.80	4.60	8.40	12.60					.302
	CONSTR. GRADE REDWOOD OR CEDAR	LF	6.85	4.60	11.45	17.18					.303
SEAT	ADD TO ABOVE RAILINGS, 18" SEAT 18" ABOVE DECK										
	PRESSURE TREATED FIR OR PINE	LF	7.47	9.45	16.92	25.38					.304
	CONSTR. GRADE REDWOOD OR CEDAR	LF	14.73	9.77	24.50	36.75					.305
STEPS TO DECK	2" X 12" STRINGERS TWO 2" X 6" PER TREAD, OPEN RISERS, CONCRETE BOTTOM TREAD ON GROUND										
	PRESSURE TREATED FIR OR PINE, 36" WIDE STEPS	PER STEP	14.25	12.50	26.75	40.13					.306
	CONSTR. GRADE REDWOOD OR CEDAR, 36" WIDE STEPS	PER STEP	30.50	12.50	43.00	64.50					.307
	PRESSURE TREATED FIR OR PINE, 48" WIDE STEPS	PER STEP	22.50	16.00	38.50	57.75					.308
	CONSTR. GRADE REDWOOD OR CEDAR, 48" WIDE STEPS	PER STEP	45.50	16.00	61.50	92.25					.309
	NOTE: STEP RAILING IS **NOT** INCLUDED IN COST										

10

10. PORCH

SPECIFICATIONS		UNIT	JOB COST			PRICE	LOCAL AREA MODIFICATION				DATA BASE ITEM NO.
			MATLS	LABOR	TOTAL		MATLS	LABOR	TOTAL	PRICE	
	NOTE: FOR PORCH FRAMING AND FINISHING ITEMS NOT SHOWN ON THIS PAGE, RE-FER TO APPROPRIATE SEC-TIONS OF THE MANUAL.										
POSTS AND COLUMNS	BUILT-UP POST, 2 X 4 COV-ERED WITH FOUR 1 X 6, UP TO 8 FEET HIGH										
	#2 PINE OR FIR	EA	28.00	20.00	48.00	72.00					.400
	CONSTR. GRADE REDWOOD	EA	38.00	20.00	58.00	87.00					.401
	ALUMINUM ROUND COLONIAL PORTICO COLUMN										
	8"	LF	13.00	13.00	26.00	39.00					.402
	10"	LF	19.00	13.00	32.00	48.00					.403
	12"	LF	40.00	14.00	54.00	81.00					.404
	FIR HOLLOW ROUND COLON-IAL PORTICO COLUMN										
	8"	LF	36.00	16.00	52.00	78.00					.405
	10"	LF	40.00	16.00	56.00	84.00					.406
	12"	LF	48.00	16.00	64.00	96.00					.407
PORCH CEILING	INCLUDES CEILING COVE AT WALL INTERSECTIONS										
	1" X 6" REDWOOD V-JOINT	SF	5.08	.79	5.87	8.81					.408
	1" X 6" KNOTTY PINE V-JOINT	SF	2.52	.79	3.31	4.97					.409
	3/8" FIR PLYWOOD CEILING INCLUDING PANEL STRIPS AND CEILING COVE	SF	.75	.55	1.30	1.95					.410
	FIR BEADED CEILING, CEN-TER BEADED ONE SIDE AND CENTER V-JOINT OTHER SIDE										
	5/8" X 4"	SF	1.53	.79	2.32	3.48					.411
PORCH FLOOR	3/4" T&G PINE PORCH FLOORING	SF	3.17	1.00	4.17	6.26					.412
HEADER TRIM	#2 PINE OR FIR, 6" SOFFIT AND TWO 12" FASCIAS	LF	4.18	2.53	6.71	10.07					.413
SCREEN	ALUMINUM SCREENING WITH 2 X 4 INTERMEDIATE FRAM-ING AND PANEL STRIPS	SF	1.42	1.34	2.76	4.14					.414
SCREEN DOOR, WOOD	WOOD FRAME SCREEN DOOR WITH ALUMINUM SCREENING, INCLUDING ALL NECESSARY HARDWARE FOR INSTALLATION	EA	72.00	36.00	108.00	162.00					.415

10

SPECIFICATIONS		UNIT	JOB COST			PRICE	LOCAL AREA MODIFICATION				DATA BASE ITEM NO.
			MATLS	LABOR	TOTAL		MATLS	LABOR	TOTAL	PRICE	
ORNAMEN-TAL IRON	ORNAMENTAL IRON PORCH AND STEP RAILING, INCLUD-ING ALL POSTS & FITTINGS REQUIRED FOR INSTALLA-TION										
	PREMIUM	LF	16.75	5.80	22.55	33.83					.416
	ECONOMY	LF	14.15	5.80	19.95	29.93					.417
	CUSTOM BUILT AND IN-STALLED IRON RAILING	LF	19.60	6.55	26.15	39.23					.418
	ORNAMENTAL IRON COLUMN, 1" SQUARE TUBING, UP TO 8 FT. HIGH										
	8"	EA	42.00	26.00	68.00	102.00					.419
	8" X 8"	EA	70.00	26.00	96.00	144.00					.420
CORRU-GATED FIBER-GLASS ROOF PANELS	NAILED TO WOOD FRAMING WITH RING SHANK NAILS WITH RUBBER WASHERS, 2-1/2" CORRUGATIONS, 8 FT., 10 FT., AND 12 FT. PANELS X 26"	SF	.88	1.35	2.23	3.35					.421

10

11. WOOD SIDING

SPECIFICATIONS		UNIT	JOB COST			PRICE	LOCAL AREA MODIFICATION				DATA BASE ITEM NO.
			MATLS	LABOR	TOTAL		MATLS	LABOR	TOTAL	PRICE	
	NOTE: ALL SIDING ITEMS ON THIS AND THE FOLLOWING PAGES INCLUDE SIDING, NAILS, PAPER, STARTER STRIPS AND ALL OTHER MATERIALS AND LABOR REQUIRED TO INSTALL ON EXISTING SHEATHED OR BRACED STUDWALL AND ON A ONE STORY BUILDING										
MORE THAN ONE STORY ABOVE GRADE	FOR INSTALLATIONS MORE THAN ONE STORY ABOVE GRADE 2 STORIES ABOVE GRADE **ADD**	SF	--	--	10%	10%					.000
	3 STORIES ABOVE GRADE **ADD**	SF	--	--	20%	20%					.001
HOUSE WRAP	SUBSTITUTE HOUSE WRAP FOR PAPER SPECIFIED IN SIDING INSTALLATIONS ON THIS AND THE FOLLOWING PAGES **ADD**	SF	.06	.08	.14	.21					.002
BEVELED SIDING	• REDWOOD OR CEDAR BEVELED SIDING • #15 FELT OR ROSIN COATED PAPER • 5/4 X 4 OUTSIDE CORNERS • 5/4 X 5/4 INSIDE CORNERS ***Cedar*** 1/2 X 4	SF	1.91	1.47	3.38	5.07					.003
	1/2 X 6	SF	1.89	1.05	2.94	4.41					.004
	1/2 X 8	SF	1.75	.97	2.72	4.08					.005
	3/4 X 8	SF	1.97	.98	2.95	4.43					.006
	3/4 X 10	SF	1.83	.90	2.73	4.10					.007
	Redwood 1/2 X 4	SF	2.84	1.47	4.31	6.47					.008
	1/2 X 6	SF	2.77	1.05	3.82	5.73					.009
	1/2 X 8	SF	2.63	.97	3.60	5.40					.010
	3/4 X 8	SF	2.94	.98	3.92	5.88					.011
	3/4 X 10	SF	2.53	.90	3.43	5.15					.012
GERMAN OR DROP SIDING	• DROP SIDING • #15 FELT OR ROSIN COATED PAPER • 3/4 X 4 OUTSIDE CORNERS • 3/4 X 3/4 INSIDE CORNERS ***Fir/Pine*** 1 X 6	SF	1.10	1.05	2.15	3.23					.013
	1 X 8	SF	1.05	1.00	2.05	3.08					.014
	Cedar 1 X 6	SF	1.83	1.05	2.88	4.32					.015
	1 X 8	SF	1.79	1.00	2.79	4.19					.016
	Redwood 1 X 6	SF	2.63	1.05	3.68	5.52					.017
	1 X 8	SF	2.42	1.00	3.42	5.13					.018

SPECIFICATIONS		UNIT	JOB COST			PRICE	LOCAL AREA MODIFICATION				DATA BASE ITEM NO.
			MATLS	LABOR	TOTAL		MATLS	LABOR	TOTAL	PRICE	
TONGUE AND GROOVE	• #15 FELT OR ROSIN COATED PAPER • 3/4 X 4 OUTSIDE CORNERS										
	Cedar										
	1 X 4	SF	3.29	1.42	4.71	7.07					.019
	1 X 6	SF	3.06	1.28	4.34	6.51					.020
	1 X 8	SF	2.97	1.16	4.13	6.20					.021
	1 X 10	SF	2.83	.98	3.81	5.72					.022
	Redwood										
	1 X 4	SF	4.88	1.42	6.30	9.45					.023
	1 X 6	SF	4.58	1.28	5.86	8.79					.024
	1 X 8	SF	4.45	1.16	5.61	8.42					.025
	1 X 10	SF	4.24	.98	5.22	7.83					.026
	Fir/Pine										
	1 X 4	SF	1.56	1.42	2.98	4.47					.027
	1 X 6	SF	1.39	1.28	2.67	4.01					.028
	1 X 8	SF	1.32	1.16	2.48	3.72					.029
	1 X 10	SF	1.25	.98	2.23	3.35					.030
BOARD AND BATTEN	• #15 FELT OR ROSIN COATED PAPER • NAILED TO EXISTING WALL SHEATHING OR 1" X 3" HORIZONTAL FURRING, 12" O.C. • 1" X 3" BATTENS										
	Cedar										
	1 X 8	SF	2.30	1.04	3.34	5.01					.031
	1 X 10	SF	2.22	.95	3.17	4.76					.032
	1 X 12	SF	2.16	.84	3.00	4.50					.033
	Redwood										
	1 X 8	SF	3.51	1.04	4.55	6.83					.034
	1 X 10	SF	3.37	.95	4.32	6.48					.035
	1 X 12	SF	3.24	.84	4.08	6.12					.036
DRIP CAP & WATER TABLE	DRIP CAP AND 1 X 4 WATER TABLE INSTALLED UNDER FIRST COURSE OF SIDING **ADD** LF = TOTAL LINEAL FEET OF COMBINATION DRIP CAP AND WATER TABLE	LF	.35	.44	.79	1.19					.037

11. CEDAR SHAKE & SHINGLE SIDING

SPECIFICATIONS		UNIT	JOB COST			PRICE	LOCAL AREA MODIFICATION				DATA BASE ITEM NO.
			MATLS	LABOR	TOTAL		MATLS	LABOR	TOTAL	PRICE	
16" CEDAR SHINGLES	• 16" CEDAR SHINGLES • #15 FELT OR ROSIN COATED PAPER • 3d RUST RESISTANT NAILS • WOVEN OUTSIDE CORNERS • WOVEN INSIDE CORNERS										
	5" EXPOSURE	SF	1.25	1.62	2.87	4.31					.100
	7-1/2" EXPOSURE	SF	1.10	1.39	2.49	3.74					.101
	SAME AS ABOVE, FIRE RETARDANT SHINGLES										
	5" EXPOSURE	SF	1.89	1.63	3.52	5.28					.102
	7-1/2 EXPOSURE	SF	1.63	1.39	3.02	4.53					.103
18" PERFECTIONS	SAME AS ABOVE, NO. 1 18" PERFECTIONS										
	5-1/2" EXPOSURE	SF	1.80	1.50	3.30	4.95					.104
	7-1/2" EXPOSURE	SF	1.32	1.26	2.58	3.87					.105
	8-1/2" EXPOSURE	SF	1.17	1.16	2.33	3.50					.106
	SAME AS ABOVE, FIRE RETARDANT SHINGLES										
	5-1/2" EXPOSURE	SF	2.60	1.50	4.10	6.15					.107
	7-1/2" EXPOSURE	SF	1.95	1.26	3.21	4.82					.108
	8-1/2" EXPOSURE	SF	1.70	1.16	2.86	4.29					.109
	SAME AS ABOVE, DOUBLE COURSING • NO. 1 18" PERFECTIONS • UNDERCOURSE NO. 3 UTILITY GRADE • BUTT NAILED										
	14" EXPOSURE	SF	.90	1.40	2.30	3.45					.110
HAND-SPLIT SHAKES	SAME AS ABOVE, 24" LONG HAND-SPLIT SHAKES										
	8-1/2" EXPOSURE	SF	1.21	1.46	2.67	4.01					.111
	10" EXPOSURE	SF	1.00	1.33	2.33	3.50					.112
	11-1/2" EXPOSURE	SF	.79	1.22	2.01	3.02					.113
	SAME AS ABOVE, FIRE RETARDANT HAND-SPLIT SHAKES										
	8-1/2" EXPOSURE	SF	1.81	1.46	3.27	4.91					.114
	10" EXPOSURE	SF	1.50	1.33	2.83	4.25					.115
	11-1/2" EXPOSURE	SF	1.19	1.22	2.41	3.62					.116

11

SPECIFICATIONS		UNIT	JOB COST			PRICE	LOCAL AREA MODIFICATION				DATA BASE ITEM NO.
			MATLS	LABOR	TOTAL		MATLS	LABOR	TOTAL	PRICE	
PANEL SID-ING	• NO. 1 RED CEDAR SHAKES OR SHINGLES BONDED TO UNDER-COURSE AND PLYWOOD • MATCHING CORNERS • MATCHING NAILS • #15 FELT PAPER										
	18" X 96" 14" EXPOSURE	SF	1.96	.50	2.46	3.69					.116
	9" X 96" 7" EXPOSURE	SF	2.01	.56	2.57	3.86					.117
CEDAR SHINGLE PANELS	• 16' LENGTHS • 7/16" THICKNESS • #15 FELT PAPER • 5/4" X 4" OUTSIDE COR-NERS • 5/4" X 5/4" INSIDE COR-NERS • CEDAR SHINGLE FINISH • PRIMED 12" WIDTH	SF	1.48	.67	2.15	3.23					.118

11

11. HARDBOARD SIDING

SPECIFICATIONS		UNIT	JOB COST			PRICE	LOCAL AREA MODIFICATION				DATA BASE ITEM NO.
			MATLS	LABOR	TOTAL		MATLS	LABOR	TOTAL	PRICE	
LAP SIDING	• 16' LENGTHS • 7/16" THICKNESS • #15 FELT OR ROSIN COATED PAPER • 5/4" X 4" OUTSIDE CORNERS • 5/4" X 5/4" INSIDE CORNERS • ROUGH SAWN CEDAR FINISH • PRIMED										
	8" WIDTH	SF	.95	.70	1.65	2.48					.200
	12" WIDTH	SF	.90	.67	1.57	2.36					.201
	• 16' LENGTHS • 7/16" THICKNESS • #15 FELT PAPER • 5/4" X 4" OUTSIDE CORNERS • 5/4" X 5/4" INSIDE CORNERS • WOODGRAIN OR SMOOTH FINISHES • PREFINISHED										
	8" WIDTH	SF	1.14	.70	1.84	2.76					.202
	12" WIDTH	SF	1.11	.67	1.78	2.67					.203

SPECIFICATIONS		UNIT	JOB COST			PRICE	LOCAL AREA MODIFICATION				DATA BASE ITEM NO.
			MATLS	LABOR	TOTAL		MATLS	LABOR	TOTAL	PRICE	
HARD-BOARD PANEL SIDING	• 4'-0" WIDTH PANELS • 7/16" THICKNESS • PRIMED • ROUGH SAWN TEXTURE • #15 FELT PAPER • 3/4" X 4" OUTSIDE CORNERS • 3/4" X 3/4" INSIDE CORNERS										
	8' LENGTH	SF	1.06	.63	1.69	2.54					.205
	9' LENGTH	SF	1.10	.63	1.73	2.60					.206
	SAME AS ABOVE, PREFINISHED										
	8' LENGTH	SF	1.22	.63	1.85	2.78					.207
	9' LENGTH	SF	1.23	.63	1.86	2.79					.208
HARD-BOARD IMITATION STUCCO SIDING	• 4'-0" WIDTH PANELS • 7/16" THICKNESS • PRIMED • STUCCO TEXTURE • #15 FELT PAPER • 3/4" X 4" OUTSIDE CORNERS • 3/4" X 3/4" INSIDE CORNERS										
	8' LENGTH	SF	1.37	.63	2.00	3.00					.209
	9' LENGTH	SF	1.40	.63	2.03	3.05					.210
	SAME AS ABOVE, PREFINISHED										
	8' LENGTH	SF	1.51	.96	2.47	3.71					.211
	9' LENGTH	SF	1.54	.96	2.50	3.75					.212
BATTENS	INSTALL BATTENS ON VERTICAL JOINTS OR TO ACHIEVE TUDOR EFFECT										
	1" X 4" BATTENS	LF	.53	.43	.96	1.44					.213
	1" X 6" BATTENS	LF	.81	.43	1.24	1.86					.214
	LF = TOTAL LENGTH OF BATTENS										

11. PLYWOOD SIDING PANELS

SPECIFICATIONS			UNIT	JOB COST			PRICE	LOCAL AREA MODIFICATION				DATA BASE ITEM NO.
				MATLS	LABOR	TOTAL		MATLS	LABOR	TOTAL	PRICE	
TEXTURE 1-11	• PLYWOOD PANELS WITH PATTERNS 4", 6" OR 8" O.C. • #15 FELT PAPER • 1" X 3" AND 1" X 4" TRIM • RUST RESISTANT NAILS											
	Cedar	3/8"	SF	1.34	.63	1.97	2.96					.300
		5/8"	SF	1.69	.63	2.32	3.48					.301
	Redwood	3/8"	SF	1.48	.63	2.11	3.17					.302
		5/8"	SF	1.74	.63	2.37	3.56					.303
	Fir	3/8"	SF	.69	.63	1.32	1.98					.304
		5/8"	SF	.89	.63	1.52	2.28					.305
	Pine	3/8"	SF	.69	.63	1.32	1.98					.306
		5/8"	SF	.91	.63	1.54	2.31					.307
NATURAL STONE CAST PANELS	• NATURAL STONE CHIPS BONDED TO PLYWOOD PANELS • RUST RESISTANT OR COLOR NAILS • #15 FELT PAPER • SEALANTS AND CAULKING AS REQUIRED											
	48" X 1/2" X 8'		SF	3.03	1.10	4.13	6.20					.308
	48" X 1/2" X 9', 10'		SF	3.33	1.10	4.43	6.65					.309

11

SPECIFICATIONS		UNIT	JOB COST			PRICE	LOCAL AREA MODIFICATION				DATA BASE ITEM NO.
			MATLS	LABOR	TOTAL		MATLS	LABOR	TOTAL	PRICE	
ALUMINUM SIDING	• .024" THICK HORIZONTAL ALUMINUM SIDING • INSULATION BOARD BACKING • #15 FELT PAPER • ALL ACCESSORIES, BUT **NOT** INCLUDING SOFFIT, FASCIA OR WINDOW AND DOOR TRIM										
	8" SMOOTH	SF	1.75	.97	2.72	4.08					.400
	8" DOUBLE 4	SF	1.78	.97	2.75	4.13					.401
	10" SMOOTH	SF	1.77	.97	2.74	4.11					.402
	10" DOUBLE 5	SF	1.88	.97	2.85	4.28					.403
	SAME AS ABOVE, 12" ALUMINUM VERTICAL BOARD AND BATTEN	SF	1.81	.87	2.68	4.02					.404
VINYL SIDING	• SOLID HORIZONTAL PVC VINYL SIDING • INSULATION BOARD BACKING • #15 FELT PAPER • ALL ACCESSORIES, BUT **NOT** INCLUDING SOFFIT, FASCIA OR WINDOW AND DOOR TRIM										
	8" SMOOTH	SF	1.30	.97	2.27	3.41					.405
	8" DOUBLE 4	SF	1.39	.97	2.36	3.54					.406
	10" SMOOTH	SF	1.34	.97	2.31	3.47					.407
	10" DOUBLE 5	SF	1.45	.97	2.42	3.63					.408
	SAME AS ABOVE, 12" VINYL VERTICAL BOARD AND BATTEN	SF	1.39	.88	2.27	3.41					.409
WITHOUT INSULATION	WHEN INSULATION BOARD BACKING IS NOT USED **DEDUCT**	SF	.16	.15	.31	.47					.410

11

12. EXTERIOR DOORS

SPECIFICATIONS		UNIT	JOB COST			PRICE	LOCAL AREA MODIFICATION				DATA BASE ITEM NO.
			MATLS	LABOR	TOTAL		MATLS	LABOR	TOTAL	PRICE	
	NOTE: ALL EXTERIOR DOORS ARE 1-3/4" THICK										
FLUSH, BIRCH	PARTICLE CORE, INCLUDING FRAME, BRICK MOULDING, 1-1/2 PAIR 4 X 4 BUTTS, INTERIOR CASINGS, ENTRANCE LOCK @ $20.00										
	2-8 X 6-8	EA	233.00	103.00	336.00	504.00					.000
	3-0 X 6-8	EA	243.00	103.00	346.00	519.00					.001
	3-0 X 7-0	EA	255.00	103.00	358.00	537.00					.002
	TO ABOVE DOORS, FOR EACH LIGHT **ADD**	EA	9.00	--	8.70	13.05					.003
COLONIAL	6 PANEL INCLUDING FRAME, TRIM AND HARDWARE, FIR										
	3-0 X 6-8	EA	361.00	103.00	464.00	696.00					.004
	3-0 X 7-0	EA	375.00	103.00	478.00	717.00					.005
	SAME AS ABOVE, PINE DOOR 3-0 X 6-8	EA	361.00	103.00	464.00	696.00					.006
	3-0 X 7-0	EA	441.00	103.00	544.00	816.00					.007
	SAME AS ABOVE, OAK DOOR 3-0 X 6-8	EA	461.00	105.00	566.00	849.00					.008
	3-0 X 7-0	EA	495.00	105.00	600.00	900.00					.009
COLONIAL WITH LIGHTS	ABOVE COLONIAL DOORS WITH 4 SMALL LIGHTS AT TOP **ADD**	EA	15.00	--	15.00	22.50					.010
DOUBLE DOORS	2 DOORS, INCLUDING FRAME, BRICK MOULDING, SILL, 3 PAIR 4 X 4 BUTTS, KEY-LOCK, BARREL BOLT, T-ASTRAGEL, CASINGS, 6-0 X 6-8										
	FLUSH BIRCH	SET	498.00	138.00	636.00	954.00					.011
	COLONIAL FIR	SET	730.00	138.00	868.00	1,302.00					.012
	COLONIAL PINE	SET	834.00	138.00	972.00	1,458.00					.013
	COLONIAL OAK	SET	936.00	138.00	1,074.00	1,611.00					
SIDE OR REAR DOOR	FIR, 3 PANELS, 4 LIGHTS, FRAME, TRIM, BRICK MOULDING, HARDWARE 2-8 X 6-8	EA	317.00	91.00	408.00	612.00					.014
REPLACE EXTERIOR DOOR	REMOVE EXTERIOR DOOR AND REPLACE WITH 6 PANEL COLONIAL FIR IN EXISTING FRAME AND CASING, INSTALL EXISTING LOCKSET	EA	252.00	82.00	334.00	501.00					.015

SPECIFICATIONS		UNIT	JOB COST			PRICE	LOCAL AREA MODIFICATION				DATA BASE ITEM NO.
			MATLS	LABOR	TOTAL		MATLS	LABOR	TOTAL	PRICE	
FRENCH	FIR, 1-3/4", 10 LIGHTS, INCLUDING FRAME, BRICK MOULDING, OAK SILL, 1-1/2 PAIR 4 X 4 BUTTS, KEYLOCK AND INTERIOR CASINGS										
	2-6 X 6-8	EA	329.00	103.00	432.00	648.00					.016
	2-8 X 6-8	EA	349.00	103.00	452.00	678.00					.017
	3-0 X 6-8	EA	365.00	103.00	468.00	702.00					.018
	2 DOORS, INCLUDING FRAME, BRICK MOULDING, SILL, 3 PAIR 4 X 4 BUTTS, KEYLOCK, BARREL BOLT, T-ASTRAGEL, CASINGS										
	5-0 X 6-8	SET	634.00	132.00	766.00	1,149.00					.019
	5-4 X 6-8	SET	749.00	133.00	882.00	1,323.00					.020
	6-0 X 6-8	SET	755.00	137.00	892.00	1,338.00					.021
DUTCH	FIR, 2 PANEL, 9 LIGHTS, WITH FRAME, BRICK MOULDING AND OAK SILL, 2 PAIR 4 X 4 BUTTS, KEYLOCK, DUTCH DOOR QUADRANT, INTERIOR CASINGS										
	2-8 X 6-8	EA	454.00	126.00	580.00	870.00					.022
	3-0 X 6-8	EA	484.00	126.00	610.00	915.00					.023
	FOR CROSSBUCK STYLE **ADD**	EA	23.00	--	23.00	34.50					.024

12

12. PRE-HUNG EXTERIOR DOORS

SPECIFICATIONS		UNIT	JOB COST			PRICE	LOCAL AREA MODIFICATION				DATA BASE ITEM NO.
			MATLS	LABOR	TOTAL		MATLS	LABOR	TOTAL	PRICE	
FLUSH, BIRCH	1-3/4", SOLID CORE, INCLUDING FRAME, BRICK MOULDING, THRESHOLD, INTERIOR CASINGS, ENTRANCE LOCK @ $20.00										
	2-8 X 6-8	EA	216.00	42.00	258.00	387.00					.025
	3-0 X 6-8	EA	226.00	44.00	270.00	405.00					.026
	3-0 X 7-0	EA	249.00	49.00	298.00	447.00					.027
COLONIAL	6 PANEL, INCLUDING FRAME, TRIM AND HARDWARE, FIR										
	3-0 X 6-8	EA	352.00	44.00	396.00	594.00					.028
	3-0 X 7-0	EA	383.00	49.00	432.00	648.00					.029
	SAME AS ABOVE, PINE DOOR										
	3-0 X 6-8	EA	400.00	44.00	444.00	666.00					.030
	3-0 X 7-0	EA	443.00	49.00	492.00	738.00					.031
	SAME AS ABOVE, OAK DOOR										
	3-0 X 6-8	EA	468.00	50.00	518.00	777.00					.032
	3-0 X 7-0	EA	498.00	50.00	548.00	822.00					.033

12

SPECIFICATIONS		UNIT	JOB COST			PRICE	LOCAL AREA MODIFICATION				DATA BASE ITEM NO.
			MATLS	LABOR	TOTAL		MATLS	LABOR	TOTAL	PRICE	
METAL PRE-HUNG	• 6 PANEL • 3-0 X 6-8 X 1-3/4" • FOAM CORE • 1-1/2 PAIR BUTTS • ALUMINUM THRESHOLD • MAGNETIC WEATHER-STRIPPING • ENTRANCE LOCK • EXTERIOR FRAME AND BRICK MOULDING • INTERIOR CASINGS	EA	222.00	44.00	266.00	399.00					.034

12

12. DOOR ACCESSORIES

SPECIFICATIONS		UNIT	JOB COST			PRICE	LOCAL AREA MODIFICATION				DATA BASE ITEM NO.
			MATLS	LABOR	TOTAL		MATLS	LABOR	TOTAL	PRICE	
	NOTE: **ADD** ORNAMENTAL ENTRANCE FRAME AND SIDE-LIGHT COSTS TO MATERIALS AND LABOR COSTS OF DOOR SELECTED.										
ORNA-MENTAL ENTRANCE FRAME & SIDELIGHTS	• ENTRANCE FRAME WITH 2 PILASTERS AND 11" HIGH PEDIMENT • SIDELIGHT, 1-2 X 6-8 X 1-3/4" WITH 7/16" INSULATED GLASS, ONE LIGHT OR ONE LIGHT AND ONE PANEL										
	SINGLE DOOR **ADD**	EA	127.00	9.00	136.00	204.00					.035
	SINGLE DOOR **ADD** WITH 1 SIDELIGHT	EA	362.00	32.00	394.00	591.00					.036
	SINGLE DOOR **ADD** WITH 2 SIDELIGHTS	EA	602.00	44.00	646.00	969.00					.037
	• ENTRANCE FRAME WITH 2 PILASTERS AND 18" HIGH PEDIMENT WITH SLOPED OR ARCHED SIDES										
	SINGLE DOOR **ADD**	EA	179.00	9.00	188.00	282.00					.038
	SINGLE DOOR **ADD** WITH 1 SIDELIGHT	EA	420.00	32.00	452.00	678.00					.039
	SINGLE DOOR **ADD** WITH 2 SIDELIGHTS	EA	650.00	44.00	694.00	1,041.00					.040
	3 OR 5 LIGHT DIVIDED LIGHT GRILLE **ADD**	EA	26.00	--	26.00	39.00					.041
DOOR WEATHER-STRIPPING	ALUMINUM THRESHOLD AND BRONZE WEATHERSTRIPPING	EA	28.00	24.00	52.00	78.00					.042
	INTERLOCKING WEATHERSTRIPPING ON 3-0 X 6-8 DOOR ON WOOD FLOOR, INCLUDING THRESHOLD	EA	60.00	42.00	102.00	153.00					.043
DEADBOLT LOCK	OUTSIDE KEY, THUMB TURN INSIDE, SINGLE CYLINDER										
	STANDARD QUALITY	EA	18.00	22.00	40.00	60.00					.044
	PREMIUM QUALITY	EA	50.00	22.00	72.00	108.00					.045

SPECIFICATIONS		UNIT	JOB COST			PRICE	LOCAL AREA MODIFICATION				DATA BASE ITEM NO.
			MATLS	LABOR	TOTAL		MATLS	LABOR	TOTAL	PRICE	
GLIDING PATIO DOOR, WOOD	WITH INSULATED GLASS, INCLUDING FRAME, HARDWARE AND INTERIOR AND EXTERIOR TRIM										
Panels	*Economy*										
2	6-0 X 6-8	EA	596.00	100.00	696.00	1,044.00					.100
2	8-0 X 6-8	EA	701.00	121.00	822.00	1,233.00					.101
3	9-0 X 6-8	EA	848.00	132.00	980.00	1,470.00					.102
3	12-0 X 6-8	EA	1071.00	169.00	1,240.00	1,860.00					.103
	Builder										
2	6-0 X 6-8	EA	696.00	100.00	796.00	1,194.00					.104
2	8-0 X 6-8	EA	817.00	121.00	938.00	1,407.00					.105
3	9-0 X 6-8	EA	958.00	132.00	1,090.00	1,635.00					.106
3	12-0 X 6-8	EA	1205.00	169.00	1,374.00	2,061.00					.107
	Premium										
2	6-0 X 6-8	EA	920.00	100.00	1,020.00	1,530.00					.108
2	8-0 X 6-8	EA	1047.00	121.00	1,168.00	1,752.00					.109
3	9-0 X 6-8	EA	1190.00	132.00	1,322.00	1,983.00					.110
3	12-0 X 6-8	EA	1565.00	169.00	1,734.00	2,601.00					.111
ATRIUM STYLE WOOD PATIO DOOR	ATRIUM STYLE WITH HINGED ACTIVE SIDE, INSULATED GLASS, FRAME, HARDWARE & EXTERIOR & INTERIOR TRIM										
Panels	*Economy*										
2	6-0 X 6-8	EA	580.00	90.00	670.00	1,005.00					.118
2	8-0 X 6-8	EA	640.00	100.00	740.00	1,110.00					.119
	Builder										
2	6-0 X 6-8	EA	678.00	90.00	768.00	1,152.00					.120
2	8-0 X 6-8	EA	750.00	100.00	850.00	1,275.00					.121
	Premium										
2	6-0 X 6-8	EA	876.00	90.00	966.00	1,449.00					.122
2	8-0 X 6-8	EA	956.00	100.00	1,056.00	1,584.00					.123
GLIDING PATIO DOOR, ALUMINUM	WITH INSULATED GLASS, INCLUDING FRAME, HARDWARE AND INTERIOR AND EXTERIOR TRIM										
Panels	*Opening*										
2	5-0 X 6-8	EA	365.00	83.00	448.00	672.00					.112
2	6-0 X 6-8	EA	402.00	88.00	490.00	735.00					.113
2	8-0 X 6-8	EA	494.00	100.00	594.00	891.00					.114
2	9-0 X 6-8	EA	573.00	105.00	678.00	1,017.00					.115
2	12-0 X 6-8	EA	603.00	121.00	724.00	1,086.00					.116
2	15-0 X 6-8	EA	635.00	139.00	774.00	1,161.00					.117

12. GARAGE DOORS

SPECIFICATIONS		UNIT	JOB COST			PRICE	LOCAL AREA MODIFICATION				DATA BASE ITEM NO.
			MATLS	LABOR	TOTAL		MATLS	LABOR	TOTAL	PRICE	
GARAGE DOOR, WOOD	4 SECTION 4 PANEL, ONE ROW OF LIGHTS (OPTIONAL) INCLUDING STOPS, HARDWARE AND EXTERIOR TRIM										
	8-0 X 7-0	EA	292.00	108.00	400.00	600.00					.124
	9-0 X 7-0	EA	314.00	112.00	426.00	639.00					.125
	16-0 X 7-0	EA	588.00	146.00	734.00	1,101.00					.126
GARAGE DOOR, METAL	SAME AS ABOVE, STEEL GARAGE DOOR										
	8-0 X 7-0	EA	310.00	108.00	418.00	627.00					.127
	9-0 X 7-0	EA	322.00	112.00	434.00	651.00					.128
	16-0 X 7-0	EA	592.00	146.00	738.00	1,107.00					.129
	SAME AS ABOVE, FIBERGLASS WITH ALUMINUM FRAME										
	8-0 X 7-0	EA	346.00	108.00	454.00	681.00					.130
	9-0 X 7-0	EA	370.00	112.00	482.00	723.00					.131
	16-0 X 7-0	EA	680.00	146.00	826.00	1,239.00					.132
GARAGE DOOR OPERATOR	AUTOMATIC CONTROL, 1/4 HP REVERSIBLE MOTOR, CHAIN DRIVE, FOR DOOR UP TO 16 X 7										
	ONE CAR SET	EA	193.00	101.00	294.00	441.00					.133
	TWO CAR SET	EA	237.00	101.00	338.00	507.00					.134
	SAME AS ABOVE WITH 1/3 HP MOTOR FOR DOOR UP TO 18 FEET										
	ONE CAR SET	EA	231.00	101.00	332.00	498.00					.135
	TWO CAR SET	EA	273.00	101.00	374.00	561.00					.136

12

SPECIFICATIONS		UNIT	JOB COST			PRICE	LOCAL AREA MODIFICATION				DATA BASE ITEM NO.
			MATLS	LABOR	TOTAL		MATLS	LABOR	TOTAL	PRICE	
ALUMINUM STORM/ SCREEN COMBINA- TION DOORS	1" ALUMINUM FRAME WITH TEMPERED GLASS & SCREEN INSERTS, WEATHERSTRIP- PED, KICK PANEL, ALL HARD- WARE, CLOSER, 6-8 OR 7-0 HIGH X 2-6, 2-8, 2-10 OR 3-0 WIDE, 2 GLASS INSERTS AND 2 SCREEN INSERTS ANODIZED	EA	124.00	48.00	172.00	258.00					.137
	WHITE	EA	132.00	48.00	180.00	270.00					.138
	SELF-STORING DOOR WITH 2 GLASS INSERTS AND 1 SCREEN INSERT ECONOMY ANODIZED	EA	126.00	48.00	174.00	261.00					.139
	PREMIUM WHITE	EA	150.00	48.00	198.00	297.00					.140
	CROSSBUCK, WITH 1 TEM- PERED GLASS AND 1 UPPER SCREEN INSERT, COLONIAL FRINGE, 3 BLACK STRAP HINGES, BLACK HANDLE, CROSSBUCK KICK PANEL ECONOMY WHITE	EA	144.00	48.00	192.00	288.00					.141
	PREMIUM WHITE	EA	168.00	48.00	216.00	324.00					.142
WOOD STORM/ SCREEN COMBINA- TION DOORS	1-1/8" THICK WOOD FRAME COMBINATION DOOR WITH 1 GLASS INSERT, INCLUDING CLOSER AND ALL HARD- WARE, 6-8 AND 7-0 HIGH X 2-6, 2-8 AND 3-0 WIDE, CUT OR PLANE TO FIT	EA	169.00	63.00	232.00	348.00					.143

12

12. PRE-HUNG INTERIOR DOOR UNITS

SPECIFICATIONS		UNIT	JOB COST			PRICE	LOCAL AREA MODIFICATION				DATA BASE ITEM NO.
			MATLS	LABOR	TOTAL		MATLS	LABOR	TOTAL	PRICE	
PRE-HUNG DOOR	• INTERIOR 1-3/8" DOOR • STAIN GRADE JAMB • 2 SIDES CASING • PRIVACY LOCK @ $10										
BIRCH FLUSH HOLLOW CORE											
	1-6 X 6-8	EA	94.00	34.00	128.00	192.00					.200
	2-0 X 6-8	EA	95.00	34.00	129.00	193.50					.201
	2-4 X 6-8	EA	96.00	34.00	130.00	195.00					.202
	2-6 X 6-8	EA	98.00	34.00	132.00	198.00					.203
	2-8 X 6-8	EA	102.00	34.00	136.00	204.00					.204
	3-0 X 6-8	EA	106.00	34.00	140.00	210.00					.205
OAK FLUSH HOLLOW CORE											
	1-6 X 6-8	EA	116.00	34.00	150.00	225.00					.206
	2-0 X 6-8	EA	118.00	34.00	152.00	228.00					.207
	2-4 X 6-8	EA	124.00	34.00	158.00	237.00					.208
	2-6 X 6-8	EA	126.00	34.00	160.00	240.00					.209
	2-8 X 6-8	EA	130.00	34.00	164.00	246.00					.210
	3-0 X 6-8	EA	134.00	34.00	168.00	252.00					.211
WALNUT FL. HOLLOW CORE											
	1-6 X 6-8	EA	186.00	34.00	220.00	330.00					.212
	2-0 X 6-8	EA	192.00	34.00	226.00	339.00					.213
	2-4 X 6-8	EA	196.00	34.00	230.00	345.00					.214
	2-6 X 6-8	EA	198.00	34.00	232.00	348.00					.215
	2-8 X 6-8	EA	200.00	34.00	234.00	351.00					.216
	3-0 X 6-8	EA	204.00	34.00	238.00	357.00					.217
PINE, 6 PANEL											
	1-6 X 6-8	EA	162.00	34.00	196.00	294.00					.218
	2-0 X 6-8	EA	172.00	34.00	206.00	309.00					.219
	2-4 X 6-8	EA	180.00	34.00	214.00	321.00					.220
	2-6 X 6-8	EA	182.00	34.00	216.00	324.00					.221
	2-8 X 6-8	EA	186.00	34.00	220.00	330.00					.222
	3-0 X 6-8	EA	198.00	34.00	232.00	348.00					.223
DOUBLE FRENCH DOORS, 15 LITE											
	1-6 X 6-8	SET	418.00	34.00	452.00	678.00					.224
	2-0 X 6-8	SET	442.00	34.00	476.00	714.00					.225
	2-4 X 6-8	SET	464.00	34.00	498.00	747.00					.226
	2-6 X 6-8	SET	518.00	34.00	552.00	828.00					.227
	2-8 X 6-8	SET	574.00	34.00	608.00	912.00					.228
	3-0 X 6-8	SET	606.00	34.00	640.00	960.00					.229

12

SPECIFICATIONS	UNIT	JOB COST			PRICE	LOCAL AREA MODIFICATION				DATA BASE ITEM NO.
		MATLS	LABOR	TOTAL		MATLS	LABOR	TOTAL	PRICE	
PRE-HUNG DOOR • INTERIOR 1-3/8" DOOR • PAINT GRADE JAMB • 2 SIDES CASING • PRIVACY LOCK @ $10										
MAHOGANY FLUSH HC										
1-6 X 6-8	EA	87.00	34.00	121.00	181.50					.300
2-0 X 6-8	EA	88.00	34.00	122.00	183.00					.301
2-4 X 6-8	EA	90.00	34.00	124.00	186.00					.302
2-6 X 6-8	EA	92.00	34.00	126.00	189.00					.303
2-8 X 6-8	EA	94.00	34.00	128.00	192.00					.304
3-0 X 6-8	EA	98.00	34.00	132.00	198.00					.305
PINE LOUVER / LOUVER										
1-6 X 6-8	EA	112.00	34.00	146.00	219.00					.306
2-0 X 6-8	EA	118.00	34.00	152.00	228.00					.307
2-4 X 6-8	EA	126.00	34.00	160.00	240.00					.308
2-6 X 6-8	EA	128.00	34.00	162.00	243.00					.309
2-8 X 6-8	EA	138.00	34.00	172.00	258.00					.310
3-0 X 6-8	EA	150.00	34.00	184.00	276.00					.311
PINE LOUVER / PANEL										
1-6 X 6-8	EA	118.00	34.00	152.00	228.00					.312
2-0 X 6-8	EA	128.00	34.00	162.00	243.00					.313
2-4 X 6-8	EA	134.00	34.00	168.00	252.00					.314
2-6 X 6-8	EA	136.00	34.00	170.00	255.00					.315
2-8 X 6-8	EA	148.00	34.00	182.00	273.00					.316
3-0 X 6-8	EA	158.00	34.00	192.00	288.00					.317
HARDBOARD FLUSH HC										
1-6 X 6-8	EA	74.00	34.00	108.00	162.00					.318
2-0 X 6-8	EA	78.00	34.00	112.00	168.00					.319
2-4 X 6-8	EA	80.00	34.00	114.00	171.00					.320
2-6 X 6-8	EA	82.00	34.00	116.00	174.00					.321
2-8 X 6-8	EA	84.00	34.00	118.00	177.00					.322
3-0 X 6-8	EA	86.00	34.00	120.00	180.00					.323
6 RAISED PANELS, HARD-BOARD FACED										
1-6 X 6-8	EA	96.00	34.00	130.00	195.00					.324
2-0 X 6-8	EA	100.00	34.00	134.00	201.00					.325
2-4 X 6-8	EA	102.00	34.00	136.00	204.00					.326
2-6 X 6-8	EA	106.00	34.00	140.00	210.00					.327
2-8 X 6-8	EA	108.00	34.00	142.00	213.00					.328
3-0 X 6-8	EA	112.00	34.00	146.00	219.00					.329

12

12. INTERIOR FLUSH DOORS

SPECIFICATIONS	UNIT	JOB COST			PRICE	LOCAL AREA MODIFICATION				DATA BASE ITEM NO.
		MATLS	LABOR	TOTAL		MATLS	LABOR	TOTAL	PRICE	
FLUSH MAHOGANY, HOLLOW CORE, 1-3/8", INCLUDING JAMBS AND CASINGS, 1 PAIR 3-1/2 X 3-1/2 BUTTS, PRIVACY LOCK										
1-4 X 6-8	EA	89.00	63.00	152.00	228.00					.400
1-6 X 6-8	EA	89.00	63.00	152.00	228.00					.401
1-8 X 6-8	EA	91.00	63.00	154.00	231.00					.402
2-0 X 6-8	EA	91.00	63.00	154.00	231.00					.403
2-4 X 6-8	EA	91.00	63.00	154.00	231.00					.404
2-6 X 6-8	EA	95.00	63.00	158.00	237.00					.405
2-8 X 6-8	EA	97.00	63.00	160.00	240.00					.406
3-0 X 6-8	EA	101.00	63.00	164.00	246.00					.407
BIRCH, HOLLOW CORE, 1-3/8", INCLUDING TRIM AND HARDWARE										
1-4 X 6-8	EA	93.00	63.00	156.00	234.00					.408
1-6 X 6-8	EA	93.00	63.00	156.00	234.00					.409
1-8 X 6-8	EA	97.00	63.00	160.00	240.00					.410
2-0 X 6-8	EA	97.00	63.00	160.00	240.00					.411
2-4 X 6-8	EA	97.00	63.00	160.00	240.00					.412
2-6 X 6-8	EA	105.00	63.00	168.00	252.00					.413
2-8 X 6-8	EA	107.00	63.00	170.00	255.00					.414
3-0 X 6-8	EA	115.00	63.00	178.00	267.00					.415
HARDBOARD, UNFINISHED HOLLOW CORE, 1-3/8" DOOR, INCLUDING TRIM AND HARDWARE										
1-4 X 6-8	EA	81.00	63.00	144.00	216.00					.416
1-6 X 6-8	EA	81.00	63.00	144.00	216.00					.417
1-8 X 6-8	EA	85.00	63.00	148.00	222.00					.418
2-0 X 6-8	EA	85.00	63.00	148.00	222.00					.419
2-4 X 6-8	EA	85.00	63.00	148.00	222.00					.420
2-6 X 6-8	EA	87.00	63.00	150.00	225.00					.421
2-8 X 6-8	EA	89.00	63.00	152.00	228.00					.422
3-0 X 6-8	EA	89.00	63.00	152.00	228.00					.423

12

SPECIFICATIONS	UNIT	JOB COST			PRICE	LOCAL AREA MODIFICATION				DATA BASE ITEM NO.
		MATLS	LABOR	TOTAL		MATLS	LABOR	TOTAL	PRICE	
SLIDING MAHOGANY FLUSH HOLLOW CORE, 1-3/8", 2 DOORS, INCLUDING JAMBS, TWO SIDES 2-1/4" CASING, ALL NECESSARY HARDWARE										
4-0 X 6-8	EA	119.00	65.00	184.00	276.00					.424
5-0 X 6-8	EA	129.00	65.00	194.00	291.00					.425
6-0 X 6-8	EA	137.00	65.00	202.00	303.00					.426
SAME AS ABOVE WITH BIRCH HOLLOW CORE										
4-0 X 6-8	EA	137.00	65.00	202.00	303.00					.427
5-0 X 6-8	EA	149.00	65.00	214.00	321.00					.428
6-0 X 6-8	EA	161.00	65.00	226.00	339.00					.429
SAME AS ABOVE WITH PINE, HEMLOCK, OR FIR LOUVER DOORS										
4-0 X 6-8	EA	219.00	65.00	284.00	426.00					.430
5-0 X 6-8	EA	245.00	65.00	310.00	465.00					.431
6-0 X 6-8	EA	285.00	65.00	350.00	525.00					.432
SAME AS ABOVE WITH OAK HOLLOW CORE										
4-0 X 6-8	EA	157.00	65.00	222.00	333.00					.433
5-0 X 6-8	EA	167.00	65.00	232.00	348.00					.434
6-0 X 6-8	EA	181.00	65.00	246.00	369.00					.435

12

12. INTERIOR DOORS

SPECIFICATIONS		UNIT	JOB COST			PRICE	LOCAL AREA MODIFICATION				DATA BASE ITEM NO.
			MATLS	LABOR	TOTAL		MATLS	LABOR	TOTAL	PRICE	
COLONIAL	HARDBOARD FACED, 6 RAISED PANELS, 1-3/8", INCLUDING JAMBS, STOPS, 2 SIDES 2-1/2" CASING, 1 PAIR 3-1/2 X 3-1/2 BUTTS, PRIVACY LOCK										
	1-6 X 6-8	EA	101.00	63.00	164.00	246.00					.500
	2-0 X 6-8	EA	111.00	63.00	174.00	261.00					.501
	2-2 X 6-8	EA	113.00	63.00	176.00	264.00					.502
	2-4 X 6-8	EA	115.00	63.00	178.00	267.00					.503
	2-6 X 6-8	EA	121.00	63.00	184.00	276.00					.504
	2-8 X 6-8	EA	125.00	63.00	188.00	282.00					.505
	3-0 X 6-8	EA	127.00	63.00	190.00	285.00					.506
	SAME AS ABOVE, WITH SOLID PINE RAISED PANELS										
	1-6 X 6-8	EA	150.00	64.00	214.00	321.00					.507
	2-0 X 6-8	EA	170.00	64.00	234.00	351.00					.508
	2-2 X 6-8	EA	170.00	64.00	234.00	351.00					.509
	2-4 X 6-8	EA	176.00	64.00	240.00	360.00					.510
	2-6 X 6-8	EA	182.00	64.00	246.00	369.00					.511
	2-8 X 6-8	EA	190.00	64.00	254.00	381.00					.512
	3-0 X 6-8	EA	202.00	64.00	266.00	399.00					.513
LOUVERED	PINE, 1-3/8", WITH OR WITHOUT CROSS RAIL, INCLUDING JAMBS, STOPS, 2 SIDES 2-1/4" CASING, 1 PAIR 3-1/2 X 3-1/2 BUTTS, PASSAGE SET, 1/4 X 1-3/8 SLATS										
	1-6 X 6-8	EA	111.00	63.00	174.00	261.00					.514
	2-0 X 6-8	EA	131.00	63.00	194.00	291.00					.515
	2-4 X 6-8	EA	137.00	63.00	200.00	300.00					.516
	2-6 X 6-8	EA	139.00	63.00	202.00	303.00					.517
	2-8 X 6-8	EA	143.00	63.00	206.00	309.00					.518
	3-0 X 6-8	EA	149.00	63.00	212.00	318.00					.519

12

SPECIFICATIONS		UNIT	JOB COST			PRICE	LOCAL AREA MODIFICATION				DATA BASE ITEM NO.
			MATLS	LABOR	TOTAL		MATLS	LABOR	TOTAL	PRICE	
POCKET DOOR	METAL REINFORCED SLIDING DOOR POCKET, INCLUDING TRACK, NYLON ROLLERS, JAMBS, CASINGS AND HARD-WARE, 1-3/8" DOOR										
	BIRCH FLUSH HOLLOW CORE										
	2-0 X 6-8	EA	158.00	96.00	254.00	381.00					.520
	2-4 X 6-8	EA	158.00	96.00	254.00	381.00					.521
	2-6 X 6-8	EA	166.00	98.00	264.00	396.00					.522
	2-8 X 6-8	EA	168.00	100.00	268.00	402.00					.523
	3-0 X 6-8	EA	176.00	100.00	276.00	414.00					.524
	HARDBOARD FLUSH HC										
	2-0 X 6-8	EA	148.00	96.00	244.00	366.00					.525
	2-4 X 6-8	EA	148.00	96.00	244.00	366.00					.526
	2-6 X 6-8	EA	150.00	98.00	248.00	372.00					.527
	2-8 X 6-8	EA	152.00	100.00	252.00	378.00					.528
	3-0 X 6-8	EA	156.00	100.00	256.00	384.00					.529
	COLONIAL WITH SOLID RAISED PANELS										
	2-0 X 6-8	EA	202.00	96.00	298.00	447.00					.530
	2-4 X 6-8	EA	204.00	96.00	300.00	450.00					.531
	2-6 X 6-8	EA	212.00	98.00	310.00	465.00					.532
	2-8 X 6-8	EA	222.00	100.00	322.00	483.00					.533
	3-0 X 6-8	EA	232.00	100.00	332.00	498.00					.534

12

12. BI-FOLD DOORS

SPECIFICATIONS		UNIT	JOB COST			PRICE	LOCAL AREA MODIFICATION				DATA BASE ITEM NO.
			MATLS	LABOR	TOTAL		MATLS	LABOR	TOTAL	PRICE	
BIRCH BI-FOLD	BIRCH HOLLOW CORE FLUSH 1-3/8" DOORS, INCLUDING JAMBS, 2 SIDES 2-1/4" CASING, BI-FOLD HARDWARE										
	Opening Size										
	TWO DOORS 2-6 X 6-8	SET	80.00	56.00	136.00	204.00					.600
	3-0 X 6-8	SET	85.00	59.00	144.00	216.00					.601
	FOUR DOORS 4-0 X 6-8	SET	108.00	62.00	170.00	255.00					.602
	5-0 X 6-8	SET	114.00	66.00	180.00	270.00					.603
	6-0 X 6-8	SET	122.00	68.00	190.00	285.00					.604
MAHOGANY BI-FOLD	SAME AS ABOVE WITH MAHOGANY HOLLOW CORE										
	TWO DOORS 2-6 X 6-8	SET	78.00	56.00	134.00	201.00					.605
	3-0 X 6-8	SET	83.00	59.00	142.00	213.00					.606
	FOUR DOORS 4-0 X 6-8	SET	104.00	62.00	166.00	249.00					.607
	5-0 X 6-8	SET	112.00	66.00	178.00	267.00					.608
	6-0 X 6-8	SET	120.00	68.00	188.00	282.00					.609
OAK BI-FOLD	SAME AS ABOVE WITH OAK HOLLOW CORE										
	TWO DOORS 2-6 X 6-8	SET	88.00	56.00	144.00	216.00					.610
	3-0 X 6-8	SET	97.00	59.00	156.00	234.00					.611
	FOUR DOORS 4-0 X 6-8	SET	120.00	62.00	182.00	273.00					.612
	5-0 X 6-8	SET	128.00	66.00	194.00	291.00					.613
	6-0 X 6-8	SET	138.00	68.00	206.00	309.00					.614
HARD-BOARD BI-FOLD	SAME AS ABOVE WITH HARD-BOARD HOLLOW CORE										
	TWO DOORS 2-6 X 6-8	SET	70.00	56.00	126.00	189.00					.615
	3-0 X 6-8	SET	73.00	59.00	132.00	198.00					.616
	FOUR DOORS 4-0 X 6-8	SET	93.00	63.00	156.00	234.00					.617
	5-0 X 6-8	SET	98.00	66.00	164.00	246.00					.618
	6-0 X 6-8	SET	108.00	68.00	176.00	264.00					.619

12

SPECIFICATIONS		UNIT	JOB COST			PRICE	LOCAL AREA MODIFICATION				DATA BASE ITEM NO.
			MATLS	LABOR	TOTAL		MATLS	LABOR	TOTAL	PRICE	
COLONIAL BI-FOLD	BIRCH HOLLOW CORE FLUSH 1-3/8" DOORS, INCLUDING JAMBS, 2 SIDES 2-1/4" CASING, BI-FOLD HARDWARE										
	Opening Size										
	TWO DOORS 2-6 X 6-8	SET	100.00	56.00	156.00	234.00					.620
	3-0 X 6-8	SET	105.00	59.00	164.00	246.00					.621
	FOUR DOORS 4-0 X 6-8	SET	134.00	62.00	196.00	294.00					.622
	5-0 X 6-8	SET	142.00	66.00	208.00	312.00					.623
	6-0 X 6-8	SET	152.00	68.00	220.00	330.00					.624
PINE LOUVERED BI-FOLD	SAME AS ABOVE WITH PINE LOUVERED DOORS										
	TWO DOORS 2-6 X 6-8	SET	118.00	56.00	174.00	261.00					.625
	3-0 X 6-8	SET	121.00	59.00	180.00	270.00					.626
	FOUR DOORS 4-0 X 6-8	SET	156.00	62.00	218.00	327.00					.627
	5-0 X 6-8	SET	164.00	66.00	230.00	345.00					.628
	6-0 X 6-8	SET	176.00	68.00	244.00	366.00					.629
MIRROR BI-FOLD	SAME AS ABOVE WITH FRAMELESS MIRROR WITH SHATTER-RESISTANT SAFETY BACKING										
	TWO DOORS 2-6 X 6-8	SET	132.00	56.00	188.00	282.00					.630
	3-0 X 6-8	SET	153.00	59.00	212.00	318.00					.631
	FOUR DOORS 4-0 X 6-8	SET	200.00	62.00	262.00	393.00					.632
	5-0 X 6-8	SET	218.00	66.00	284.00	426.00					.633
	6-0 X 6-8	SET	264.00	68.00	332.00	498.00					.634

12

13. SINGLE GLAZED WINDOWS

SPECIFICATIONS		UNIT	JOB COST			PRICE	LOCAL AREA MODIFICATION				DATA BASE ITEM NO.
			MATLS	LABOR	TOTAL		MATLS	LABOR	TOTAL	PRICE	
	NOTE: ALL *WINDOW SIZES* SHOWN ON THIS AND THE FOLLOWING PAGES ARE *SASH SIZES* (1-8 X 3-2 = 20" WIDE X 38" HIGH)										
WOOD DOUBLE-HUNG	• FRAME & BRICK MOULD • SINGLE GLAZED SASH • SASH LOCK • INTERIOR TRIM, STOPS • STOOL AND APRON										
	2 LIGHTS 1-8 X 3-2	EA	59.00	55.00	114.00	171.00					.000
	3-10	EA	63.00	55.00	118.00	177.00					.001
	4-2	EA	69.00	55.00	124.00	186.00					.002
	4-6	EA	73.00	55.00	128.00	192.00					.003
	4 LIGHTS 1-8 X 3-2	EA	73.00	55.00	128.00	192.00					.004
	3-10	EA	75.00	55.00	130.00	195.00					.005
	4-2	EA	79.00	55.00	134.00	201.00					.006
	4-6	EA	83.00	55.00	138.00	207.00					.007
	6 LIGHTS 2-0 X 3-2	EA	77.00	55.00	132.00	198.00					.008
	3-10	EA	79.00	55.00	134.00	201.00					.009
	4-2	EA	83.00	55.00	138.00	207.00					.010
	4-6	EA	85.00	55.00	140.00	210.00					.011
	2-4 X 3-2	EA	79.00	55.00	134.00	201.00					.012
	3-10	EA	83.00	55.00	138.00	207.00					.013
	4-2	EA	85.00	55.00	140.00	210.00					.014
	4-6	EA	93.00	55.00	148.00	222.00					.015
	2-8 X 3-2	EA	85.00	55.00	140.00	210.00					.016
	3-10	EA	89.00	55.00	144.00	216.00					.017
	4-2	EA	93.00	55.00	148.00	222.00					.018
	4-6	EA	103.00	55.00	158.00	237.00					.019
	5-2	EA	117.00	55.00	172.00	258.00					.020
	5-6	EA	125.00	55.00	180.00	270.00					.021
	8 LIGHTS 3-0 X 3-2	EA	89.00	55.00	144.00	216.00					.022
	3-10	EA	93.00	55.00	148.00	222.00					.023
	4-2	EA	107.00	55.00	162.00	243.00					.024
	4-6	EA	113.00	55.00	168.00	252.00					.025
	5-2	EA	127.00	55.00	182.00	273.00					.026
	5-6	EA	133.00	55.00	188.00	282.00					.027
TWIN UNITS	DOUBLE THE SINGLE UNIT AMOUNT AND **ADD**	EA	7.00	9.00	16.00	24.00					.028
TRIPLE UNITS	TRIPLE THE SINGLE UNIT AMOUNT AND **ADD**	EA	9.00	13.00	22.00	33.00					.029

SPECIFICATIONS		UNIT	JOB COST			PRICE	LOCAL AREA MODIFICATION				DATA BASE ITEM NO.
			MATLS	LABOR	TOTAL		MATLS	LABOR	TOTAL	PRICE	
ONE LIGHT PICTURE SASH AND FRAME	WOOD, WITH 1/4" PLATE GLASS, INCLUDING EX-TERIOR BRICK MOULD, FRAME, SASH, INTERIOR CASINGS, STOPS, STOOL										
	4-0 X 3-10	EA	175.00	55.00	230.00	345.00					.030
	5-0 X 3-10	EA	207.00	55.00	262.00	393.00					.031
	6-0 X 3-10	EA	229.00	55.00	284.00	426.00					.032
	4-0 X 4-2	EA	185.00	55.00	240.00	360.00					.033
	5-0 X 4-2	EA	217.00	55.00	272.00	408.00					.034
	6-0 X 4-2	EA	265.00	55.00	320.00	480.00					.035
	4-0 X 4-6	EA	205.00	55.00	260.00	390.00					.036
	5-0 X 4-6	EA	243.00	55.00	298.00	447.00					.037
	6-0 X 4-6	EA	255.00	55.00	310.00	465.00					.038
	4-0 X 5-2	EA	229.00	55.00	284.00	426.00					.039
	5-0 X 5-2	EA	261.00	55.00	316.00	474.00					.040
	6-0 X 5-2	EA	309.00	55.00	364.00	546.00					.041
	SAME AS ABOVE WITH 1/2" INSULATED GLASS PICTURE SASH **ADD**	EA	70%	--	50%	50%					.042
PICTURE WINDOW	SINGLE GLAZED PICTURE WINDOW WITH FLANKING DOUBLE-HUNG FRAMES AND SASH										
	Side Center Side Height 2-0 4-0 2-0 4-6	EA	388.00	66.00	454.00	681.00					.043
	5-0	EA	424.00	66.00	490.00	735.00					.044
	6-0	EA	440.00	66.00	506.00	759.00					.045
	3-0 4-0 3-0 5-2	EA	498.00	66.00	564.00	846.00					.046
	5-0	EA	532.00	66.00	598.00	897.00					.047
	6-0	EA	580.00	66.00	646.00	969.00					.048
	SAME AS ABOVE USING 1/2" INSULATED GLASS IN CEN-TER SECTION **ADD**	EA	65%	--	60%	60%					.049

13

13. DOUBLE GLAZED WINDOWS -- BUILDER QUALITY

SPECIFICATIONS			UNIT	JOB COST			PRICE	LOCAL AREA MODIFICATION				DATA BASE ITEM NO.
				MATLS	LABOR	TOTAL		MATLS	LABOR	TOTAL	PRICE	
WOOD DOUBLE HUNG	• FRAME, SASH, BRICK MOULD • PRIMED SASH WITH 1/2" INSULATING GLASS • SASH LOCK • INTERIOR TRIM, STOPS • STOOL AND APRON • SCREEN											
	2 LIGHTS 1-8 X	4-2	EA	161.00	55.00	216.00	324.00					.100
		4-6	EA	161.00	55.00	216.00	324.00					.101
	2-0 X	3-2	EA	149.00	55.00	204.00	306.00					.102
		4-2	EA	159.00	55.00	214.00	321.00					.103
		4-6	EA	169.00	55.00	224.00	336.00					.104
		5-2	EA	173.00	55.00	228.00	342.00					.105
	2-4 X	2-10	EA	139.00	59.00	198.00	297.00					.106
		3-2	EA	141.00	59.00	200.00	300.00					.107
		3-10	EA	153.00	59.00	212.00	318.00					.108
		4-2	EA	159.00	59.00	218.00	327.00					.109
		4-6	EA	163.00	59.00	222.00	333.00					.110
		5-2	EA	173.00	59.00	232.00	348.00					.111
	2-8 X	2-10	EA	159.00	59.00	218.00	327.00					.112
		3-2	EA	163.00	59.00	222.00	333.00					.113
		3-10	EA	173.00	59.00	232.00	348.00					.114
		4-2	EA	183.00	59.00	242.00	363.00					.115
		4-6	EA	193.00	59.00	252.00	378.00					.116
		5-2	EA	201.00	59.00	260.00	390.00					.117
		5-6	EA	211.00	59.00	270.00	405.00					.118
		6-2	EA	221.00	59.00	280.00	420.00					.119
	3-0 X	2-10	EA	179.00	59.00	238.00	357.00					.120
		3-2	EA	187.00	59.00	246.00	369.00					.121
		3-10	EA	195.00	59.00	254.00	381.00					.122
		4-2	EA	205.00	59.00	264.00	396.00					.123
		4-6	EA	211.00	59.00	270.00	405.00					.124
		5-2	EA	221.00	59.00	280.00	420.00					.125
		5-6	EA	233.00	59.00	292.00	438.00					.126
		6-2	EA	241.00	59.00	300.00	450.00					.127
	3-4 X	3-2	EA	203.00	61.00	264.00	396.00					.128
		3-10	EA	215.00	61.00	276.00	414.00					.129
		4-2	EA	225.00	61.00	286.00	429.00					.130
		4-6	EA	235.00	61.00	296.00	444.00					.131
		5-2	EA	245.00	61.00	306.00	459.00					.132
		5-6	EA	255.00	61.00	316.00	474.00					.133
	3-8 X	4-6	EA	247.00	61.00	308.00	462.00					.134
TWIN UNITS	DOUBLE THE SINGLE UNIT AMOUNT AND **ADD**		EA	7.00	9.00	16.00	24.00					.135
TRIPLE UNITS	TRIPLE THE SINGLE UNIT AMOUNT AND **ADD**		EA	9.00	13.00	22.00	33.00					.136

13

SPECIFICATIONS		UNIT	JOB COST			PRICE	LOCAL AREA MODIFICATION				DATA BASE ITEM NO.
			MATLS	LABOR	TOTAL		MATLS	LABOR	TOTAL	PRICE	
DIVIDED LIGHT GRILLE	DIVIDED LIGHT GRILLES FOR WINDOWS **ADD**	SF	2.30	--	2.30	3.45					.137
SCREENS	FULL LENGTH INSECT SCREENS FOR WINDOWS **ADD**	SF	1.73	--	1.73	2.60					.138
ONE LIGHT PICTURE SASH & FRAME	• FRAME, SASH AND BRICK MOULDING • PRIMED SASH WITH 1/2" INSULATING GLASS • INTERIOR TRIM, STOPS, STOOL AND APRON										
	3-0 X 4-2	EA	226.00	56.00	282.00	423.00					.139
	4-6	EA	242.00	56.00	298.00	447.00					.140
	4-0 X 4-2	EA	258.00	56.00	314.00	471.00					.141
	4-6	EA	272.00	56.00	328.00	492.00					.142
	5-2	EA	304.00	56.00	360.00	540.00					.143
	5-4 X 4-6	EA	334.00	56.00	390.00	585.00					.144
PICTURE WINDOW WITH FLANKERS	PICTURE WINDOW ABOVE WITH DOUBLE HUNG FRAMES AND SASH, INCLUDING ALL TRIM, ETC. TOTAL THE COSTS FOR DOUBLE HUNG AND PICTURE WINDOW AND **ADD** EA = TOTAL UNIT	EA	11.00	--	11.00	16.50					.145

13

13. DOUBLE GLAZED WINDOWS -- PREMIUM QUALITY

SPECIFICATIONS			UNIT	JOB COST			PRICE	LOCAL AREA MODIFICATION				DATA BASE ITEM NO.
				MATLS	LABOR	TOTAL		MATLS	LABOR	TOTAL	PRICE	
WOOD DOUBLE HUNG	• VINYL COVERED FRAME AND BRICK MOULDING • PRIMED SASH WITH 1/2" INSULATING GLASS • SASH LOCK • INTERIOR TRIM, STOPS • STOOL AND APRON • SCREEN											
	2 LIGHTS	1-8 X 4-2	EA	202.00	56.00	258.00	387.00					.200
		4-6	EA	202.00	56.00	258.00	387.00					.201
		2-0 X 3-2	EA	178.00	56.00	234.00	351.00					.202
		4-2	EA	180.00	56.00	236.00	354.00					.203
		4-6	EA	194.00	56.00	250.00	375.00					.204
		5-2	EA	202.00	56.00	258.00	387.00					.205
		2-4 X 2-10	EA	187.00	59.00	246.00	369.00					.206
		3-2	EA	201.00	59.00	260.00	390.00					.207
		3-10	EA	215.00	59.00	274.00	411.00					.208
		4-2	EA	221.00	59.00	280.00	420.00					.209
		4-6	EA	227.00	59.00	286.00	429.00					.210
		5-2	EA	229.00	59.00	288.00	432.00					.211
		2-8 X 2-10	EA	201.00	59.00	260.00	390.00					.212
		3-2	EA	207.00	59.00	266.00	399.00					.213
		3-10	EA	219.00	59.00	278.00	417.00					.214
		4-2	EA	235.00	59.00	294.00	441.00					.215
		4-6	EA	243.00	59.00	302.00	453.00					.216
		5-2	EA	255.00	59.00	314.00	471.00					.217
		5-6	EA	267.00	59.00	326.00	489.00					.218
		6-2	EA	279.00	59.00	338.00	507.00					.219
		3-0 X 2-10	EA	227.00	59.00	286.00	429.00					.220
		3-2	EA	237.00	59.00	296.00	444.00					.221
		3-10	EA	245.00	59.00	304.00	456.00					.222
		4-2	EA	261.00	59.00	320.00	480.00					.223
		4-6	EA	267.00	59.00	326.00	489.00					.224
		5-2	EA	279.00	59.00	338.00	507.00					.225
		5-6	EA	293.00	59.00	352.00	528.00					.226
		6-2	EA	305.00	59.00	364.00	546.00					.227
		3-4 X 3-2	EA	257.00	61.00	318.00	477.00					.228
		3-10	EA	269.00	61.00	330.00	495.00					.229
		4-2	EA	285.00	61.00	346.00	519.00					.230
		4-6	EA	295.00	61.00	356.00	534.00					.231
		5-2	EA	309.00	61.00	370.00	555.00					.232
		5-6	EA	321.00	61.00	382.00	573.00					.233
		3-8 X 4-6	EA	313.00	61.00	374.00	561.00					.234
TWIN UNITS	DOUBLE THE SINGLE UNIT AMOUNT AND **ADD**		EA	7.00	9.00	16.00	24.00					.235
TRIPLE UNITS	TRIPLE THE SINGLE UNIT AMOUNT AND **ADD**		EA	11.00	13.00	24.00	36.00					.236

13

13. DOUBLE GLAZED WINDOWS -- PREMIUM QUALITY

SPECIFICATIONS		UNIT	JOB COST			PRICE	LOCAL AREA MODIFICATION				DATA BASE ITEM NO.
			MATLS	LABOR	TOTAL		MATLS	LABOR	TOTAL	PRICE	
DIVIDED LIGHT GRILLE	DIVIDED LIGHT GRILLES FOR WINDOWS **ADD**	SF	2.30	--	2.30	3.45					.237
SCREENS	FULL LENGTH INSECT SCREENS FOR WINDOWS **ADD**	SF	1.73	--	1.73	2.60					.238
ONE LIGHT PICTURE SASH & FRAME	• VINYL COVERED FRAME AND BRICK MOULDING • PRIMED SASH WITH 1/2" INSULATING GLASS • INTERIOR TRIM, STOPS, STOOL AND APRON										
	3-0 X 4-2	EA	262.00	56.00	318.00	477.00					.239
	4-6	EA	280.00	56.00	336.00	504.00					.240
	4-0 X 4-2	EA	298.00	56.00	354.00	531.00					.241
	4-6	EA	314.00	56.00	370.00	555.00					.242
	5-2	EA	350.00	56.00	406.00	609.00					.243
	5-4 X 4-6	EA	388.00	56.00	444.00	666.00					.244
PICTURE WINDOW WITH FLANKERS	PICTURE WINDOW ABOVE WITH DOUBLE HUNG FRAMES AND SASH, INCLUDING ALL TRIM, ETC. TOTAL THE COSTS FOR DOUBLE HUNG AND PICTURE WINDOW AND **ADD** EA = TOTAL UNIT	EA	12.00	--	12.00	18.00					.245

13

13. WOOD AWNING WINDOWS

SPECIFICATIONS			UNIT	JOB COST			PRICE	LOCAL AREA MODIFICATION				DATA BASE ITEM NO.
				MATLS	LABOR	TOTAL		MATLS	LABOR	TOTAL	PRICE	
AWNING WINDOW, SINGLE GLAZED	INCLUDING WEATHERSTRIPPING, EXTERIOR BRICK MOULD, FRAME, SASH, DRIP CAP, SCREEN, HARDWARE, INTERIOR TRIM											
	(1)	2-8 X 1-8	EA	141.00	37.00	178.00	267.00					.500
	(2)	3-4	EA	208.00	48.00	256.00	384.00					.501
	(3)	5-0	EA	286.00	48.00	334.00	501.00					.502
	(1)	2-4	EA	159.00	37.00	196.00	294.00					.503
	(2)	4-8	EA	230.00	48.00	278.00	417.00					.504
	(3)	7-0	EA	318.00	48.00	366.00	549.00					.505
	(1)	4-0 X 1-8	EA	174.00	48.00	222.00	333.00					.506
	(2)	3-4	EA	278.00	48.00	326.00	489.00					.507
	(3)	5-0	EA	360.00	56.00	416.00	624.00					.508
	(1)	2-4	EA	180.00	56.00	236.00	354.00					.509
	(2)	4-8	EA	278.00	56.00	334.00	501.00					.510
	(3)	7-0	EA	380.00	56.00	436.00	654.00					.511
AWNING WINDOW, INSULATED GLASS	INSULATED GLASS, INCLUDING WEATHERSTRIPPING, EXTERIOR BRICK MOULD, FRAME, SASH, DRIP CAP, SCREEN, HARDWARE, INTERIOR TRIM											
	(1)	2-8 X 1-8	EA	157.00	37.00	194.00	291.00					.512
	(2)	3-4	EA	232.00	48.00	280.00	420.00					.513
	(3)	5-0	EA	318.00	48.00	366.00	549.00					.514
	(1)	2-4	EA	177.00	37.00	214.00	321.00					.515
	(2)	4-8	EA	256.00	48.00	304.00	456.00					.516
	(3)	7-0	EA	352.00	48.00	400.00	600.00					.517
	(1)	4-0 X 1-8	EA	194.00	48.00	242.00	363.00					.518
	(2)	3-4	EA	296.00	48.00	344.00	516.00					.519
	(3)	5-0	EA	400.00	56.00	456.00	684.00					.520
	(1)	2-4	EA	200.00	56.00	256.00	384.00					.521
	(2)	4-8	EA	308.00	56.00	364.00	546.00					.522
	(3)	7-0	EA	424.00	56.00	480.00	720.00					.523

13

SPECIFICATIONS		UNIT	JOB COST			PRICE	LOCAL AREA MODIFICATION				DATA BASE ITEM NO.
			MATLS	LABOR	TOTAL		MATLS	LABOR	TOTAL	PRICE	
VENTING, SINGLE GLAZED	• STANDARD GLASS • EXTERIOR FRAME AND BRICK MOULDING • SASH • INTERIOR CASING • STOOL CAP AND APRON • ROTO OPERATOR • HARDWARE • SCREEN • ONE PANEL **Sash Size** 1-11 X 2-6	EA	102.00	48.00	150.00	225.00					.524
	3-1	EA	114.00	48.00	162.00	243.00					.525
	3-6	EA	126.00	48.00	174.00	261.00					.526
	4-6	EA	136.00	48.00	184.00	276.00					.527
	5-6	EA	152.00	48.00	200.00	300.00					.528
	2-4 X 2-6	EA	114.00	48.00	162.00	243.00					.529
	3-1	EA	128.00	48.00	176.00	264.00					.530
	3-6	EA	130.00	48.00	178.00	267.00					.531
	4-6	EA	142.00	48.00	190.00	285.00					.532
VENTING, INSULATED GLASS, BUILDER QUALITY	SAME AS ABOVE WITH IN-SULATED GLASS, MEDIUM QUALITY **Sash Size** 1-11 X 2-6	EA	134.00	48.00	182.00	273.00					.533
	3-1	EA	140.00	48.00	188.00	282.00					.534
	4-6	EA	176.00	48.00	224.00	336.00					.535
	5-6	EA	206.00	48.00	254.00	381.00					.536
	2-4 X 2-6	EA	148.00	48.00	196.00	294.00					.537
	3-1	EA	156.00	48.00	204.00	306.00					.538
	3-6	EA	168.00	48.00	216.00	324.00					.539
	4-6	EA	174.00	48.00	222.00	333.00					.540
	5-6	EA	210.00	48.00	258.00	387.00					.541
VENTING, INSULATED GLASS, PREMIUM QUALITY	• INSULATED GLASS • VINYL COVERED SASH AND FRAME • INTERIOR CASING • STOOL CAP AND APRON • ROTO OPERATOR • HARDWARE • SCREEN • ONE PANEL **Sash Size** 1-11 X 2-6	EA	177.00	53.00	230.00	345.00					.542
	3-1	EA	185.00	53.00	238.00	357.00					.543
	4-6	EA	233.00	53.00	286.00	429.00					.544
	5-6	EA	271.00	53.00	324.00	486.00					.545
	2-4 X 2-6	EA	195.00	53.00	248.00	372.00					.546
	3-0	EA	205.00	53.00	258.00	387.00					.547
	4-6	EA	229.00	53.00	282.00	423.00					.548
	5-6	EA	277.00	53.00	330.00	495.00					.549
STATION-ARY	SAME AS ABOVE WITHOUT ROTO OPERATOR **DEDUCT**	EA	7.00	--	7.00	10.50					

13

13. WOOD SLIDING WINDOWS

SPECIFICATIONS		UNIT	JOB COST			PRICE	LOCAL AREA MODIFICATION				DATA BASE ITEM NO.
			MATLS	LABOR	TOTAL		MATLS	LABOR	TOTAL	PRICE	
SINGLE GLAZED	• BUILDER QUALITY SASH AND FRAME • SINGLE GLAZED • SCREEN • APRON, STOOL AND TRIM • HARDWARE										
	3-0 X 3-0	EA	144.00	42.00	186.00	279.00					.600
	3-6	EA	150.00	42.00	192.00	288.00					.601
	4-0 X 3-6	EA	174.00	42.00	216.00	324.00					.602
	4-0	EA	178.00	42.00	220.00	330.00					.603
	5-0 X 3-6	EA	200.00	42.00	242.00	363.00					.604
	4-0	EA	206.00	42.00	248.00	372.00					.605
	5-0	EA	218.00	42.00	260.00	390.00					.606
	6-0 X 4-0	EA	286.00	42.00	328.00	492.00					.607
	5-0	EA	298.00	42.00	340.00	510.00					.608
INSULATED GLASS	• BUILDER QUALITY SASH AND FRAME • INSULATED GLASS • SCREEN • APRON, STOOL AND TRIM • HARDWARE										
	3-0 X 3-0	EA	202.00	42.00	244.00	366.00					.609
	3-6	EA	208.00	42.00	250.00	375.00					.610
	4-0 X 3-6	EA	244.00	42.00	286.00	429.00					.611
	4-0	EA	250.00	42.00	292.00	438.00					.612
	5-0 X 3-6	EA	280.00	42.00	322.00	483.00					.613
	4-0	EA	290.00	42.00	332.00	498.00					.614
	5-0	EA	306.00	42.00	348.00	522.00					.615
	6-0 X 4-0	EA	400.00	42.00	442.00	663.00					.616
	5-0	EA	416.00	42.00	458.00	687.00					.617
PREMIUM GRADE	• VINYL COVERED SASH AND FRAME • INSULATED GLASS • SCREEN • APRON, STOOL AND TRIM • HARDWARE										
	3-0 X 3-0	EA	274.00	42.00	316.00	474.00					.618
	3-6	EA	284.00	42.00	326.00	489.00					.619
	4-0 X 3-6	EA	330.00	42.00	372.00	558.00					.620
	4-0	EA	340.00	42.00	382.00	573.00					.621
	5-0 X 3-6	EA	380.00	42.00	422.00	633.00					.622
	4-0	EA	394.00	42.00	436.00	654.00					.623
	5-0	EA	414.00	42.00	456.00	684.00					.624
	6-0 X 4-0	EA	542.00	42.00	584.00	876.00					.625
	5-0	EA	564.00	42.00	606.00	909.00					.626

13

SPECIFICATIONS		UNIT	JOB COST			PRICE	LOCAL AREA MODIFICATION				DATA BASE ITEM NO.
			MATLS	LABOR	TOTAL		MATLS	LABOR	TOTAL	PRICE	
BUILDER GRADE	• PRIMED WOOD EXTERIOR SASH, FRAME AND TRIM • INSULATED GLASS • 2-1/4" INTERIOR ARCHED CASING • CURVED EXTENSION JAMBS • STOOL CAP										
	Width **Height** 2-2 x 1-4	EA	192.00	58.00	250.00	375.00					.627
	2-6 x 1-4	EA	228.00	58.00	286.00	429.00					.628
	2-10 x 1-4	EA	242.00	58.00	300.00	450.00					.629
	3-2 x 1-4	EA	258.00	58.00	316.00	474.00					.630
	3-6 x 2-0	EA	274.00	58.00	332.00	498.00					.631
PREMIUM GRADE	• VINYL CLAD EXTERIOR SASH, FRAME AND TRIM • INSULATED GLASS • 2-1/4" INTERIOR ARCHED CASING • CURVED EXTENSION JAMBS • STOOL CAP										
	Width **Height** 2-2 x 1-4	EA	274.00	58.00	332.00	498.00					.632
	2-6 x 1-4	EA	326.00	58.00	384.00	576.00					.633
	2-10 x 1-4	EA	348.00	58.00	406.00	609.00					.634
	3-2 x 1-4	EA	370.00	58.00	428.00	642.00					.635
	3-6 x 2-0	EA	400.00	58.00	458.00	687.00					.636
DIVIDED LIGHT GRILLES	DIVIDED LIGHT GRILLES FOR ABOVE UNITS										
	Width **Height** 2-2 x 1-4	EA	18.00	--	18.00	27.00					.637
	2-6 x 1-4	EA	20.00	--	20.00	30.00					.638
	2-10 x 1-4	EA	22.00	--	22.00	33.00					.639
	3-2 x 1-4	EA	26.00	--	26.00	39.00					.640
	3-6 x 2-0	EA	28.00	--	28.00	42.00					.641

13

13. BAY WINDOW

SPECIFICATIONS		UNIT	JOB COST			PRICE	LOCAL AREA MODIFICATION				DATA BASE ITEM NO.
			MATLS	LABOR	TOTAL		MATLS	LABOR	TOTAL	PRICE	
BAY WINDOW, PREMIUM QUALITY	PRE-FABRICATED 30° BAY WINDOW IN EXIST'G OPENING • TWO 1-8 X 4-6 WOOD DOUBLE HUNG, ONE STATIONARY 4-4 X 4-6 • HIGH PERFORMANCE INSULATED GLASS • VINYL CLAD • PROJECTION 14" • STOPS • INTERIOR CASINGS • EXTERIOR BRICK MOULDING	EA	1248.00	204.00	1,452.00	2,178.00					.700
	PRE-FABRICATED 45° BAY WINDOW IN EXIST'G OPENING • TWO 1-8 X 4-6 WOOD DOUBLE HUNG, ONE STATIONARY 4-4 X 4-6 • HIGH PERFORMANCE INSULATED GLASS • VINYL CLAD • PROJECTION 20" • STOPS • INTERIOR CASINGS • EXTERIOR BRICK MOULDING	EA	1324.00	204.00	1,528.00	2,292.00					.701
	PRE-FABRICATED 90° BOX BAY WINDOW IN EXISTING OPENING • TWO 1-8 X 4-6 WOOD DOUBLE HUNG, ONE STATIONARY 4-4 X 4-6 • HIGH PERFORMANCE INSULATED GLASS • VINYL CLAD • PROJECTION 23" • STOPS • INTERIOR CASINGS • EXTERIOR BRICK MOULDING	EA	1348.00	204.00	1,552.00	2,328.00					.702
	BUILD ROOF OVER PROJECTION WITH 2 X 4 FRAMING, 1/2" PLYWOOD ROOF SHEATHING, 3/4" X 3/4" COVE FASCIA	EA	57.00	217.00	274.00	411.00					.703
	ON EXISTING SHEATHING, INSTALL FELT PAPER, FLASHING AND ASPHALT SHINGLES	EA	18.00	144.00	162.00	243.00					.704
	3/4" PLYWOOD HEADBOARD, TWO 3/4" PLYWOOD SEAT BOARDS AND TWO KNEE BRACES EA = ALL THREE ITEMS	EA	137.00	109.00	246.00	369.00					.705

13

SPECIFICATIONS		UNIT	JOB COST			PRICE	LOCAL AREA MODIFICATION				DATA BASE ITEM NO.
			MATLS	LABOR	TOTAL		MATLS	LABOR	TOTAL	PRICE	
BOW WINDOW PREMIUM QUALITY	PRE-FABRICATED BOW WINDOW IN EXISTING OPENING • TWO 2-0 X 5-6 VENTING WOOD CASEMENT SASH AND TWO STATIONARY • HIGH PERFORMANCE INSULATED GLASS • REMOVABLE MUNTINS • ROLL-UP HARDWARE • PROJECTION 10" • INTERIOR CASINGS • EXTERIOR BRICK MOULDING • SCREEN	EA	1480.00	204.00	1,684.00	2,526.00					.706
	SAME AS ABOVE, ALUMINUM CLAD WINDOW	EA	1780.00	204.00	1,984.00	2,976.00					.707
	BUILD ROOF OVER PROJECTION WITH 2 X 4 FRAMING, 1/2" PLYWOOD ROOF SHEATHING, 3/4" X 3/4" COVE FASCIA	EA	58.00	114.00	172.00	258.00					.708
	ON EXISTING SHEATHING, INSTALL FELT PAPER, FLASHING AND ASPHALT SHINGLES	EA	18.00	138.00	156.00	234.00					.709
	3/4" PLYWOOD HEADBOARD, TWO 3/4" PLYWOOD SEAT BOARDS AND TWO KNEE BRACES EA = ALL THREE ITEMS	EA	137.00	109.00	246.00	369.00					.710
GREEN-HOUSE WINDOW	• BUMP-OUT WINDOW • 1/2" INSULATED GLASS • ADJUSTABLE SHELVES • ALUMINUM FRAME • PROJECTS OUT 14" • BRONZE OR WHITE										
	2-0 X 3-0	EA	328.00	62.00	390.00	585.00					.711
	4-0	EA	392.00	64.00	456.00	684.00					.712
	5-0	EA	458.00	66.00	524.00	786.00					.713
	3-0 X 3-0	EA	393.00	65.00	458.00	687.00					.714
	4-0	EA	464.00	68.00	532.00	798.00					.715
	5-0	EA	523.00	71.00	594.00	891.00					.716
	6-0	EA	591.00	75.00	666.00	999.00					.717
	4-0 X 3-0	EA	460.00	68.00	528.00	792.00					.718
	4-0	EA	526.00	72.00	598.00	897.00					.719
	5-0	EA	596.00	78.00	674.00	1,011.00					.720
	6-0	EA	658.00	80.00	738.00	1,107.00					.721

13. ROOF WINDOW

SPECIFICATIONS	UNIT	JOB COST			PRICE	LOCAL AREA MODIFICATION				DATA BASE ITEM NO.
		MATLS	LABOR	TOTAL		MATLS	LABOR	TOTAL	PRICE	
NOTE: SEE PAGE 71 FOR ROOF WINDOW INSTALLATION IN EXISTING ROOF										
ROOF WINDOW — INSTALL DURING NEW CONSTRUCTION										
• CENTER PIVOT • ALUMINUM CLAD WOOD FRAME • INSULATED GLASS • EXTERIOR AWNING • INSECT SCREEN • BUILD CURB ON EXISTING FRAMED OPENING • INTERIOR WELL & TRIM										
Outside Frame										
30-3/4" X 38-1/2"	EA	542.00	94.00	636.00	954.00					.722
30-3/4" X 55"	EA	634.00	94.00	728.00	1,092.00					.723
36-7/8" X 62-7/8"	EA	772.00	94.00	866.00	1,299.00					.724
44-7/8" X 46-1/2"	EA	740.00	94.00	834.00	1,251.00					.725
27-1/2" X 46-1/2"	EA	564.00	94.00	658.00	987.00					.726
21-5/8" X 38-1/2"	EA	460.00	94.00	554.00	831.00					.727
52-3/4" X 38-1/2"	EA	770.00	94.00	864.00	1,296.00					.728
52-3/4" X 55"	EA	880.00	94.00	974.00	1,461.00					.729
21-5/8" X 27-1/2"	EA	428.00	94.00	522.00	783.00					.730
SAME AS ABOVE, TOP-HUNG										
30-3/4" X 38-1/2"	EA	570.00	94.00	664.00	996.00					.731
30-3/4" X 55"	EA	664.00	94.00	758.00	1,137.00					.732
44-7/8" X 46-1/2"	EA	774.00	94.00	868.00	1,302.00					.733
21-5/8" X 38-1/2"	EA	484.00	94.00	578.00	867.00					.734

13

SPECIFICATIONS	UNIT	JOB COST			PRICE	LOCAL AREA MODIFICATION				DATA BASE ITEM NO.
		MATLS	LABOR	TOTAL		MATLS	LABOR	TOTAL	PRICE	
NOTE: SEE PAGE 70 FOR SKYLIGHT INSTALLATION IN EXISTING ROOF										
HATCH TYPE SKYLIGHT • INSULATED GLASS • 1/8" TEMPERED OVER 1/4" LAMINATED GLASS WITH LOW E COATING • DOUBLE ALUMINUM WINDOW FRAME • COPPER FLASHING COVERS • MARINE GRADE PLYWOOD LINER BOX										
FIXED SKYLIGHT *Rough Opening Sizes*										
22" X 30"	EA	366.00	78.00	444.00	666.00					.735
X 45-1/2"	EA	430.00	78.00	508.00	762.00					.736
30" X 22"	EA	366.00	78.00	444.00	666.00					.737
X 30"	EA	404.00	80.00	484.00	726.00					.738
X 45-1/2"	EA	492.00	82.00	574.00	861.00					.739
45-1/2" X 30"	EA	492.00	82.00	574.00	861.00					.740
X 45-1/2"	EA	595.00	83.00	678.00	1,017.00					.741
VENTILATING SKYLIGHT W/INSECT SCREEN *Rough Opening Sizes*										
22" X 30"	EA	407.00	83.00	490.00	735.00					.742
X 45-1/2"	EA	479.00	83.00	562.00	843.00					.743
30" X 22"	EA	407.00	83.00	490.00	735.00					.744
X 30"	EA	447.00	83.00	530.00	795.00					.745
X 45-1/2"	EA	548.00	88.00	636.00	954.00					.746
45-1/2" X 30"	EA	548.00	88.00	636.00	954.00					.747
X 45-1/2"	EA	658.00	88.00	746.00	1,119.00					.748

13

13. ALUMINUM SINGLE HUNG WINDOWS

SPECIFICATIONS			UNIT	JOB COST			PRICE	LOCAL AREA MODIFICATION				DATA BASE ITEM NO.
				MATLS	LABOR	TOTAL		MATLS	LABOR	TOTAL	PRICE	
ALUMINUM SINGLE HUNG WINDOW, DOUBLE GLAZED	INCL. ALUMINUM FRAME, WEATHERSTRIPPING, HARD-WARE AND SCREEN, DOUBLE GLAZED											
	2-0 X	3-0	EA	60.00	34.00	89.00	133.50					.800
		4-0	EA	64.00	34.00	93.00	139.50					.801
		5-0	EA	86.00	34.00	113.00	169.50					.802
		6-0	EA	92.00	34.00	119.00	178.50					.803
	2-8 X	3-0	EA	68.00	34.00	97.00	145.50					.804
		4-0	EA	76.00	34.00	104.00	156.00					.805
		5-0	EA	94.00	34.00	122.00	183.00					.806
		6-0	EA	106.00	34.00	133.00	199.50					.807
	3-0 X	3-0	EA	80.00	34.00	109.00	163.50					.808
		4-0	EA	92.00	34.00	120.00	180.00					.809
		5-0	EA	96.00	34.00	124.00	186.00					.810
		6-0	EA	110.00	34.00	137.00	205.50					.811
	4-0 X	3-0	EA	88.00	34.00	115.00	172.50					.812
		4-0	EA	98.00	34.00	125.00	187.50					.813
		5-0	EA	114.00	34.00	141.00	211.50					.814
		6-0	EA	138.00	34.00	164.00	246.00					.815
SINGLE GLAZED	SAME AS ABOVE, SINGLE GLAZED											
	2-0 X	2-4	EA	34.00	34.00	64.00	96.00					.816
		3-0	EA	38.00	34.00	68.00	102.00					.817
		5-0	EA	58.00	34.00	87.00	130.50					.818
	2-8 X	3-0	EA	48.00	34.00	77.00	115.50					.819
		4-0	EA	56.00	34.00	86.00	129.00					.820
		4-4	EA	58.00	34.00	87.00	130.50					.821
		5-0	EA	64.00	34.00	92.00	138.00					.822
	3-0 X	3-0	EA	50.00	34.00	80.00	120.00					.823
		4-0	EA	60.00	34.00	90.00	135.00					.824
		4-4	EA	62.00	34.00	91.00	136.50					.825
		6-0	EA	66.00	34.00	95.00	142.50					.826

13. ALUMINUM SLIDING AND AWNING WINDOWS

SPECIFICATIONS			UNIT	JOB COST			PRICE	LOCAL AREA MODIFICATION				DATA BASE ITEM NO.
				MATLS	LABOR	TOTAL		MATLS	LABOR	TOTAL	PRICE	
ALUMINUM SLIDING WINDOW, SINGLE GLAZED	INCL. ALUMINUM FRAME, WEATHERSTRIPPING, HARD-WARE, SCREEN & INTERIOR TRIM, SINGLE GLAZED											
	2-0 X	3-0	EA	67.00	39.00	106.00	159.00					.827
	3-0 X	3-0	EA	73.00	39.00	112.00	168.00					.828
		4-0	EA	85.00	39.00	124.00	186.00					.829
	4-0 X	3-0	EA	79.00	39.00	118.00	177.00					.830
		3-6	EA	85.00	39.00	124.00	186.00					.831
		4-0	EA	91.00	39.00	130.00	195.00					.832
	5-0 X	3-6	EA	95.00	39.00	134.00	201.00					.833
		4-0	EA	105.00	39.00	144.00	216.00					.834
	6-0 X	4-0	EA	113.00	39.00	152.00	228.00					.835
ALUMINUM SLIDING WINDOW, INSULATED GLASS	INCL. ALUMINUM FRAME, WEATHERSTRIPPING, HARD-WARE, SCREEN & INTERIOR TRIM, INSULATED GLASS											
	2-0 X	3-0	EA	87.00	39.00	126.00	189.00					.836
	3-0 X	3-0	EA	103.00	39.00	142.00	213.00					.837
		4-0	EA	125.00	39.00	164.00	246.00					.838
	4-0 X	3-0	EA	117.00	39.00	156.00	234.00					.839
		3-6	EA	131.00	39.00	170.00	255.00					.840
		4-0	EA	143.00	39.00	182.00	273.00					.841
	5-0 X	3-6	EA	155.00	39.00	194.00	291.00					.842
		4-0	EA	171.00	39.00	210.00	315.00					.843
	6-0 X	4-0	EA	193.00	39.00	232.00	348.00					.844
ALUMINUM AWNING	INCL. ALUMINUM FRAME, SCREEN AND INTERIOR TRIM											
	Panels											
	1 2-1 X	2-1	EA	79.00	39.00	118.00	177.00					.845
	2	3-1	EA	87.00	39.00	126.00	189.00					.846
	3	4-1	EA	113.00	39.00	152.00	228.00					.847
	4	5-1	EA	121.00	39.00	160.00	240.00					.848
	5	6-1	EA	135.00	39.00	174.00	261.00					.849
	1 3-1 X	2-1	EA	83.00	39.00	122.00	183.00					.850
	2	3-1	EA	101.00	39.00	140.00	210.00					.851
	3	4-1	EA	121.00	39.00	160.00	240.00					.852
	4	5-1	EA	145.00	39.00	184.00	276.00					.853
	5	6-1	EA	165.00	39.00	204.00	306.00					.854
	1 4-1 X	2-1	EA	101.00	39.00	140.00	210.00					.855
	2	3-1	EA	115.00	39.00	154.00	231.00					.856
	3	4-1	EA	145.00	39.00	184.00	276.00					.857
	4	5-1	EA	157.00	39.00	196.00	294.00					.858
	5	6-1	EA	199.00	39.00	238.00	357.00					.859

13

13. DOUBLE HUNG REPLACEMENT WINDOWS

SPECIFICATIONS		UNIT	JOB COST			PRICE	LOCAL AREA MODIFICATION				DATA BASE ITEM NO.
			MATLS	LABOR	TOTAL		MATLS	LABOR	TOTAL	PRICE	
	THE REPLACEMENT WINDOW COSTS SHOWN ON THIS AND THE FOLLOWING PAGES ARE FOR INSTALLATION IN OPENINGS APPROXIMATELY THE SAME SIZE AS THE REPLACEMENT WINDOWS										
WINDOW OPENING SIZE ADJUSTMENT	FOR OPENINGS UP TO 2-1/2" LARGER IN HEIGHT OR WIDTH THAN REPLACEMENT WINDOW, FRAME OPENING TO WINDOW SIZE WITH FILLER STRIPS AND APPLY WIDER CASINGS AND TRIM **ADD**	EA	7.00	25.00	32.00	48.00					.316
	TO REDUCE OPENING SIZE IN STUCCO WALL, APPLY STUCCO ON EXTERIOR, AND DRYWALL OR TRIM ON INTERIOR **ADD**	EA	16.00	50.00	66.00	99.00					.317
WOOD DOUBLE HUNG	REPLACE EXISTING WOOD DOUBLE HUNG WINDOW WITH NEW WOOD DOUBLE HUNG WINDOW IN WOOD OR BRICK VENEER EXTERIOR WALL • REMOVE INTERIOR CASING AND EXTERIOR WINDOW TRIM • REMOVE EXISTING WINDOW AND FRAME FROM WALL WITH DRYWALL OR PLASTER INTERIOR • INSTALL NEW WOOD DOUBLE HUNG TWO LIGHT WINDOW AND FRAME WITH INSULATED GLASS AND SCREEN • REPLACE INTERIOR AND EXTERIOR TRIM AND CASING • TOUCH UP PAINT TO MATCH EXISTING										
	UP TO 72 UI	EA	192.00	62.00	254.00	381.00					.300
	73 TO 82 UI	EA	234.00	66.00	300.00	450.00					.301
	83 TO 92 UI	EA	272.00	72.00	344.00	516.00					.302
	93 TO 101 UI	EA	310.00	78.00	388.00	582.00					.303
	UI = ADD TOGETHER THE WIDTH IN INCHES AND THE HEIGHT IN INCHES										
VINYL DOUBLE HUNG WINDOW	SAME AS ABOVE EXCEPT REPLACEMENT WINDOW IS VINYL WITH INSULATED GLASS AND SCREEN										
	UP TO 72 UI	EA	164.00	54.00	218.00	327.00					.305
	73 TO 82 UI	EA	190.00	58.00	248.00	372.00					.306
	83 TO 92 UI	EA	215.00	62.00	277.00	415.50					.307
	93 TO 101 UI	EA	234.00	68.00	302.00	453.00					.308
WINDOWS IN MASONRY WALL	ANY OF THE ABOVE WINDOWS INSTALLED IN BRICK OR BRICK & BLOCK WALL **ADD**	EA	--	16.00	16.00	24.00					.310
DIVIDED LIGHT GRILLE	DIVIDED LIGHT GRILLE FOR ABOVE WINDOWS **ADD** SF= OVERALL SIZE OF WINDOW SASH	SF	2.30	--	2.30	3.45					.311

SPECIFICATIONS		UNIT	JOB COST			PRICE	LOCAL AREA MODIFICATION				DATA BASE ITEM NO.
			MATLS	LABOR	TOTAL		MATLS	LABOR	TOTAL	PRICE	
WOOD CASEMENT WINDOW	REPLACE EXISTING WOOD DOUBLE HUNG OR CASEMENT WINDOW WITH NEW WOOD CASEMENT WINDOW IN WOOD OR BRICK VENEER EXTERIOR WALL • REMOVE INTERIOR CASING AND EXTERIOR WINDOW TRIM • REMOVE EXISTING WINDOW AND FRAME FROM WALL • INSTALL NEW WOOD CASEMENT WINDOW WITH INSULATED GLASS AND SCREEN • REPLACE INTERIOR AND EXTERIOR TRIM AND CASING • TOUCH UP PAINT TO MATCH EXISTING										
	UP TO 72 UI	EA	200.00	62.00	262.00	393.00					.300
	73 TO 82 UI	EA	243.00	66.00	309.00	463.50					.301
	83 TO 92 UI	EA	281.00	72.00	353.00	529.50					.302
	93 TO 101 UI	EA	320.00	78.00	398.00	597.00					.303
	UI = ADD TOGETHER THE WIDTH IN INCHES AND THE HEIGHT IN INCHES										
VINYL CASEMENT WINDOW	SAME AS ABOVE EXCEPT REPLACEMENT WINDOW IS VINYL WITH INSULATED GLASS AND SCREEN										
	UP TO 72 UI	EA	160.00	54.00	214.00	321.00					.305
	73 TO 82 UI	EA	183.00	58.00	241.00	361.50					.306
	83 TO 92 UI	EA	210.00	62.00	272.00	408.00					.307
	93 TO 101 UI	EA	240.00	68.00	308.00	462.00					.308
WINDOWS IN MASONRY WALL	ANY OF THE ABOVE WINDOWS INSTALLED IN BRICK OR BRICK & BLOCK WALL **ADD**	EA	--	16.00	16.00	24.00					.310
DIVIDED LIGHT GRILLE	DIVIDED LIGHT GRILLE FOR ABOVE WINDOWS **ADD** SF= OVERALL SIZE OF WINDOW SASH	SF	2.30	--	2.30	3.45					.311
REMOVE STEEL CASEMENT AND REPLACE	REMOVE EXISTING STEEL CASEMENT WINDOW SET IN MASONRY WALL AND REPLACE WITH ANY OF ABOVE WINDOWS **ADD**	EA	--	10.00	10.00	15.00					
WOOD SASH REPLACEMENT ONLY	DOUBLE HUNG REPLACEMENT SASH AND JAMB LINERS • 1/2" INSULATING GLASS • 2 LIGHTS PER OPENING										
	UP TO 72 UI	EA	114.00	25.00	139.00	208.50					.312
	73 TO 82 UI	EA	137.00	25.00	162.00	243.00					.313
	83 TO 92 UI	EA	163.00	25.00	188.00	282.00					.314
	93 TO 101 UI	EA	185.00	25.00	210.00	315.00					.315

13

13. VINYL SLIDER & BASEMENT REPLACEMENT WINDOW

SPECIFICATIONS		UNIT	JOB COST			PRICE	LOCAL AREA MODIFICATION				DATA BASE ITEM NO.
			MATLS	LABOR	TOTAL		MATLS	LABOR	TOTAL	PRICE	
VINYL SLIDER	REPLACE EXISTING WOOD OR METAL WINDOW WITH VINYL BYPASS SLIDER UNIT • REMOVE INTERIOR CASING AND EXTERIOR WINDOW TRIM • REMOVE EXISTING WINDOW AND FRAME FROM WOOD OR BRICK VENEER EXTERIOR WALL • INSTALL NEW VINYL BYPASS SLIDER UNIT WITH INSULATED GLASS AND SCREEN • REPLACE INTERIOR AND EXTERIOR TRIM AND CASING • TOUCH UP PAINT TO MATCH EXISTING										
	UP TO 72 UI	EA	170.00	64.00	234.00	351.00					.400
	73 TO 82 UI	EA	195.00	68.00	263.00	394.50					.401
	83 TO 92 UI	EA	220.00	72.00	292.00	438.00					.402
	93 TO 101 UI	EA	235.00	78.00	313.00	469.50					.403
	102 TO 112 UI	EA	260.00	82.00	342.00	513.00					.404
	UI = ADD TOGETHER THE WIDTH IN INCHES AND THE HEIGHT IN INCHES										
WINDOWS IN MASONRY WALL	ANY OF THE ABOVE WINDOWS INSTALLED IN BRICK OR BRICK & BLOCK WALL **ADD**	EA	--	16.00	16.00	24.00					.405
BASEMENT WINDOW	REPLACE EXISTING AWNING STYLE METAL BASEMENT WINDOW WITH AWNING STYLE VINYL BASEMENT WINDOW • REMOVE EXISTING WINDOW FROM MASONRY WALL • INSTALL NEW AWNING STYLE VINYL BASEMENT WINDOW WITH INSULATED GLASS AND SCREEN										
	UP TO 72 UI	EA	126.00	48.00	174.00	261.00					.406

13

SPECIFICATIONS		UNIT	JOB COST			PRICE	LOCAL AREA MODIFICATION				DATA BASE ITEM NO.
			MATLS	LABOR	TOTAL		MATLS	LABOR	TOTAL	PRICE	
ALUMINUM COMBINA- TION WINDOWS	TRIPLE TRACK, SELF STOR- ING 2 GLASS INSERTS AND 1 SCREEN INSERT WEATHERSTRIPPING MILL FINISH UP TO 101 UNITED INCHES (ADD HEIGHT OF WINDOW PLUS WIDTH OF WINDOW)										
	ECONOMY	EA	54.00	16.00	70.00	105.00					.407
	MEDIUM	EA	60.00	16.00	76.00	114.00					.408
	PREMIUM	EA	74.00	16.00	90.00	135.00					.409
	FOR WINDOWS OVER 101 UNITED INCHES **ADD**	PER INCH	.70	--	.70	1.05					.410
	WHITE FINISH **ADD**	EA	2.00	--	2.00	3.00					.411
	COLOR FINISH **ADD**	EA	3.00	--	3.00	4.50					.412
CASEMENT	16-3/4" X 23-5/8"	EA	27.00	23.00	50.00	75.00					.413
	36-1/8"	EA	29.00	23.00	52.00	78.00					.414
	48-3/8"	EA	35.00	23.00	58.00	87.00					.415

13

14. PLUMBING

SPECIFICATIONS		UNIT	JOB COST			PRICE	LOCAL AREA MODIFICATION				DATA BASE ITEM NO.
			MATLS	LABOR	TOTAL		MATLS	LABOR	TOTAL	PRICE	
	THE LABOR COSTS SHOWN IN THIS SECTION INCLUDE A PLUMBING SUBCONTRACTOR'S OVERHEAD AND PROFIT										
HOSE BIBB	NEW OUTSIDE FROST-FREE SPIGOT (FAUCET)	EA	38.00	72.00	110.00	165.00					.000
	RE-LOCATE EXISTING HOSE BIBB UP TO 20 FEET	EA	28.00	72.00	100.00	150.00					.001
CUTOFF	INSTALL CUTOFF VALVES ON SINK LINES	EA	4.00	24.00	28.00	42.00					.002
GAS LINE	EXTEND GAS LINE INSIDE BUILDING WITHOUT BREAKING WALLS	LF	3.00	11.00	14.00	21.00					.003
REPLACE WATER PIPES	REPLACE HORIZONTAL GALVANIZED WATER PIPES IN BASEMENT WITH 3/4" COPPER PIPES, OPEN CEILING	EA PLUS	186.00	450.00	636.00	954.00					.004
	FULL BASEMENT	SF	.18	.27	.45	.68					.005
	CRAWL SPACE	SF	.18	.47	.65	.98					.006
	SF = FLOOR AREA										
	REPLACE VERTICAL GALVANIZED WATER PIPES TO EACH BATH OR KITCHEN WITH 3/4" COPPER, OPEN FLOORS AND WALLS										
	BASEMENT TO FIRST FLOOR	EA	38.00	360.00	398.00	597.00					.007
	BASEM'T TO SECOND FLOOR	EA	56.00	490.00	546.00	819.00					.008
VENT	RAISE EXISTING VENT THROUGH ROOF UP TO 10 FEET	EA	20.00	92.00	112.00	168.00					.009
	RUN NEW VENT THROUGH EXISTING ROOF	LF	3.00	11.00	14.00	21.00					.010
STACK	NEW STACK RUNNING FROM BASEMENT OR CRAWL SPACE UP AND VENTING THROUGH ROOF										
	ONE STORY	EA	36.00	270.00	306.00	459.00					.011
	TWO STORY	EA	76.00	360.00	436.00	654.00					.012
WASTE PIPE	EXTEND WASTE LINE FROM NEW STACK IN BASEMENT OR CRAWL SPACE										
	HANGING WASTE LINE	LF	8.00	26.00	34.00	51.00					.013
	BELOW GROUND WASTE LINE, INCLUDING EXCAVATION AND BACKFILL	LF	7.00	31.00	38.00	57.00					.014
	BELOW GROUND WASTE LINE, INCLUDING BREAKING CONCRETE FLOOR AND REPLACING	LF	12.00	28.00	40.00	60.00					.015
SEWER EJECTION PUMP	STANDARD SEWER EJECTION PUMP FOR BASEMENT BATH, WITH ADEQUATE ACCESS	EA	361.00	415.00	776.00	1,164.00					.016
GRAY BOX	INSTALL WALL MOUNTED SUPPLY AND WASTE "GRAY BOX" IN LAUNDRY AREA	EA	49.00	185.00	234.00	351.00					.017

14

SPECIFICATIONS		UNIT	JOB COST			PRICE	LOCAL AREA MODIFICATION				DATA BASE ITEM NO.
			MATLS	LABOR	TOTAL		MATLS	LABOR	TOTAL	PRICE	
OVER 5 FEET FROM STACK	*NOTE:* THE FOLLOWING IN-STALLATIONS ARE FOR NEW WORK LOCATED WITHIN 5 FEET OF EXISTING STACK AND INCLUDE COST OF ROUGH-IN, FIXTURES AND FITTINGS, AND COMPLETE INSTALLATION. IF EXISTING STACK IS OVER 5 FEET FROM THE NEW W.C. INSTALLATION, A NEW STACK WILL PROBABLY BE RE-QUIRED, RUNNING FROM BASEMENT UP & THROUGH EXISTING ROOF. SEE PAGE 202.										
INSTALL TWO FIXTURES	ANY TWO FIXTURES ABOVE, ADD TOTAL COST OF BOTH AND **DEDUCT**	SET	5%	15%	10%	10%					.100
INSTALL THREE OR MORE FIXTURES	THREE OR MORE OF ABOVE FIXTURES, ADD TOTAL COST AND **DEDUCT**	SET	5%	25%	15%	15%					.101
WATER CLOSET	2-PIECE FLOOR MOUNTED VITREOUS CHINA WATER CLOSET WITHIN 5 FEET OF EXISTING STACK										
	WHITE	EA	166.00	700.00	866.00	1,299.00					.102
	COLOR	EA	206.00	700.00	906.00	1,359.00					.103
	1-PIECE FLOOR MOUNTED VITREOUS CHINA WATER CLOSET WITHIN 5 FEET OF EXISTING STACK										
	WHITE	EA	376.00	800.00	1,176.00	1,764.00					.104
	COLOR	EA	486.00	800.00	1,286.00	1,929.00					.105

14

14. BATHTUB, SHOWER, BIDET -- NEW WORK

SPECIFICATIONS		UNIT	JOB COST			PRICE	LOCAL AREA MODIFICATION				DATA BASE ITEM NO.
			MATLS	LABOR	710.00		MATLS	LABOR	TOTAL	PRICE	
	NOTE: THE FOLLOWING IN-STALLATIONS ARE FOR NEW WORK LOCATED WITHIN 5 FEET OF EXISTING STACK AND INCLUDE COST OF ROUGH-IN, FIXTURES AND FITTINGS, AND COMPLETE INSTALLATION.										
OVER 5 FEET FROM STACK	GOING AWAY FROM STACK OVER 5 FEET IN INSTALLA-TIONS ON THIS AND THE PREVIOUS PAGE **ADD** LF = TOTAL DISTANCE FROM EXISTING STACK	LF	16.00	20.00	36.00	54.00					.106
BATHTUB	5-FOOT BATHTUB, FAUCET @ $60, DRAIN ASSEMBLY @ $75										
	CAST IRON @ $250	EA	386.00	740.00	1,126.00	1,689.00					.107
	STEEL @ $150	EA	300.00	686.00	986.00	1,479.00					.108
	FIBERGLASS @ $170	EA	340.00	686.00	1,026.00	1,539.00					.109
	ENAMELED CAST IRON 5-FOOT BATHTUB WITH BUILT-IN WHIRLPOOL, IN-CLUDING ELECTRICAL CON-NECTION	EA	1760.00	1286.00	3,046.00	4,569.00					.110
FIBER-GLASS WHIRL-POOL	FIBERGLASS WHIRLPOOL TUB, 5-FOOT X 32" X 18" DEEP, 3 JETS, INCLUDING SKIRT, INCLUDING ELECTRI-CAL CONNECTION	EA	1600.00	1250.00	2,850.00	4,275.00					.111
STALL SHOWER	ROUGH AND INSTALL PLUMB-ING FOR STALL SHOWER, INCLUDING VINYL OR RUB-BER PAN, READY FOR WALL AND FLOOR TILE, FAUCET @ $45, DRAIN ASSEMBLY @ $65	EA	135.00	575.00	710.00	1,065.00					.112
	ROUGH AND INSTALL FIBER-GLASS SHOWER STALL @ $300, FAUCET @ $45, DRAIN ASSEMBLY @ $65	EA	436.00	670.00	1,106.00	1,659.00					.113
	ROUGH AND INSTALL FIBER-GLASS TUB/SHOWER UNIT @ $337, FAUCET @ $55, DRAIN ASSEMBLY @ $65	EA	470.00	696.00	1,166.00	1,749.00					.114
BIDET	INSTALL FLOOR MOUNTED BIDET, FITTINGS										
	WHITE	EA	606.00	750.00	1,356.00	2,034.00					.115
	COLOR	EA	736.00	750.00	1,486.00	2,229.00					.116

SPECIFICATIONS		UNIT	JOB COST			PRICE	LOCAL AREA MODIFICATION				DATA BASE ITEM NO.
			MATLS	LABOR	TOTAL		MATLS	LABOR	TOTAL	PRICE	
LAVATORY, WALL HUNG	WALL HUNG LAVATORY, PORCELAIN ENAMEL ON CAST IRON, 20" X 18"										
	WHITE	EA	216.00	490.00	706.00	1,059.00					.117
	COLOR	EA	240.00	490.00	730.00	1,095.00					.118
	VITREOUS CHINA, 19" X 17"										
	WHITE	EA	190.00	476.00	666.00	999.00					.119
	COLOR	EA	220.00	476.00	696.00	1,044.00					.120
PEDESTAL SINK	PEDESTAL MOUNTED, VITREOUS CHINA, 27" X 23"										
	WHITE	EA	310.00	580.00	890.00	1,335.00					.121
	COLOR	EA	350.00	580.00	930.00	1,395.00					.122
LAVATORY SET IN VANITY BASE	PORCELAIN ENAMEL ON CAST IRON, 20" X 10"										
	WHITE	EA	166.00	580.00	746.00	1,119.00					.123
	COLOR	EA	196.00	580.00	776.00	1,164.00					.124
	ENAMELED STEEL, 10" X 17"										
	WHITE	EA	120.00	546.00	666.00	999.00					.125
	COLOR	EA	150.00	546.00	696.00	1,044.00					.126
	VITREOUS CHINA, 20" X 16"										
	WHITE	EA	140.00	560.00	700.00	1,050.00					.127
	COLOR	EA	170.00	560.00	730.00	1,095.00					.128
VANITY BASE	PRE-FINISHED VANITY BASE, BUILDER QUALITY, 21" DEEP										
	Doors Drawers Width 2 0 24"	EA	236.00	24.00	260.00	390.00					.129
	2 0 30"	EA	290.00	24.00	314.00	471.00					.130
	2 0 36"	EA	375.00	31.00	406.00	609.00					.131
	3 3 42"	EA	420.00	40.00	460.00	690.00					.132
	3 3 48"	EA	480.00	46.00	526.00	789.00					.133
	3 3 60"	EA	615.00	61.00	676.00	1,014.00					.134
VANITY TOPS	CULTURED MARBLE TOP AND 4" SPLASH	LF	25.00	9.00	34.00	51.00					.135

14

14. REPLACE BATHROOM FIXTURES

SPECIFICATIONS		UNIT	JOB COST			PRICE	LOCAL AREA MODIFICATION				DATA BASE ITEM NO.
			MATLS	LABOR	TOTAL		MATLS	LABOR	TOTAL	PRICE	
	NOTE: REMOVE EXISTING BATHROOM FIXTURES AND REPLACE WITH NEW FIXTURES IN SAME LOCATION										
REPLACE WATER CLOSET	2-PIECE FLOOR MOUNTED VITREOUS CHINA WATER CLOSET										
	WHITE	EA	146.00	110.00	256.00	384.00					.200
	COLOR	EA	196.00	110.00	306.00	459.00					.201
	1-PIECE FLOOR MOUNTED VITREOUS CHINA WATER CLOSET										
	WHITE	EA	350.00	206.00	556.00	834.00					.202
	COLOR	EA	476.00	206.00	682.00	1,023.00					.203
REPLACE BIDET	FLOOR MOUNTED BIDET, BRASS FITTINGS										
	WHITE	EA	600.00	370.00	970.00	1,455.00					.204
	COLOR	EA	686.00	370.00	1,056.00	1,584.00					.205
REPLACE BATHTUB	5-FOOT BATHTUB, INCL. FAUCET @ $60, DRAIN ASSEMBLY @ $75										
	CAST IRON @ $250	EA	385.00	475.00	860.00	1,290.00					.206
	STEEL @ $150	EA	286.00	370.00	656.00	984.00					.207
	FIBERGLASS @ $170	EA	310.00	400.00	710.00	1,065.00					.208
	ENAMELED CAST IRON 5-FOOT BATHTUB WITH BUILT-IN WHIRLPOOL, INCLUDING ELECTRICAL CONNECTION	EA	1650.00	736.00	2,386.00	3,579.00					.209
	FIBERGLASS WHIRLPOOL TUB, 5-FOOT X 32" X 18" DEEP, 3 JETS, INCLUDING SKIRT, INCLUDING ELECTRICAL CONNECTION	EA	1476.00	680.00	2,156.00	3,234.00					.210
STALL SHOWER	RE-ROUGH AND INSTALL PLUMBING FOR STALL SHOWER, INCLUDING VINYL OR RUBBER PAN, READY FOR WALL AND FLOOR TILE, FAUCET @ $45, DRAIN ASSEMBLY @ $65	EA	136.00	350.00	486.00	729.00					.211
REPLACE SHOWER PAN	REMOVE EXISTING DEFECTIVE SHOWER PAN (AFTER WALL AND FLOOR TILE HAVE BEEN REMOVED BY OTHERS), AND INSTALL NEW SHOWER PAN WITH DRAIN ASSEMBLY AND DRAIN COVER. FOR COMPLETE JOB REPLACEMENT OF SHOWER PAN, SEE PAGE 64.	EA	80.00	400.00	480.00	720.00					.212

14

SPECIFICATIONS		UNIT	JOB COST			PRICE	LOCAL AREA MODIFICATION				DATA BASE ITEM NO.
			MATLS	LABOR	TOTAL		MATLS	LABOR	TOTAL	PRICE	
REPLACE LAVATORY	WALL HUNG LAVATORY, PORCELAIN ENAMEL ON CAST IRON, 20" X 18"										
	WHITE	EA	180.00	160.00	340.00	510.00					.213
	COLOR	EA	210.00	160.00	370.00	555.00					.214
	VITREOUS CHINA, 19" X 17"										
	WHITE	EA	160.00	150.00	310.00	465.00					.215
	COLOR	EA	190.00	150.00	340.00	510.00					.216
	PEDESTAL MOUNTED, VITREOUS CHINA, 27" X 23"										
	WHITE	EA	300.00	190.00	490.00	735.00					.217
	COLOR	EA	330.00	190.00	520.00	780.00					.218
REPLACE LAVATORY SET IN VANITY BASE	PORCELAIN ENAMEL ON CAST IRON, 20" X 10"										
	WHITE	EA	165.00	139.00	304.00	456.00					.219
	COLOR	EA	195.00	139.00	334.00	501.00					.220
	ENAMELED STEEL, 10" X 17"										
	WHITE	EA	120.00	138.00	258.00	387.00					.221
	COLOR	EA	150.00	138.00	288.00	432.00					.222
	VITREOUS CHINA, 20" X 16"										
	WHITE	EA	140.00	140.00	280.00	420.00					.223
	COLOR	EA	170.00	140.00	310.00	465.00					.224
REPLACE VANITY BASE	PRE-FINISHED VANITY BASE, BUILDER QUALITY, 21" DEEP										
	Doors Drawers Width 2 0 24"	EA	235.00	45.00	280.00	420.00					.225
	2 0 30"	EA	291.00	45.00	336.00	504.00					.226
	2 0 36"	EA	376.00	50.00	426.00	639.00					.227
	3 3 42"	EA	420.00	60.00	480.00	720.00					.228
	3 3 48"	EA	480.00	68.00	548.00	822.00					.229
	3 3 60"	EA	616.00	80.00	696.00	1,044.00					.230
REPLACE VANITY TOP	CULTURED MARBLE TOP AND 4" SPLASH	LF	24.00	10.00	34.00	51.00					.231
REMOVE AND REPLACE FIXTURE	DISCONNECT AND REMOVE BATHROOM FIXTURE AND REPLACE IN SAME LOCATION AT LATER TIME										.232
	WATER CLOSET	EA	10.00	110.00	120.00	180.00					
	BIDET	EA	15.00	325.00	340.00	510.00					.233
	WALL HUNG OR PEDESTAL LAVATORY	EA	15.00	221.00	236.00	354.00					.234
	LAVATORY & VANITY	EA	15.00	151.00	166.00	249.00					.235

14

14. KITCHEN PLUMBING AND APPLIANCES -- NEW WORK

SPECIFICATIONS		UNIT	JOB COST			PRICE	LOCAL AREA MODIFICATION				DATA BASE ITEM NO.
			MATLS	LABOR	TOTAL		MATLS	LABOR	TOTAL	PRICE	
	THE FOLLOWING COSTS INCLUDE FIXTURES AND ALL MATERIALS AND LABOR FOR ROUGHING IN AND INSTALLATION. ELECTRICAL COSTS ARE **NOT** INCLUDED.										
KITCHEN SINK	ROUGH AND INSTALL KITCHEN SINK WITHIN 5 FEET FROM EXISTING STACK, INCLUDING FAUCET, SPRAY AND TWO STRAINERS	EA	315.00	515.00	830.00	1,245.00					.300
BAR SINK	15" X 15" STAINLESS STEEL BAR SINK WITH FAUCET AND STRAINER, WITHIN 5 FEET OF EXISTING STACK	EA	196.00	230.00	426.00	639.00					.301
DISPOSER	1/2 HP DISPOSER INSTALLED AT EXISTING SINK LOCATION	EA	106.00	90.00	196.00	294.00					.302
DISH-WASHER	INSTALL DISHWASHER NEXT TO EXISTING SINK										
	ECONOMY	EA	280.00	240.00	520.00	780.00					.303
	BUILDER	EA	354.00	240.00	594.00	891.00					.304
	PREMIUM	EA	444.00	240.00	684.00	1,026.00					.305
DROP WASTE	DROP WASTE FOR ABOVE DISPOSER OR DISHWASHER	EA	25.00	105.00	130.00	195.00					.306
LAUNDRY TUB	POLYPROPYLENE 20 GAL. SINK, INCLUDING RUBBER STOPPER, RAISED SOAP DISH AND DRAIN ASSEMBLY WITH FAUCET	EA	126.00	490.00	616.00	924.00					.307
CLOTHES WASHER	INSTALL CLOTHES WASHER										
	ECONOMY	EA	379.00	247.00	626.00	939.00					.308
	BUILDER	EA	467.00	247.00	714.00	1,071.00					.309
	PREMIUM	EA	547.00	247.00	794.00	1,191.00					.310
GAS DRYER	INSTALL GAS DRYER										
	ECONOMY	EA	358.00	190.00	548.00	822.00					.311
	BUILDER	EA	478.00	190.00	668.00	1,002.00					.312
	PREMIUM	EA	526.00	190.00	716.00	1,074.00					.313
GAS RANGE	GAS COOKTOP 28" X 20"	EA	168.00	174.00	342.00	513.00					.314
	35" X 20"	EA	314.00	174.00	488.00	732.00					.315
GAS OVEN	GAS WALL OVEN, SINGLE	EA	258.00	190.00	448.00	672.00					.316
	DOUBLE	EA	320.00	190.00	510.00	765.00					.317
	FREE STANDING GAS RANGE AND OVEN										
	ECONOMY	EA	337.00	205.00	542.00	813.00					.318
	BUILDER	EA	453.00	205.00	658.00	987.00					.319
	PREMIUM	EA	735.00	205.00	940.00	1,410.00					.320
INSTANT HOT	INSTALL INSTANT HOT NEXT TO SINK	EA	96.00	220.00	316.00	474.00					.321
ICEMAKER	RUN WATER LINE FOR REFRIGERATOR ICEMAKER WITHIN 5 FEET OF SINK	EA	25.00	75.00	100.00	150.00					.322
	SAME AS ABOVE, MORE THAN 5 FEET FROM SINK **ADD** LF = TOTAL DISTANCE FROM EXISTING SINK	LF	.50	1.10	1.60	2.40					.323

14

14. REPLACE KITCHEN FIXTURES AND APPLIANCES

SPECIFICATIONS		UNIT	JOB COST			PRICE	LOCAL AREA MODIFICATION				DATA BASE ITEM NO.
			MATLS	LABOR	TOTAL		MATLS	LABOR	TOTAL	PRICE	
	THE FOLLOWING COSTS INCLUDE FIXTURES AND ALL MATERIALS AND LABOR FOR ROUGHING IN AND INSTALLATION. ELECTRICAL COSTS ARE **NOT** INCLUDED.										
	IF MORE THAN ONE FIXTURE OR APPLIANCE BELOW IS REMOVED AND REPLACED AT THE SAME TIME **DEDUCT** EA = EACH FIXTURE	EA	--	10%	5%	5%					
REPLACE KITCHEN SINK	REPLACE KITCHEN SINK IN SAME LOCATION	EA	310.00	140.00	450.00	675.00					.400
RELOCATE SINK	RE-LOCATE EXISTING KITCHEN SINK ON ANOTHER WALL IN KITCHEN, RE-ROUGH DRAIN AND VENT	EA	46.00	590.00	636.00	954.00					.401
REPLACE BAR SINK	REPLACE BAR SINK WITHIN 5 FEET OF EXISTING LINE	EA	180.00	140.00	320.00	480.00					.402
REPLACE DISPOSER	REPLACE DISPOSER WITH NEW 1/2 HP DISPOSER	EA	106.00	80.00	186.00	279.00					.403
REPLACE DISH-WASHER	REPLACE DISHWASHER										
	ECONOMY	EA	280.00	146.00	426.00	639.00					.404
	BUILDER	EA	354.00	146.00	500.00	750.00					.405
	PREMIUM	EA	444.00	146.00	590.00	885.00					.406
REPLACE LAUNDRY TUB	REPLACE LAUNDRY TUB WITH NEW POLYPROPYLENE 20 GAL. SINK	EA	100.00	136.00	236.00	354.00					.407
REPLACE CLOTHES WASHER	REPLACE CLOTHES WASHER										
	ECONOMY	EA	360.00	76.00	436.00	654.00					.408
	BUILDER	EA	456.00	76.00	532.00	798.00					.409
	PREMIUM	EA	520.00	80.00	600.00	900.00					.410
REPLACE GAS DRYER	REPLACE GAS DRYER										
	ECONOMY	EA	340.00	76.00	416.00	624.00					.411
	BUILDER	EA	456.00	76.00	532.00	798.00					.412
	PREMIUM	EA	500.00	86.00	586.00	879.00					.413
REPLACE GAS APPLIANCES	REPLACE GAS COOKTOP										
	28" X 20"	EA	110.00	30.00	140.00	210.00					.414
	35" X 20"	EA	258.00	40.00	298.00	447.00					.415
	REPLACE GAS WALL OVEN										
	SINGLE	EA	230.00	30.00	260.00	390.00					.416
	DOUBLE	EA	270.00	40.00	310.00	465.00					.417
	REPLACE FREE STANDING GAS RANGE AND OVEN										
	ECONOMY	EA	320.00	50.00	370.00	555.00					.418
	BUILDER	EA	430.00	50.00	480.00	720.00					.419
	PREMIUM	EA	700.00	60.00	760.00	1,140.00					.420
REMOVE AND REPLACE SAME FIXTURES	DISCONNECT AND REMOVE KITCHEN FIXTURES AND APPLIANCES AND REPLACE SAME FIXTURES IN SAME LOCATION AT LATER TIME										
	SINK	EA	16.00	150.00	166.00	249.00					.421
	DISPOSER	EA	8.00	60.00	68.00	102.00					.422
	DISHWASHER	EA	20.00	116.00	136.00	204.00					.423
	GAS RANGE	EA	--	90.00	90.00	135.00					.424
	REFRIGERATOR W/ICE MAKER	EA	--	66.00	66.00	99.00					.425
	LAUNDRY TUB	EA	16.00	110.00	126.00	189.00					.426

14

14. HOT WATER HEATER

SPECIFICATIONS		UNIT	JOB COST			PRICE	LOCAL AREA MODIFICATION				DATA BASE ITEM NO.
			MATLS	LABOR	TOTAL		MATLS	LABOR	TOTAL	PRICE	
HOT WATER HEATER, NEW WORK	GAS HOT WATER HEATER, INCLUDING 2" FOAM INSULATION, FLUE AND ALL LABOR AND MATERIALS TO CONNECT WITH WATER AND GAS SUPPLY										
	30 GAL	EA	182.00	198.00	380.00	570.00					.500
	40 GAL	EA	224.00	204.00	428.00	642.00					.501
	50 GAL	EA	298.00	204.00	502.00	753.00					.502
	ELECTRIC WATER HEATER, INCLUDING ALL LABOR AND MATERIALS TO CONNECT TO WATER SUPPLY AND 220 VOLT LINE FROM ELECTRICAL PANEL BOX										
	30 GAL	EA	170.00	172.00	342.00	513.00					.503
	40 GAL	EA	214.00	176.00	390.00	585.00					.504
	52 GAL	EA	285.00	178.00	463.00	694.50					.505
	82 GAL	EA	486.00	198.00	684.00	1,026.00					.506
REPLACEMENT	REPLACE EXISTING GAS WATER HEATER WITH SAME SIZE NEW HEATER, **NO** PLUMBING MODIFICATIONS INCLUDED										
	30 GAL	EA	182.00	100.00	282.00	423.00					.507
	40 GAL	EA	224.00	106.00	330.00	495.00					.508
	50 GAL	EA	298.00	116.00	414.00	621.00					.509
	REPLACE EXISTING ELECTRIC WATER HEATER WITH SAME SIZE NEW HEATER, **NO** PLUMBING MODIFICATIONS INCLUDED										
	30 GAL	EA	170.00	100.00	270.00	405.00					.510
	40 GAL	EA	214.00	106.00	320.00	480.00					.511
	52 GAL	EA	285.00	116.00	401.00	601.50					.512
	82 GAL	EA	486.00	138.00	624.00	936.00					.513
REMOVE AND REPLACE WATER HEATER	DISCONNECT AND REMOVE WATER HEATER AND REPLACE SAME HEATER IN SAME LOCATION AT A LATER TIME	EA	5.00	115.00	120.00	180.00					.514

SPECIFICATIONS		UNIT	JOB COST			PRICE	LOCAL AREA MODIFICATION				DATA BASE ITEM NO.
			MATLS	LABOR	TOTAL		MATLS	LABOR	TOTAL	PRICE	
	THE HEATING AND AIR CONDITIONING LABOR COSTS SHOWN IN THIS SECTION INCLUDE A SUBCONTRACTOR'S OVERHEAD AND PROFIT										
CONVECTOR	CONVECTORS ON EXISTING SYSTEM										
	FIRST FLOOR	EA	210.00	236.00	446.00	669.00					.000
	SECOND FLOOR	EA	295.00	309.00	604.00	906.00					.001
CAST IRON	NEW CAST IRON RADIATORS ON EXISTING SYSTEM										
	FIRST FLOOR	EA	345.00	231.00	576.00	864.00					.002
	SECOND FLOOR	EA	410.00	326.00	736.00	1,104.00					.003
	USED RADIATOR ON EXISTING SYSTEM										
	FIRST FLOOR	EA	170.00	230.00	400.00	600.00					.004
	SECOND FLOOR	EA	195.00	325.00	520.00	780.00					.005
	REMOVE RADIATOR, RUN CAST IRON BASEBOARD HEAT ON SAME WALL, 8-FOOT STRETCH	EA	280.00	316.00	596.00	894.00					.006
	RE-LOCATE RADIATOR WITHIN 10 FEET OF SAME LOCATION, FLOOR, WALL AND CEILING PATCHING **NOT** INCLUDED	EA	35.00	305.00	340.00	510.00					.008
	EXTEND HEATING LINES TO NEW ADDITION, INSTALL ONE RADIATOR	EA	310.00	346.00	656.00	984.00					.009
BASEBOARD	ON EXISTING SYSTEM, 8-FOOT STRETCH										
	COPPER	EA	270.00	90.00	360.00	540.00					.010
	CAST IRON	EA	375.00	195.00	570.00	855.00					.011
	SAME AS ABOVE, 16-FOOT STRETCH										
	COPPER	EA	420.00	156.00	576.00	864.00					.012
	CAST IRON	EA	590.00	180.00	770.00	1,155.00					.013

15

15. HOT WATER HEAT, OIL FIRED HEAT

SPECIFICATIONS		UNIT	JOB COST			PRICE	LOCAL AREA MODIFICATION				DATA BASE ITEM NO.
			MATLS	LABOR	TOTAL		MATLS	LABOR	TOTAL	PRICE	
CIRCULAT-ING PUMP	INSTALL CIRCULATING PUMP ON EXISTING SYSTEM	EA	196.00	140.00	336.00	504.00					.014
SEPARATE ZONE	INSTALL SEPARATE ZONE AND THERMOSTAT, INCLUD-ING BY-PASS VALVE AND CIR-CULATING PUMP	EA	384.00	210.00	594.00	891.00					.015
HEAT RISERS	EXTEND HEAT RISERS FROM FIRST TO SECOND FLOOR	EA	75.00	215.00	290.00	435.00					.016
DRAIN SYSTEM	DRAIN SYSTEM, DISCONNECT AND CAP OFF RADIATOR, FILL SYSTEM	EA	8.00	120.00	128.00	192.00					.017
BOILER	RE-BUILD CHAMBER IN EXISTING BOILER										
	MINIMUM	EA	116.00	160.00	276.00	414.00					.018
	MAXIMUM	EA	370.00	380.00	750.00	1,125.00					.019
	NEW BOILER, GAS FIRED										
	100,000 BTU	EA	2090.00	190.00	2,280.00	3,420.00					.020
	125,000 BTU	EA	2400.00	210.00	2,610.00	3,915.00					.021
OIL BURNER	NEW OIL BURNER ON EXIST-ING BOILER	EA	392.00	180.00	572.00	858.00					.100
OIL TANK	INDOOR, 275 GALLONS	EA	640.00	176.00	816.00	1,224.00					.101
	OUTDOOR, BURIED FIBER-GLASS TANK AND PIPING, 550 GALLONS	EA	770.00	678.00	1,448.00	2,172.00					.102
FURNACE	OIL FURNACE ON EXISTING DUCT SYSTEM AND FLUE, NO ELECTRICAL HOOKUP OR TANK INCLUDED										
	85,000 BTU	EA	1985.00	115.00	2,100.00	3,150.00					.103
	100,000 BTU	EA	2090.00	230.00	2,320.00	3,480.00					.104
	125,000 BTU	EA	2370.00	270.00	2,640.00	3,960.00					.105
	NEW BOILER, OIL FIRED										
	100,000 BTU	EA	2250.00	210.00	2,460.00	3,690.00					.106
	125,000 BTU	EA	2520.00	260.00	2,780.00	4,170.00					.107

15

SPECIFICATIONS		UNIT	JOB COST			PRICE	LOCAL AREA MODIFICATION				DATA BASE ITEM NO.
			MATLS	LABOR	TOTAL		MATLS	LABOR	TOTAL	PRICE	
ELECTRIC FURNACE	ELECTRIC FURNACE ON EXISTING DUCT SYSTEM, INCLUDING ALL LABOR AND MATERIALS, ON EXISTING CIRCUIT										
	10 KW	EA	590.00	150.00	740.00	1,110.00					.200
	15 KW	EA	850.00	150.00	1,000.00	1,500.00					.201
	20 KW	EA	1035.00	195.00	1,230.00	1,845.00					.202
	25 KW	EA	1510.00	230.00	1,740.00	2,610.00					.203
	30 KW	EA	1665.00	265.00	1,930.00	2,895.00					.204
ELECTRIC HEAT PUMP	ELECTRIC HEAT PUMP FOR HEAT AND AIR CONDITIONING, ON EXISTING SYSTEM. DETERMINE SIZE PER A/C LOAD, ON EXISTING CIRCUIT										
	10 KW, 2 TON	EA	3760.00	340.00	4,100.00	6,150.00					.205
	15 KW, 2.5 TON	EA	4180.00	410.00	4,590.00	6,885.00					.206
	19 KW, 3 TON	EA	4790.00	520.00	5,310.00	7,965.00					.207
	21 KW, 3.5 TON	EA	5695.00	585.00	6,280.00	9,420.00					.208
	28 KW, 4 TON	EA	5940.00	650.00	6,590.00	9,885.00					.209
	28 KW, 5 TON	EA	6560.00	760.00	7,320.00	11,964.					.210
GAS FURNACE	GAS FURNACE, USING EXISTING DUCT SYSTEM & FLUE, INCLUDING GAS PIPE HOOKUP IF GAS LINE AT FURNACE, ON EXISTING CIRCUIT										
	Approximate Heating Area										
60,000 BTU	750 SF	EA	1395.00	181.00	1,576.00	2,364.00					.211
80,000 BTU	1000 SF	EA	1495.00	231.00	1,726.00	2,589.00					.212
100,000 BTU	1200 SF	EA	1555.00	245.00	1,800.00	2,700.00					.213
125,000 BTU	1350 SF	EA	1665.00	275.00	1,940.00	2,910.00					.214
150,000 BTU	1500 SF	EA	1890.00	300.00	2,190.00	3,285.00					.215
175,000 BTU	2000 SF	EA	1990.00	316.00	2,306.00	3,459.00					.216
200,000 BTU	2500 SF	EA	2465.00	315.00	2,780.00	4,170.00					.217
	GAS WALL FURNACE, RECESSED	EA	540.00	316.00	856.00	1,284.00					.233

15

15. ELECTRIC HEAT

SPECIFICATIONS		UNIT	JOB COST			PRICE	LOCAL AREA MODIFICATION				DATA BASE ITEM NO.
			MATLS	LABOR	TOTAL		MATLS	LABOR	TOTAL	PRICE	
BASE-BOARD ELECTRIC, OPEN WALLS	INCLUDING BUILT-IN THER-MOSTAT ON EXISTING SYS-TEM WITHIN 10 LF OF BOX										
	4'-0" (1000 WATT)	EA	105.00	75.00	180.00	270.00					.218
	6'-0" (1500 WATT)	EA	120.00	86.00	206.00	309.00					.219
	8'-0" (2000 WATT)	EA	130.00	90.00	220.00	330.00					.220
	10'-0" (2500 WATT)	EA	145.00	101.00	246.00	369.00					.221
	SAME AS ABOVE, FISHED										
	4'-0" (1000 WATT)	EA	105.00	93.00	198.00	297.00					.222
	6'-0" (1500 WATT)	EA	120.00	96.00	216.00	324.00					.223
	8'-0" (2000 WATT)	EA	130.00	116.00	246.00	369.00					.224
	10'-0" (2500 WATT)	EA	145.00	123.00	268.00	402.00					.225
	OVER 10 LF FROM BOX										
	4'-0" (1000 WATT)	EA	112.00	106.00	218.00	327.00					.226
	6'-0" (1500 WATT)	EA	130.00	110.00	240.00	360.00					.227
	8'-0" (2000 WATT)	EA	142.00	130.00	272.00	408.00					.228
	10'-0" (2500 WATT)	EA	160.00	138.00	298.00	447.00					.229
WALL HEATER	WALL HEATER WITH BUILT-IN THERMOSTAT AND OUTLET, 12" X 12", 1250 WATTS	EA	115.00	75.00	190.00	285.00					.230
CEILING HEATER	CEILING HEATER WITH RE-MOTE THERMOSTAT AND SWITCH, 1250 WATTS	EA	150.00	86.00	236.00	354.00					.231
INFRA-RED HEATER	INFRA-RED HEATER IN CEIL-ING ON SEPARATE SWITCH	EA	165.00	65.00	230.00	345.00					.232

15

SPECIFICATIONS	UNIT	JOB COST			PRICE	LOCAL AREA MODIFICATION				DATA BASE ITEM NO.
		MATLS	LABOR	TOTAL		MATLS	LABOR	TOTAL	PRICE	
AIR CONDITIONER INSTALL SEPARATE A/C SYSTEM EACH FLOOR WITH AIR HANDLER INSIDE AND COMPRESSOR OUTSIDE										
Approximate Cooling Area										
2 TON 750 SF	EA	2185.00	315.00	2,500.00	3,750.00					.300
2.5 TON 1500 SF	EA	2435.00	345.00	2,780.00	4,170.00					.301
3 TON 2500 SF	EA	2770.00	380.00	3,150.00	4,725.00					.302
INSTALL A/C SYSTEM ON EXISTING FURNACE, COMPRESSOR ON GROUND										
Approximate Cooling Area										
2 TON 750 SF	EA	1495.00	225.00	1,720.00	2,580.00					.303
2.5 TON 1500 SF	EA	1765.00	241.00	2,006.00	3,009.00					.304
3 TON 2500 SF	EA	1925.00	265.00	2,190.00	3,285.00					.305
3.5 TON 3000 SF	EA	2155.00	275.00	2,430.00	3,645.00					.306
4 TON 3500 SF	EA	2310.00	300.00	2,610.00	3,915.00					.307
5 TON 4500 SF	EA	2915.00	315.00	3,230.00	4,845.00					.308
FOR COMPRESSOR ON ROOF (INCLUDING CRANE COST) **ADD**	EA	250.00	650.00	900.00	1,350.00					.309
INCREASE BLOWER MOTOR CAPACITY TO ADD A/C	EA	40.00	90.00	130.00	195.00					.310
CONDENSATE PUMP TO REMOVE EVAPORATOR WATER FROM A/C	EA	35.00	95.00	130.00	195.00					.311
WINDOW UNITS HIGH EFFICIENCY AIR CONDITIONING AND COOLING UNITS ON EXISTING SERVICE										
Approximate Cooling Area										
5,600 BTU 190 SF	EA	390.00	90.00	480.00	720.00					.319
9,000 BTU 420 SF	EA	480.00	90.00	570.00	855.00					.320
13,000 BTU 720 SF	EA	630.00	90.00	720.00	1,080.00					.321
15,000 BTU 875 SF	EA	680.00	90.00	770.00	1,155.00					.322

15

15. DUCTWORK

SPECIFICATIONS		UNIT	JOB COST			PRICE	LOCAL AREA MODIFICATION				DATA BASE ITEM NO.
			MATLS	LABOR	TOTAL		MATLS	LABOR	TOTAL	PRICE	
REGISTER	CUT OUT AND INSTALL REGISTER IN EXISTING WARM AIR OR RETURN DUCT	EA	30.00	90.00	120.00	180.00					.312
DUCTWORK	EXTEND SUPPLY OR RETURN DUCT TO ROOM ADDITION WITHIN 10 FEET	EA	75.00	287.00	362.00	543.00					.313
	EXTEND DUCTWORK IN BASEMENT	LF	2.50	8.30	10.80	16.20					.314
	CONVERT EXISTING LOW DUCT TO HIGH-LOW DUCT AND REGISTER FOR CENTRAL AIR CONDITIONING, **NOT** INCLUDING WALL REPAIR OR PLASTERING	EA	60.00	136.00	196.00	294.00					.315
	RUN DUCTWORK FROM WALL TO BASE OF EITHER KITCHEN CABINET OR VANITY WITH NEW REGISTER	EA	30.00	116.00	146.00	219.00					.316
	EXTEND HIGH REGISTER OUT TO SOFFIT IN KITCHEN	EA	25.00	85.00	110.00	165.00					.317
HUMIDIFIER	INSTALL FLUSH TYPE HUMIDIFIER ON EXISTING FURNACE DUCT SYSTEM	EA	250.00	116.00	366.00	549.00					.318

15

SPECIFICATIONS		UNIT	JOB COST			PRICE	LOCAL AREA MODIFICATION				DATA BASE ITEM NO.
			MATLS	LABOR	TOTAL		MATLS	LABOR	TOTAL	PRICE	
	THE LABOR COSTS SHOWN IN THIS SECTION INCLUDE AN ELECTRICAL SUBCONTRACTOR'S OVERHEAD AND PROFIT										
SERVICE	INCREASE SERVICE WITH CIRCUIT BREAKERS										
	TO 150 AMPS	EA	176.00	360.00	536.00	804.00					.000
	TO 200 AMPS	EA	246.00	360.00	606.00	909.00					.001
	INCREASE SERVICE WITH FUSE BOX										
	TO 150 AMPS	EA	155.00	225.00	380.00	570.00					.002
	TO 200 AMPS	EA	235.00	225.00	460.00	690.00					.003
	RE-LOCATE EXISTING SERVICE ON SAME SERVICE PANEL	EA	70.00	270.00	340.00	510.00					.004
FUSE	INSTALL FUSE BOX										
	2-CIRCUIT	EA	33.00	53.00	86.00	129.00					.005
	4-CIRCUIT	EA	45.00	65.00	110.00	165.00					.006
	6-CIRCUIT	EA	48.00	80.00	128.00	192.00					.007
	8-CIRCUIT	EA	56.00	88.00	144.00	216.00					.008
SPECIAL WIRING	SINGLE HEAVY-DUTY LINE FOR APPLIANCE OR WORK-SHOP										
	BASEMENT	EA	12.00	74.00	86.00	129.00					.009
	FIRST FLOOR	EA	16.00	90.00	106.00	159.00					.010
	SECOND FLOOR	EA	22.00	114.00	136.00	204.00					.011
	THIRD FLOOR	EA	34.00	152.00	186.00	279.00					.012
	REMOVE AND RE-INSTALL CEILING LIGHT FIXTURE ON ONE TRIP	EA	--	28.00	28.00	42.00					.013
	REWIRE WITH:										
	ROMEX CABLE	EA	12.00	60.00	72.00	108.00					.014
	BX ARMORED CABLE	EA	18.00	72.00	90.00	135.00					.015
	EA = EACH OUTLET										

16

16. ELECTRICAL OUTLETS

SPECIFICATIONS		UNIT	JOB COST			PRICE	LOCAL AREA MODIFICATION				DATA BASE ITEM NO.
			MATLS	LABOR	TOTAL		MATLS	LABOR	TOTAL	PRICE	
	NOTE: ALL ELECTRICAL OUTLETS SHOWN HERE ARE FOR NEW WORK (IN OPEN WALLS AND CEILINGS) AND INCLUDE CIRCUITS AS REQUIRED										
DUPLEX OUTLETS	DUPLEX OUTLETS	EA	13.00	25.00	38.00	57.00					.100
	WEATHERPROOF OUTLET	EA	25.00	45.00	70.00	105.00					.103
APPLIANCE OUTLETS	DISPOSAL OUTLET WITH SWITCH	EA	28.00	60.00	88.00	132.00					.104
	DISHWASHER OUTLET	EA	23.00	45.00	68.00	102.00					.105
	HOOD AND FAN OUTLET	EA	25.00	45.00	70.00	105.00					.106
	APPLIANCE OUTLET	EA	22.00	40.00	62.00	93.00					.107
GFIC	GROUND FAULT OUTLET	EA	33.00	45.00	78.00	117.00					.108
	GROUND FAULT BREAKER	EA	35.00	25.00	60.00	90.00					.109
	TRASH COMPACTOR OUTLET	EA	22.00	40.00	62.00	93.00					.110
220-VOLT OUTLET	220-VOLT OUTLET FOR A/C, CLOTHES DRYER, RANGE OR OVEN										
	BASEMENT	EA	29.00	75.00	104.00	156.00					.111
	FIRST FLOOR	EA	37.00	105.00	142.00	213.00					.112
	SECOND FLOOR	EA	45.00	125.00	170.00	255.00					.113
	THIRD FLOOR	EA	57.00	145.00	202.00	303.00					.114
SWITCH	SINGLE-POLE SWITCH	EA	13.00	25.00	38.00	57.00					.115
	THREE-WAY SWITCH, SET	EA	29.00	45.00	74.00	111.00					.116
	REPLACE EXISTING SWITCH WITH DIMMER SWITCH	EA	13.00	25.00	38.00	57.00					.117
FISH OUTLETS OR SWITCHES	FISH OUTLETS OR SWITCHES IN CLOSED WALLS AND CEILINGS, **ADD**										
	FRAME WALL	EA	--	26.00	26.00	39.00					.118
	MASONRY WALL	EA	--	50.00	50.00	75.00					.119

16

SPECIFICATIONS		UNIT	JOB COST			PRICE	LOCAL AREA MODIFICATION				DATA BASE ITEM NO.
			MATLS	LABOR	TOTAL		MATLS	LABOR	TOTAL	PRICE	
CEILING FIXTURE	CEILING OUTLET AND SWITCH (IN ADDITION TO COST OF FIXTURE)	EA	24.00	50.00	74.00	111.00					.120
FLUOR-ESCENT BATH FIXTURE	BATH FIXTURE OVER MEDI-CINE CABINET, ONE TUBE AND ONE PLUG, INCLUDING OUTLET, FIXTURE @ $40 AND SWITCH	EA	47.00	85.00	132.00	198.00					.121
RECESSED FIXTURE	RECESSED 6" ROUND FIX-TURE @ $30, INCLUDING FIX-TURE, OUTLET AND SWITCH	EA	46.00	90.00	136.00	204.00					.122
FLUOR-ESCENT FIXTURE	FLUORESCENT LIGHT, INCLUDING FIXTURE, OUTLET AND SWITCH, OPEN FRAMING										
	4'-0" (2 TUBE) @ $50	EA	57.00	85.00	142.00	213.00					.123
	4'-0" (4 TUBE) @ $75	EA	81.00	85.00	166.00	249.00					.124
	INSTALL FLUORESCENT 12" FIXTURE @ $20 UNDER CAB-INET IN KITCHEN WITH SWITCH ON FIXTURE	EA	21.00	45.00	66.00	99.00					.125
SPOTLIGHT	INSTALL SWITCH AND OUT-SIDE SPOTLIGHTS @ $25, TWO SPOTLIGHTS ON EAVES OF HOUSE										
	FIRST FLOOR	EA	68.00	110.00	178.00	267.00					.126
	SECOND FLOOR	EA	80.00	124.00	204.00	306.00					.127
EXTERIOR FIXTURE	INSTALL SWITCH AND OUT-SIDE OUTLET OVER DOOR OR AT SIDE OF DOOR (IN AD-DITION TO COST OF FIX-TURE)	EA	10.00	90.00	100.00	150.00					.128

16

16. KITCHEN & BATHROOM FANS AND HOODS

SPECIFICATIONS		UNIT	JOB COST			PRICE	LOCAL AREA MODIFICATION				DATA BASE ITEM NO.
			MATLS	LABOR	TOTAL		MATLS	LABOR	TOTAL	PRICE	
	NOTE: ALL FANS & HOODS SHOWN HERE ARE FOR NEW WORK (IN OPEN WALLS AND CEILINGS) AND INCLUDE CIRCUITS AS REQUIRED										
BATHROOM FANS	EXHAUST FAN FOR BATHROOM @ $40, INCL. SWITCH	EA	46.00	90.00	136.00	204.00					.200
	1,000 WATT COMBINATION FAN & HEATER @ $95 IN BATHROOM, INCL. SWITCH	EA	104.00	90.00	194.00	291.00					.201
	COMBINATION FAN, HEATER AND LIGHT @ $120 IN BATHROOM, INCL. SWITCH	EA	126.00	90.00	216.00	324.00					.202
	EXHAUST FAN AND LIGHT @ $55 IN BATHROOM, INCL. SWITCH	EA	64.00	90.00	154.00	231.00					.203
RANGE HOODS	DUCT TYPE 30" @ $70	EA	96.00	90.00	186.00	279.00					.204
	36" @ $75	EA	100.00	90.00	190.00	285.00					.205
	42" @ $77	EA	102.00	90.00	192.00	288.00					.206
	DUCTLESS 30" @ $98	EA	108.00	90.00	198.00	297.00					.207
	36" @ $106	EA	116.00	90.00	206.00	309.00					.208
	42" @ $109	EA	120.00	90.00	210.00	315.00					.209
	COOKTOP DOWNDRAFT DUCTED EXHAUST FAN	EA	61.00	125.00	186.00	279.00					.210
DUCTWORK	METAL DUCTWORK FOR HOOD OR FAN INSTALLATION, **NO** WALL BREAKTHROUGH	LF	4.00	10.00	14.00	21.00					.211
	BREAK THROUGH WALL AND VENT FAN FRAME WALL	EA	--	28.00	28.00	42.00					.212
	MASONRY WALL	EA	--	52.00	52.00	78.00					.213
	VENT FAN THROUGH ROOF OF SINGLE-STORY HOUSE	EA	--	46.00	46.00	69.00					.214
CEILING FAN	INSTALL CEILING FAN USING EXISTING WIRING	EA	136.00	60.00	196.00	294.00					.215
	INSTALL LIGHT KIT FOR CEILING FAN	EA	20.00	20.00	40.00	60.00					.216

16

16. HOUSE EXHAUST FANS, CENTRAL VACUUM, SMOKE DETECTOR

SPECIFICATIONS		UNIT	JOB COST			PRICE	LOCAL AREA MODIFICATION				DATA BASE ITEM NO.
			MATLS	LABOR	TOTAL		MATLS	LABOR	TOTAL	PRICE	
ATTIC EXHAUST FAN	• 1100-1500 CFM • INCL. ELECTRICAL WIRING • GABLE VENT MOUNTED ATTIC EXHAUST FAN WITH AUTOMATIC THERMOSTAT @ $64	EA	72.00	120.00	192.00	288.00					.217

	SAME AS ABOVE, ROOF MOUNTED FAN AND ROOF HOOD
	  TYPE OF ROOF
	
	    BUILT-UP
		EA	140.00	190.00	330.00	495.00					.218

	ASPHALT SHINGLES, ROLL ROOFING OR SELVAGE	EA	120.00	110.00	230.00	345.00					.219
	ASBESTOS, SLATE OR METAL	EA	120.00	170.00	290.00	435.00					.220
	CEDAR SHINGLES OR SHAKES	EA	120.00	130.00	250.00	375.00					.221
	WHOLE HOUSE CEILING MOUNTED EXHAUST FAN	EA	216.00	110.00	326.00	489.00					.225
CENTRAL VACUUM SYSTEM	INCLUDING POWER UNIT, HOSE, CLEANING TOOLS, TUBING INSTALLATION EA = CENTRAL UNIT AND THREE STATIONS	EA	590.00	446.00	1,036.00	1,554.00					.222
	EACH ADDITIONAL STATION **ADD**	EA	75.00	105.00	180.00	270.00					.223
SMOKE DETECTOR	HARD-WIRED AUTOMATIC FIRE AND SMOKE DETECTOR	EA	45.00	65.00	110.00	165.00					.224

16

17. INSULATION, NEW WORK

SPECIFICATIONS		UNIT	JOB COST			PRICE	LOCAL AREA MODIFICATION				DATA BASE ITEM NO.
			MATLS	LABOR	TOTAL		MATLS	LABOR	TOTAL	PRICE	
	NOTE: THE COSTS ON THIS PAGE ARE FOR NEW WORK ONLY										
FIBER-GLASS BLANKET	STAPLED TO OPEN FRAMING OR LAID FLAT BETWEEN CEILING JOISTS ON IN-STALLED DRYWALL										
	UNFACED 3-1/2"	SF	.24	.20	.44	.66					.000
	6"	SF	.36	.20	.56	.84					.001
	9"	SF	.56	.21	.77	1.16					.002
	FOILBACK ONE FACE										
	3-1/2	SF	.28	.20	.48	.72					.003
	6"	SF	.41	.20	.61	.92					.004
	9"	SF	.58	.21	.79	1.19					.005
	KRAFTBACK ONE FACE										
	3-1/2"	SF	.25	.20	.45	.68					.006
	6"	SF	.37	.20	.57	.86					.007
	9"	SF	.54	.21	.75	1.13					.008
	STAPLED TO FURRING STRIPS										
	UNFACED 1"	SF	.17	.19	.36	.54					.009
	FOILBACK 1" ONE FACE	SF	.20	.19	.39	.59					.010
	STAPLED TO BASEMENT CEILING										
	UNFACED 6"	SF	.36	.20	.56	.84					.011
	FOILBACK 3-1/2" ONE FACE	SF	.27	.20	.47	.71					.012
	RIGID INSULATION STAPLED TO ROOF RAFTERS OF CA-THEDRAL CEILING										
	FOIL FACED 3-1/2"	SF	.28	.20	.48	.72					.013
	6"	SF	.41	.20	.61	.92					.014
	1/2" FOAM INSULATION BOARD	SF	.41	.21	.62	.93					.015
PERIMETER INSULA-TION	FIBERBOARD, STYROFOAM OR RUBBERBOARD, SIDE OF EXTERIOR SLAB AND 18" IN FROM EDGE AT BOTTOM										
	3/4"	LF	.34	.60	.94	1.41					.016
	1"	LF	.42	.60	1.02	1.53					.017
	2"	LF	.96	.60	1.56	2.34					.018
	LF = PERIMETER OF SLAB										

17

SPECIFICATIONS		UNIT	JOB COST			PRICE	LOCAL AREA MODIFICATION				DATA BASE ITEM NO.
			MATLS	LABOR	TOTAL		MATLS	LABOR	TOTAL	PRICE	
	NOTE: THE COSTS ON THIS PAGE ARE FOR INSTALLATIONS IN OLD WORK										
FIBER-GLASS BLANKET	STAPLED TO EXISTING KNEE-WALLS IN ATTIC										
	UNFACED (R-11) 3-1/2	SF	.24	.26	.50	.75					.019
	FOIL FACE (INSIDE) 3-1/2	SF	.27	.26	.53	.80					.020
	KRAFT FACED 3-1/2"	SF	.25	.26	.51	.77					.021
	FIBERGLASS BLANKET TO FILL SOFFITS AND DROPPED CEILINGS IN ATTICS CF = CUBIC FEET	CF	.96	.65	1.61	2.42					.022
	STAPLE 3-1/2" FIBERGLASS BLANKET TO TRAP DOOR IN ATTIC OR TO BACK OF PULL-DOWN STAIRS	EA	6.00	2.50	8.50	12.75					.023
FIBER-GLASS BATTS	UNFACED, INSTALLED OVER EXISTING INSULATION BEHIND KNEEWALLS										
	(R-8) 2-1/2"	SF	.20	.21	.41	.62					.024
BAFFLE	INSTALL BAFFLE AROUND RECESSED LIGHTS										
	OPEN CEILING	EA	3.00	9.50	12.50	18.75					.025
	RESTRICTED CEILING	EA	3.00	12.50	15.50	23.25					.026
	INSTALL BAFFLE AROUND CHIMNEY	EA	6.00	16.00	22.00	33.00					.027
	INSTALL BAFFLE AROUND FAN	EA	3.00	9.50	12.50	18.75					.028
	INSTALL VENTILATION BAFFLES IN RAFTER SPACES	LF	.23	.30	.53	.80					.029

17

17. INSULATION, BLOWN IN

SPECIFICATIONS		UNIT	JOB COST			PRICE	LOCAL AREA MODIFICATION				DATA BASE ITEM NO.
			MATLS	LABOR	TOTAL		MATLS	LABOR	TOTAL	PRICE	
	THE LABOR COSTS SHOWN ON THIS PAGE INCLUDE AN INSULATION SUBCONTRACTOR'S OVERHEAD AND PROFIT										
BLOWN-IN FIBER-GLASS	BEHIND ASBESTOS, ALUMINUM OR WOOD SIDING, INCLUDING DRILLING AND PLUGGING, **NO** PAINTING, FROM OUTSIDE	SF	.95	1.11	2.06	3.09					.100
	BEHIND STUCCO OR FRAME WALL, FROM OUTSIDE	SF	.95	1.25	2.20	3.30					.102
	BEHIND DRYWALL OR PLASTER WALL, FROM INSIDE	SF	.95	1.50	2.45	3.68					.104
	BLOWN IN BETWEEN OPEN JOISTS, UNRESTRICTED										
	R-11	SF	.20	.28	.48	.72					.107
	R-19	SF	.35	.45	.80	1.20					.108
	R-30	SF	.55	.65	1.20	1.80					.109
	R-38	SF	.75	.85	1.60	2.40					.110
BLOWN-IN CELLULOSE	FIRE-RETARDANT CELLULOSE BLOWN IN BETWEEN OPEN JOISTS, UNRESTRICTED										
	R-11	SF	.15	.20	.35	.53					.111
	R-19	SF	.25	.30	.55	.83					.112
	R-30	SF	.40	.50	.90	1.35					.113
	R-38	SF	.75	.50	1.25	1.88					.114

17

SPECIFICATIONS		UNIT	JOB COST			PRICE	LOCAL AREA MODIFICATION				DATA BASE ITEM NO.
			MATLS	LABOR	TOTAL		MATLS	LABOR	TOTAL	PRICE	
RESTRICT-ED AREAS	IN RESTRICTED AREAS, FOR DIFFICULT ACCESS **ADD**	SF	--	.25	.25	.38					.115
MISCELLA-NEOUS	LAY 4-MIL VAPOR BARRIER ON GROUND IN CRAWL SPACE	SF	.04	.09	.13	.20					.116
	WRAP HEATING/COOLING PIPES OR DUCTS IN CRAWL SPACE OR ATTIC WITH INSU-LATION										
	1/2" ON PIPES	LF	.30	.30	.60	.90					.117
	1-1/2" ON DUCTS	LF	.38	.32	.70	1.05					.118
RADIANT BARRIERS	REINFORCED FOIL HEAT SHIELD (FOIL FACES DOWN)	SF	.10	.11	.21	.32					.119

17

18. GYPSUM DRYWALL, NEW WORK

SPECIFICATIONS		UNIT	JOB COST			PRICE	LOCAL AREA MODIFICATION				DATA BASE ITEM NO.
			MATLS	LABOR	TOTAL		MATLS	LABOR	TOTAL	PRICE	
GYPSUM DRYWALL ON NEW WALL – COMPLETE JOB	UP TO 300 SF OF WALL, NAILED, TAPED, FINISHED AND SANDED, 3 COATS										
	3/8"	EA PLUS	--	100.00	100.00	150.00					.000
		SF	.18	.33	.51	.77					.001
	1/2"	EA PLUS	--	100.00	100.00	150.00					.000
		SF	.19	.35	.54	.81					.002
	1/2" FIRECODE	EA PLUS	--	100.00	100.00	150.00					.000
		SF	.22	.37	.59	.89					.003
	5/8" FIRECODE	EA PLUS	--	100.00	100.00	150.00					.000
		SF	.23	.40	.63	.95					.004
	1/2" MOISTURE RESISTANT	EA PLUS	--	100.00	100.00	150.00					.000
EA = EACH JOB (NOT EACH SHEET)		SF	.23	.35	.58	.87					.005
	SAME AS ABOVE, WHEN TOTAL JOB IS OVER 300 SF										
	3/8"	SF	.18	.66	.84	1.26					.006
	1/2"	SF	.19	.68	.87	1.31					.007
	1/2" FIRECODE	SF	.22	.70	.92	1.38					.008
	5/8" FIRECODE	SF	.23	.74	.97	1.46					.009
	1/2" MOISTURE RESISTANT	SF	.23	.68	.91	1.37					.010
NAILED ONLY	NAILED ONLY, APPLIED TO STUDS OR FURRING										
	3/8"	SF	.16	.18	.34	.51					.011
	1/2"	SF	.17	.20	.37	.56					.012
	1/2" FIRECODE	SF	.20	.22	.42	.63					.013
	5/8" FIRECODE	SF	.21	.25	.46	.69					.014
	1/2" MOISTURE RESISTANT	SF	.21	.20	.41	.62					.015
TAPE AND FINISH ONLY	TAPE JOINTS, FINISH AND SAND ONLY										
	UP TO 300 SF OF WALL	EA	4.30	57.70	62.00	93.00					.016
	OVER 300 SF OF WALL	SF	.03	.48	.51	.77					.017
	1/2" BLUEBOARD	SF	.19	.33	.52	.78					.018
	SKIM COAT PLASTER	SF	.09	.44	.53	.80					.019

18

SPECIFICATIONS		UNIT	JOB COST			PRICE	LOCAL AREA MODIFICATION				DATA BASE ITEM NO.
			MATLS	LABOR	TOTAL		MATLS	LABOR	TOTAL	PRICE	
GYPSUM DRYWALL OVER EXISTING WALL – COMPLETE JOB	*UP TO 300 SF* OF WALL, NAILED WITH NAILS OF REQUIRED LENGTH THROUGH EXISTING WALL FINISH AND INTO STUDS, TAPED, FINISHED AND SANDED										
	3/8"	EA	--	74.00	74.00	111.00					.100
		PLUS SF	.18	.51	.69	1.04					.101
	1/2"	EA	--	74.00	74.00	111.00					.100
		PLUS SF	.19	.53	.72	1.08					.102
	5/8" FIRECODE	EA	--	74.00	74.00	111.00					.100
		PLUS SF	.23	.60	.83	1.25					.103
	EA = EACH JOB (**NOT** EACH SHEET)										
	SAME AS ABOVE, WHEN TOTAL JOB IS *OVER 300 SF*										
	3/8"	SF	.18	.49	.67	1.01					.104
	1/2"	SF	.19	.54	.73	1.10					.105
	5/8" FIRECODE	SF	.23	.63	.86	1.29					.106
TEXTURE SPACKLE FINISH	SPRAY FINISHED DRYWALL WITH SPACKLE OR TEXTURE SPRAY, *UP TO 300 SF* OF WALL	EA	--	42.00	42.00	63.00					.107
		PLUS SF	.15	.17	.32	.48					.108
	SAME AS ABOVE WHEN JOB IS *OVER 300 SF*	SF	.15	.30	.45	.68					.109
NAILED OVER EXISTING WALL	DRYWALL OVER EXISTING PLASTERED WALL WITH NAILS OF REQUIRED LENGTH THROUGH EXISTING PLASTER & LATH & INTO STUDS, NAILED ONLY										
	3/8"	SF	.16	.51	.67	1.01					.110
	1/2"	SF	.17	.53	.70	1.05					.111
	5/8" FIRECODE	SF	.21	.60	.81	1.22					.112
VINYL SURFACE DRYWALL	VINYL SURFACE GYPSUM PANELS WITH COLOR PINS	SF	.67	.60	1.27	1.91					.113

18

18. LATHING

SPECIFICATIONS		UNIT	JOB COST			PRICE	LOCAL AREA MODIFICATION				DATA BASE ITEM NO.
			MATLS	LABOR	TOTAL		MATLS	LABOR	TOTAL	PRICE	
GYPSUM LATH	GYPSUM LATH NAILED TO WOOD STUDS, INCLUDING CORNER BEADS AND CORNERITE										
	3/8"	SF	.29	.27	.56	.84					.114
	1/2"	SF	.30	.29	.59	.89					.115
METAL LATH	3.4 DIAMOND MESH NAILED TO WOOD STUDS	SF	.45	.24	.69	1.04					.116
WOOD LATH	3/8" X 1-1/2" X 48" LATH NAILED TO STUDS (3/8" SPACED)	SF	.70	.32	1.02	1.53					.117
GROUNDS	WOOD GROUNDS FOR PLAS-TERING										
	1" X 1"	LF	.03	.56	.59	.89					.118
	1" X 3"	LF	.09	.56	.65	.98					.119

18

SPECIFICATIONS		UNIT	JOB COST			PRICE	LOCAL AREA MODIFICATION				DATA BASE ITEM NO.
			MATLS	LABOR	TOTAL		MATLS	LABOR	TOTAL	PRICE	
	THE PLASTERING LABOR COSTS SHOWN ON THIS PAGE INCLUDE A PLASTERING SUBCONTRACTOR'S OVERHEAD AND PROFIT										
PLASTER	GYPSUM PLASTER OVER EXISTING GYPSUM LATH										
	2 COATS	SF	.35	.95	1.30	1.95					.120
	3 COATS	SF	.38	1.33	1.71	2.57					.121
	3 COATS GYPSUM PLASTER										
	OVER MASONRY	SF	.41	1.66	2.07	3.11					.122
	OVER METAL LATH		.61	1.26	1.87	2.81					.123
	OVER WOOD LATH	SF	.41	1.36	1.77	2.66					.124
SKIM-COAT PLASTERING	TAPE GYPSUM BOARD (SHEETROCK) AND APPLY LIME PUTTY PLASTER FINISH	SF	.07	.76	.83	1.25					.125

18

18. CERAMIC TILE

SPECIFICATIONS		UNIT	JOB COST			PRICE	LOCAL AREA MODIFICATION				DATA BASE ITEM NO.
			MATLS	LABOR	TOTAL		MATLS	LABOR	TOTAL	PRICE	
IN MUD	INCLUDING METAL LATHING AND SCRATCH COAT, WITH AVERAGE AMOUNT OF COVE AND BULLNOSE, 4-1/4" X 4-1/4"	SF	3.15	9.00	12.15	18.23					.200
WONDER-BOARD OR DUROCK	CERAMIC TILE BACKER BOARD, 1/2" 3' X 5' SHEETS NOTE: CERAMIC TILE MAY ALSO BE INSTALLED OVER MOISTURE-RESISTANT DRYWALL; SEE PAGE 224	SF	2.06	2.34	4.40	6.60					.210
THIN-SET OR MASTIC	CERAMIC TILE WITH AVERAGE AMOUNT OF COVE AND BULLNOSE, 4-1/4" X 4-1/4", 6" X 6", AND 8" X 8", OVER EXISTING BACKER BOARD OR MOISTURE RESISTANT DRYWALL	SF	2.10	5.60	7.70	11.55					.201
FIXTURES	CERAMIC TILE BATHROOM FIXTURES: TOWEL BAR, TOOTHBRUSH HOLDER, PAPER HOLDER, SOAP AND GRAB	SET	26.00	16.00	42.00	63.00					.202
EXTEND TILE	EXTEND CERAMIC TILE ABOVE EXISTING TILE WALL IN MUD	SF	4.05	8.50	12.55	18.83					.203
	IN MASTIC	SF	4.43	7.50	11.93	17.90					.204

18

SPECIFICATIONS		UNIT	JOB COST			PRICE	LOCAL AREA MODIFICATION				DATA BASE ITEM NO.
			MATLS	LABOR	TOTAL		MATLS	LABOR	TOTAL	PRICE	
PLYWOOD PANELING	PREFINISHED, 1/4" X 4 X 8 SHEETS NAILED TO STUDS OR FURRING. **NO** MOULDING INCLUDED *Retail Price Per 4 X 8 Sheet*										
	10.00	SF	.32	.58	.90	1.35					.300
	11.00	SF	.35	.58	.93	1.40					.301
	12.00	SF	.39	.58	.97	1.46					.302
	13.00	SF	.42	.58	1.00	1.50					.303
	14.00	SF	.45	.58	1.03	1.55					.304
	15.00	SF	.48	.58	1.06	1.59					.305
	16.00	SF	.51	.58	1.09	1.64					.306
	17.00	SF	.54	.58	1.12	1.68					.307
	18.00	SF	.57	.58	1.15	1.73					.308
	19.00	SF	.60	.58	1.18	1.77					.309
	20.00	SF	.63	.58	1.21	1.82					.310
	25.00	SF	.78	.58	1.36	2.04					.311
	30.00	SF	.94	.60	1.54	2.31					.312
	35.00	SF	1.09	.61	1.70	2.55					.313
	40.00	SF	1.25	.62	1.87	2.81					.314
	45.00	SF	1.41	.64	2.05	3.08					.315
V-JOINT, SOLID WOOD	UNFINISHED, V-JOINT 3/4" THICK AND RANDOM WIDTHS TO 10" INCLUDING INSTALLATION OF 1 X 4 BASE OF SAME MATERIAL TO WHICH THE PANELING IS BUTTED										
	KNOTTY PINE	SF	2.13	.92	3.05	4.58					.316
	CEDAR	SF	2.16	.92	3.08	4.62					.317
	REDWOOD	SF	2.79	1.01	3.80	5.70					.318
CEDAR CLOSET LINING	3/8" T&G AND END MATCHED, APPLIED OVER EXISTING SOLID WALL	SF	1.85	1.29	3.14	4.71					.319
	1/4" CEDAR PARTICLEBOARD, 4-0 X 8-0 SHEETS	SF	.86	.90	1.76	2.64					.320

18

18. PATCHING WALLS

SPECIFICATIONS		UNIT	JOB COST			PRICE	LOCAL AREA MODIFICATION				DATA BASE ITEM NO.
			MATLS	LABOR	TOTAL		MATLS	LABOR	TOTAL	PRICE	
	THE LABOR COSTS SHOWN ON THIS PAGE INCLUDE CERAMIC TILE, DRYWALL AND PLASTERING SUBCONTRACTORS' OVERHEAD AND PROFIT										
DRYWALL PATCHING	PATCH HOLES AND/OR CRACKS IN DRYWALL	EA PLUS	--	120.00	120.00	180.00					.400
	EA = EACH JOB	SF	.20	1.30	1.50	2.25					.401
CORRECT NAIL POPS	CORRECT NAIL POPS IN EXISTING WALL, **NO** PAINT TOUCH-UP INCLUDED EA = UP TO ONE DOZEN NAIL POPS	EA	--	18.00	18.00	27.00					.402
PLASTER PATCHING	PATCH HOLES AND/OR CRACKS IN EXISTING PLASTER, INCLUDING LATHING WHERE REQUIRED	EA PLUS	--	190.00	190.00	285.00					.403
	EA = EACH JOB	SF	.30	1.30	1.60	2.40					.404
PATCHING	PATCH CERAMIC TILE WALL, INCLUDING REMOVING DEFECTIVE OR BROKEN TILES	EA PLUS	20.00	46.00	66.00	99.00					.405
	EA = EACH JOB	SF	4.30	9.00	13.30	19.95					.406
	PATCH CERAMIC TILE COVE, BASE OR BULLNOSE	LF	3.77	5.00	8.77	13.16					.407
	COVER EXISTING CERAMIC TILE WALL WITH URETHANE FINISH	SF	2.00	3.50	5.50	8.25					.408

18

SPECIFICATIONS		UNIT	JOB COST			PRICE	LOCAL AREA MODIFICATION				DATA BASE ITEM NO.
			MATLS	LABOR	TOTAL		MATLS	LABOR	TOTAL	PRICE	
GYPSUM DRYWALL ON NEW CEILING	UP TO 300 SF CEILING, APPLIED TO JOISTS OR FURRING, NAILED, TAPED, FINISHED AND SANDED, 3 COATS										
	3/8"	EA	--	100.00	100.00	150.00					.000
	PLUS	SF	.18	.37	.55	.83					.001
	1/2"	EA	--	100.00	100.00	150.00					.000
	PLUS	SF	.19	.39	.58	.87					.002
	1/2" FIRECODE	EA	--	100.00	100.00	150.00					.000
	PLUS	SF	.21	.39	.60	.90					.003
	5/8" FIRECODE	EA	--	100.00	100.00	150.00					.000
	PLUS	SF	.23	.44	.67	1.01					.004
	1/2" MOISTURE RESISTANT	EA	--	100.00	100.00	150.00					.000
EA = EACH JOB (**NOT** EACH SHEET)	PLUS	SF	.23	.39	.62	.93					.005
	SAME AS ABOVE, WHEN TOTAL JOB IS OVER 300 SF										
	3/8"	SF	.18	.70	.88	1.32					.006
	1/2"	SF	.19	.72	.91	1.37					.007
	1/2" FIRECODE	SF	.21	.72	.93	1.40					.008
	5/8" FIRECODE	SF	.23	.78	1.01	1.52					.009
	1/2" MOISTURE RESISTANT	SF	.23	.72	.95	1.43					.010
	FOR CEILINGS OVER 8'-0" HIGH **ADD**	SF	--	.11	.11	.17					.011
NAILED ONLY	NAILED ONLY, APPLIED TO JOISTS OR FURRING										
	3/8"	SF	.16	.20	.36	.54					.012
	1/2"	SF	.17	.22	.39	.59					.013
	5/8" FIRECODE	SF	.21	.27	.48	.72					.014
	FOR CEILINGS OVER 8'-0" HIGH **ADD**	SF	--	.05	.05	.08					.015
TAPE AND FINISH ONLY	TAPE JOINTS, FINISH AND SAND ONLY										
	UP TO 300 SF CEILING	EA	4.30	57.70	62.00	93.00					.016
	OVER 300 SF CEILING	SF	.03	.50	.53	.80					.017
	FOR CEILINGS OVER 8'-0" HIGH **ADD**	SF	--	.07	.07	.11					.018
	1/2" BLUEBOARD	SF	.19	.33	.52	.78					.019
	SKIM COAT PLASTER	SF	.09	.44	.53	.80					.020

19

19. DRYWALL CEILING, OVER EXISTING

SPECIFICATIONS		UNIT	JOB COST			PRICE	LOCAL AREA MODIFICATION				DATA BASE ITEM NO.
			MATLS	LABOR	TOTAL		MATLS	LABOR	TOTAL	PRICE	
GYPSUM DRYWALL OVER EXISTING CEILING	UP TO 300 SF CEILING, INSTALL DRYWALL OVER EXISTING PLASTERED CEILING WITH NAILS OF REQUIRED LENGTH THROUGH EXISTING PLASTER & LATH & INTO JOISTS, TAPED, FINISHED AND SANDED, 3 COATS										
	3/8"	EA PLUS	--	100.00	100.00	150.00					.021
		SF	.19	.56	.75	1.13					.022
	1/2"	EA PLUS	--	100.00	100.00	150.00					.021
		SF	.20	.60	.80	1.20					.023
	5/8" FIRECODE	EA PLUS	--	100.00	100.00	150.00					.021
	EA = EACH JOB (NOT EACH SHEET)	SF	.24	.68	.92	1.38					.024
	SAME AS ABOVE, WHEN TOTAL JOB IS OVER 300 SF										
	3/8"	SF	.19	.89	1.08	1.62					.025
	1/2"	SF	.20	.93	1.13	1.70					.026
	5/8" FIRECODE	SF	.24	1.02	1.26	1.89					.027
	FOR CEILINGS OVER 8'-0" HIGH **ADD**	SF	--	.11	.11	.17					.028
TEXTURE SPACKLE FINISH	SPRAY FINISHED DRYWALL WITH SPACKLE OR TEXTURE SPRAY, UP TO 300 SF OF CEILING	EA PLUS	--	42.00	42.00	63.00					.029
		SF	.16	.17	.33	.50					.030
	SAME AS ABOVE WHEN JOB IS OVER 300 SF	SF	.16	.30	.46	.69					.031
CORRECT NAIL POPS	CORRECT NAIL POPS IN EXISTING CEILING. NO PAINT TOUCH-UP INCLUDED EA = DOZEN NAIL POPS	EA	--	18.00	18.00	27.00					.032
DRYWALL PATCHING	PATCH HOLES AND/OR CRACKS IN DRYWALL EA = EACH JOB	EA PLUS	--	120.00	120.00	180.00					.033
		SF	.20	1.30	1.50	2.25					.034

19

SPECIFICATIONS		UNIT	JOB COST			PRICE	LOCAL AREA MODIFICATION				DATA BASE ITEM NO.
			MATLS	LABOR	TOTAL		MATLS	LABOR	TOTAL	PRICE	
GYPSUM LATH	GYPSUM LATH NAILED TO WOOD STUDS, INCLUDING CORNERITE										
	3/8"	SF	.29	.27	.56	.84					.100
	1/2"	SF	.30	.29	.59	.89					.101
METAL LATH	3.4 DIAMOND MESH NAILED TO JOISTS OR FURRING	SF	.44	.24	.68	1.02					.102
WOOD LATH	3/8" X 1-1/2" X 48" WOOD LATH NAILED TO JOISTS (3/8" SPACED)	SF	.77	.32	1.09	1.64					.103

19

19. PLASTER CEILING

SPECIFICATIONS		UNIT	JOB COST			PRICE	LOCAL AREA MODIFICATION				DATA BASE ITEM NO.
			MATLS	LABOR	TOTAL		MATLS	LABOR	TOTAL	PRICE	
PLASTER	GYPSUM PLASTER OVER EXISTING GYPSUM LATH										
	2 COATS	SF	.35	.95	1.30	1.95					.104
	3 COATS	SF	.38	1.33	1.71	2.57					.105
	PLASTER, 3 COATS										
	OVER METAL LATH	SF	.61	1.26	1.87	2.81					.106
	OVER WOOD LATH	SF	.41	1.36	1.77	2.66					.107
PATCHING	PATCH HOLES AND/OR CRACKS IN EXISTING PLASTER, INCLUDING LATHING WHERE REQUIRED	EA PLUS	—	190.00	190.00	285.00					.108
	EA = EACH JOB	SF	.30	1.30	1.60	2.40					.109
	PATCH CEILING WITH DRY-WALL OR PLASTER WHERE WALL BELOW HAS BEEN REMOVED	EA PLUS	—	20.00	20.00	30.00					.110
	EA = EACH JOB	LF	.30	2.70	3.00	4.50					.111
	LF = LENGTH OF WALL REMOVED										

19

SPECIFICATIONS	UNIT	JOB COST			PRICE	LOCAL AREA MODIFICATION				DATA BASE ITEM NO.
		MATLS	LABOR	TOTAL		MATLS	LABOR	TOTAL	PRICE	
12" X 12" OR 12" X 24" TILE OVER EXISTING CEILING OR FURRING STRIPS WITH STAPLES OR ADHESIVE										
Tile ***SF Retail Cost***										
.35	SF	.38	.56	.94	1.41					.200
.40	SF	.44	.56	1.00	1.50					.201
.45	SF	.49	.56	1.05	1.58					.202
.50	SF	.54	.56	1.10	1.65					.203
.55	SF	.59	.56	1.15	1.73					.204
.60	SF	.65	.56	1.21	1.82					.205
.70	SF	.75	.56	1.31	1.97					.206
.80	SF	.86	.56	1.42	2.13					.207
FURRING INSTALL 1 X 3 FURRING STRIPS 12" O.C., READY FOR INSTALLATION OF CEILING TILE	SF	.17	.60	.77	1.16					.208
IF FURRING NAILED THROUGH EXISTING PLASTERED CEILING	SF	.17	.53	.70	1.05					.209
SUSPEND 1 X 3 FURRING STRIPS MORE THAN 4" BELOW JOISTS	SF	.32	1.47	1.79	2.69					.210
GRID SYSTEM SUSPEND CEILING WITH STEEL GRID SYSTEM USING 24" X 48" SECTIONS, INCLUDING ENTIRE GRID SYSTEM AND CEILING PANELS										
Ceiling Panels ***SF Retail Cost***										
.50	SF	1.15	1.18	2.33	3.50					.211
.55	SF	1.20	1.18	2.38	3.57					.212
.60	SF	1.25	1.18	2.43	3.65					.213
.65	SF	1.30	1.18	2.48	3.72					.214
.70	SF	1.35	1.18	2.53	3.80					.215
.75	SF	1.40	1.18	2.58	3.87					.216
.80	SF	1.45	1.18	2.63	3.95					.217

19

20. INTERIOR MOULDINGS

SPECIFICATIONS		UNIT	JOB COST			PRICE	LOCAL AREA MODIFICATION				DATA BASE ITEM NO.
			MATLS	LABOR	TOTAL		MATLS	LABOR	TOTAL	PRICE	
BASE	PINE CLAM, RANCH OR COLONIAL BASE										
	2"	LF	.61	.51	1.12	1.68					.000
	2-1/4"	LF	.79	.53	1.32	1.98					.001
	3-1/4"	LF	1.09	.56	1.65	2.48					.002
	4-1/4"	LF	1.58	.58	2.16	3.24					.003
	5-1/4"	LF	1.94	.65	2.59	3.89					.004
	8"	LF	3.41	1.08	4.49	6.74					.005
	OAK BASE, 3-1/4"	LF	2.17	1.03	3.20	4.80					.006
SHOE MOULD	1/2" X 3/4" HARDWOOD	LF	.79	.46	1.25	1.88					.007
	PINE	LF	.34	.39	.73	1.10					.008
3-PIECE BASE	• TOP MOULD, 11/16" X 1-1/8" • OAK SHOE 1/2" X 3/4" • BASE 11/16" X 3-1/2" LF = ALL THREE PIECES	LF	2.22	1.40	3.62	5.43					.009
CEILING MOULDING	BED, COVE, CROWN OR PIC- TURE MOULDS										
	3/4"	LF	.42	.75	1.17	1.76					.010
	1-5/8"	LF	.73	.79	1.52	2.28					.011
	2-1/4"	LF	.98	.82	1.80	2.70					.012
	3-1/4"	LF	1.55	.97	2.52	3.78					.013
	4-1/4"	LF	2.00	.88	2.88	4.32					.014
	5-1/4"	LF	2.48	.91	3.39	5.09					.015

20°

SPECIFICATIONS		UNIT	JOB COST			PRICE	LOCAL AREA MODIFICATION				DATA BASE ITEM NO.
			MATLS	LABOR	TOTAL		MATLS	LABOR	TOTAL	PRICE	
3-MEMBER WOOD CORNICE	INSTALLED ON PERIMETER OF ROOM OVER TRUE WALLS 5/8 X 1-3/4" AND 3/4" X 4-3/4" AND 5/8" X 1-3/4" LF = INCLUDES ALL THREE PIECES	LF	2.90	3.13	6.03	9.05					.016
CHAIR RAIL	ONE PIECE										
	5/8" X 2-1/2"	LF	1.24	.68	1.92	2.88					.017
	5/8" X 3-1/2"	LF	1.80	.68	2.48	3.72					.018
	TWO-PIECE ROUND-EDGE CHAIR RAIL WITH CAP TRIM	LF	2.37	1.22	3.59	5.39					.019
PANEL STRIPS	PANEL STRIPS NAILED TO EXISTING PLASTERED DRY-WALL OR PLYWOOD WALL FOR PANEL EFFECT, ALL JOINTS MITERED										
	5/8" X 7/8"	LF	.57	.62	1.19	1.79					.020
	1-1/16" X 2-1/4"	LF	.74	.88	1.62	2.43					.021

20

20. SPECIAL MOULDINGS AND PLASTER MOULD

SPECIFICATIONS		UNIT	JOB COST			PRICE	LOCAL AREA MODIFICATION				DATA BASE ITEM NO.
			MATLS	LABOR	TOTAL		MATLS	LABOR	TOTAL	PRICE	
SPECIAL WOOD MOULD-INGS	CUSTOM MOULDINGS MILLED TO ORDER IN SMALL QUAN-TITIES. MATERIALS COSTS INCLUDE SETTING UP AND SHAPING IN SHOP, & LABOR COST IS FOR INSTALLATION ON JOB	EA PLUS	76.00	--	76.00	114.00					.100
	3/4" X 1-1/8"	LF	.74	.64	1.38	2.07					.101
	2-1/4"	LF	1.03	.75	1.78	2.67					.102
	3-1/2"	LF	1.21	.90	2.11	3.17					.103
	4-1/2"	LF	1.50	.95	2.45	3.68					.104
	5-1/2"	LF	1.58	1.08	2.66	3.99					.105
	FOR SPECIAL MOULDINGS REQUIRING NEW BIT **ADD**	EA	56.00	--	56.00	84.00					.106
PLASTER MOULD	ORNAMENTAL PLASTER MOULDING, STOCK OR SCULPTED. INSTALLED ON JOB BY ARCHITECTURAL SCULPTORS										
	Projection Depth										
	2" 4"	LF	.40	7.80	8.20	12.30					.107
	2" 7"	LF	.40	8.35	8.75	13.13					.108
	3" 4"	LF	.40	8.55	8.95	13.43					.109
	3" 8"	LF	.40	9.00	9.40	14.10					.110
	4" 5"	LF	.40	11.05	11.45	17.18					.111
	5" 7"	LF	.40	18.50	18.90	28.35					.112
	SAME AS ABOVE, ON CURVED WALLS **ADD**	LF	--	--	25%	25%					.113

20

SPECIFICATIONS	UNIT	JOB COST			PRICE	LOCAL AREA MODIFICATION				DATA BASE ITEM NO.
		MATLS	LABOR	TOTAL		MATLS	LABOR	TOTAL	PRICE	
FIREPLACE MANTEL — PINE WITH PLAIN OR FLUTED SIDES, PRE-ASSEMBLED										
OVERALL WIDTH UP TO 72", SHELF SIZE UP TO 12", WOOD OPENING UP TO 53" IN WIDTH										
ECONOMY	EA	162.00	66.00	228.00	342.00					.114
MEDIUM	EA	360.00	88.00	448.00	672.00					.115
PREMIUM	EA	670.00	114.00	784.00	1,176.00					.116
PLAIN PINE SURROUND MOULDING WITH NO SHELF, SURROUND MOULDING UP TO 1-5/8" X 5-3/4"	EA	67.00	39.00	106.00	159.00					.117
CHINA CORNER CASE — PINE, PRE-ASSEMBLED, OVERALL HEIGHT TO 7-4, OVERALL WIDTH TO 2-6										
ECONOMY	EA	252.00	52.00	304.00	456.00					.118
MEDIUM	EA	308.00	52.00	360.00	540.00					.119
PREMIUM	EA	392.00	52.00	444.00	666.00					.120
CASED OPENING — JAMBS AND TWO SIDES OF 2-1/2" DOOR TRIM										
3-0 X 6-8	EA	37.00	39.00	76.00	114.00					.121
6-0 X 6-8	EA	44.00	42.00	86.00	129.00					.122
TRIM EXISTING DOOR OPENING — ONE SIDE ONLY, 2-1/2" DOOR TRIM ON EXISTING JAMBS										
3-0 X 6-8	EA	13.00	15.00	28.00	42.00					.123
6-0 X 6-8	EA	14.00	16.00	30.00	45.00					.124
TRIM EXISTING WINDOW — INTERIOR TRIM ONLY • 2-1/2" WINDOW TRIM • APRON • STOOL CAP • WINDOW STOPS										
Window Size										
2-0 X 3-2	EA	15.00	27.00	42.00	63.00					.125
3-0 X 5-2	EA	21.00	29.00	50.00	75.00					.126

20

20. BOOKSHELVES

SPECIFICATIONS		UNIT	JOB COST			PRICE	LOCAL AREA MODIFICATION				DATA BASE ITEM NO.
			MATLS	LABOR	TOTAL		MATLS	LABOR	TOTAL	PRICE	
BUILT-IN BOOKCASE	• NAILED AND GLUED ENDS • VERTICAL SUPPORTS 30" O.C. AND AT EACH END • KICKBOARD • TOP RAIL • SHELVES SPACED 10" O.C. VERTICALLY MEASURED • BUILD AND INSTALL SF = FRONT SQUARE FOOTAGE *TIGHT KNOT PINE SHELVING*										
	8"	SF	2.53	2.48	5.01	7.52					.200
	10"	SF	3.17	2.54	5.71	8.57					.201
	12"	SF	3.82	2.61	6.43	9.65					.202
	CLEAR WHITE PINE										
	8"	SF	7.57	2.59	10.16	15.24					.203
	10"	SF	9.51	2.65	12.16	18.24					.204
	12"	SF	11.47	2.73	14.20	21.30					.205
	OAK										
	8"	SF	8.75	2.70	11.45	17.18					.206
	10"	SF	11.03	2.78	13.81	20.72					.207
	12"	SF	13.31	2.85	16.16	24.24					.208
	BIRCH PLYWOOD (3/4")										
	8"	SF	3.42	3.22	6.64	9.96					.209
	10"	SF	4.26	3.43	7.69	11.54					.210
	12"	SF	5.11	3.64	8.75	13.13					.211
	16"	SF	6.84	4.05	10.89	16.34					.212
SHELF MOULDING	MOULDING AT SHELF EDGE	LF	.39	.88	1.27	1.91					.213
STILES	1 X 2 CWP STILES ALONG VERTICAL SUPPORTS OF BOOKCASE	LF	.46	.50	.96	1.44					.214
BACKING	1/4" PLYWOOD OR HARD-BOARD	SF	.84	.56	1.40	2.10					.215

20

SPECIFICATIONS		UNIT	JOB COST			PRICE	LOCAL AREA MODIFICATION				DATA BASE ITEM NO.
			MATLS	LABOR	TOTAL		MATLS	LABOR	TOTAL	PRICE	
PIN TYPE SHELF SUPPORTS	• DRILL 1/4" HOLES IN SIDES OF VERTICAL SUPPORTS • INSERT METAL PIN TYPE SHELF SUPPORTS • INSTALL SHELVES ON SHELF SUPPORTS WITHOUT NAILING **ADD**	SF	.42	--	.42	.63					.216
ADJUSTABLE STANDARDS AND 3/4" SUPPORTS	• FASTEN TWO 1/2" ADJUSTABLE STANDARDS ALONG SIDES OF VERTICAL SUPPORTS • INSERT 3/4" SUPPORTS IN SLOTS IN ADJUSTABLE STANDARDS • INSTALL SHELVES ON SHELF SUPPORTS WITHOUT NAILING **ADD**	SF	.78	--	.78	1.17					.217
OPEN SHELVES ON BRACKETS	• FASTEN SLOTTED STANDARDS TO EXISTING WALL • INSERT SHELF BRACKETS UP TO 12" LONG INTO SLOTTED STANDARDS • INSTALL SHELVES ON BRACKETS WITHOUT NAILING • **NO** VERTICAL WOOD SUPPORTS										
	#2 PINE 10"	LF	3.42	2.39	5.81	8.72					.218
	12"	LF	3.76	2.73	6.49	9.74					.219
	CLEAR WHITE PINE 10"	LF	6.12	2.39	8.51	12.77					.220
	12"	LF	6.89	2.73	9.62	14.43					.221
	OAK 10"	LF	5.79	2.39	8.18	12.27					.222
	12"	LF	6.54	2.73	9.27	13.91					.223
	BIRCH PLYWOOD 10"	LF	4.20	2.79	6.99	10.49					.224
	12"	LF	4.73	2.90	7.63	11.45					.225
	LF = TOTAL LENGTH OF ALL SHELVES										

20

20. CLOSET TRIM

SPECIFICATIONS		UNIT	JOB COST			PRICE	LOCAL AREA MODIFICATION				DATA BASE ITEM NO.
			MATLS	LABOR	TOTAL		MATLS	LABOR	TOTAL	PRICE	
CLOTHES CLOSET	24" TO 30" DEEP, FULLY TRIMMED AS FOLLOWS: #2 PINE BASE AND SHOE #2 PINE HOOKSTRIP 12" PARTICLEBOARD SHELF 1-3/8" CLOTHES POLE AND CLOTHES POLE SOCKETS **NO** FRAMING, WALL COVERING, DOOR OR DOOR TRIM LF = WIDTH OF CLOSET	LF	9.18	6.60	15.78	23.67					.300
WALK-IN CLOSET	SAME AS ABOVE, ADDING ONE POLE AND ONE SHELF	SF	2.25	1.92	4.17	6.26					.301
CLOSET SHELVES	ADDITIONAL STORAGE SHELVES IN ABOVE CLOSETS, #2 PINE OR PLYWOOD SHELVES SET ON WOOD CLEATS										
	12"	LF	2.33	.90	3.23	4.85					.302
	16"	LF	2.79	1.08	3.87	5.81					.303
LINEN CLOSET	24" DEEP WITH 24" PLYWOOD OR PARTICLEBOARD SHELVES 12" O.C., VERTICALLY MEASURED, #2 PINE BASE AND #2 PINE SHELF CLEATS	LF	45.54	15.65	61.19	91.79					.304
BASEMENT CLOSET	BUILD CLOSET UNDER EXISTING STAIRWAY, 36" X 36" • FRAME DOORWAY • TRIM DOOR INSIDE & OUT • CLOTHES CLOSET TRIM • DRYWALL, FINISHED	EA	240.00	162.00	402.00	603.00					.305

20

SPECIFICATIONS		UNIT	JOB COST			PRICE	LOCAL AREA MODIFICATION				DATA BASE ITEM NO.
			MATLS	LABOR	TOTAL		MATLS	LABOR	TOTAL	PRICE	
ATTIC STAIRWAY	BETWEEN TWO WALLS, HOUSED OUT PINE STRINGERS, 1-1/8" PINE TREADS, 3/4" PINE RISERS, FIR HANDRAIL AND TWO METAL HANDRAIL BRACKETS, SHOP BUILT AND DELIVERED TO JOB BY STAIRBUILDER	EA	375.00	95.00	470.00	705.00					.306
	SAME AS ABOVE WITH RIGHT ANGLE TURN AT TOP OR BOTTOM	EA	400.00	140.00	540.00	810.00					.307
CELLAR STAIRWAY	STRINGERS AND TREADS OF 2 X 10 FIR, OPEN RISERS, 2 X 4 HANDRAIL	EA	167.00	211.00	378.00	567.00					.308
MAIN STAIRWAY	STARTING NEWEL, BIRCH HANDRAIL WITH EASEMENT AND NEWEL CAP, PINE BALUSTERS, OAK TREADS AND RISERS, SHOP BUILT AND INSTALLED ON JOB BY STAIRBUILDER										
	ONE SIDE OPEN	EA	1461.00	245.00	1,706.00	2,559.00					.309
	TWO SIDES OPEN	EA	1620.00	380.00	2,000.00	3,000.00					.310
	SAME SPECIFICATIONS AS ABOVE WITH PLATFORM AND RIGHT ANGLE TURN AT TOP OR BOTTOM **ADD**	EA	176.00	90.00	266.00	399.00					.311
WROUGHT IRON HANDRAIL	SUBSTITUTE WROUGHT IRON HANDRAIL IN ABOVE STAIRWAYS **ADD**										
	ONE SIDE	EA	74.00	70.00	144.00	216.00					.312
	TWO SIDES	EA	140.00	90.00	230.00	345.00					.313
WOOD HANDRAIL	2 X 2 PINE OR FIR HANDRAIL WITH STANDARD HARDWARE	LF	3.05	2.05	5.10	7.65					.328

20

20. SPIRAL/FOLDING STAIRS

SPECIFICATIONS		UNIT	JOB COST			PRICE	LOCAL AREA MODIFICATION				DATA BASE ITEM NO.
			MATLS	LABOR	TOTAL		MATLS	LABOR	TOTAL	PRICE	
SPIRAL STAIRWAY, WOOD	HARDWOOD TREADS AND PLATFORM, LAMINATED HANDRAIL AND CENTER POLE, BALUSTERS, UP TO 9-0 FLOOR TO FLOOR										
	Diameter										
	4-0	EA	2558.00	338.00	2,896.00	4,344.00					.314
	4-6	EA	2795.00	357.00	3,152.00	4,728.00					.315
	SAME AS ABOVE, 9-0 TO 11-0 FLOOR TO FLOOR										
	Diameter										
	4-0	EA	3109.00	447.00	3,556.00	5,334.00					.316
	4-6	EA	3316.00	462.00	3,778.00	5,667.00					.317
SPIRAL STAIRWAY, METAL	UNPAINTED METAL STAIR-WAY										
	Diameter										
	5-0	EA	902.00	196.00	1,098.00	1,647.00					.318
	6-0	EA	1026.00	196.00	1,222.00	1,833.00					.319
FOLDING STAIRWAY, WOOD	"DISAPPEARING STAIRWAY", WOOD, UP TO 10-0 FLOOR TO FLOOR X 26" WIDE, IN EXISTING FRAMED OPENING										
	ECONOMY	EA	119.00	95.00	214.00	321.00					.320
	MEDIUM	EA	207.00	95.00	302.00	453.00					.321
	PREMIUM	EA	313.00	99.00	412.00	618.00					.322
	OPEN CEILING AND HEAD OFF TO INSTALL FOLDING STAIRS, CUTTING ONE CEILING JOIST	EA	19.00	63.00	82.00	123.00					.323

20

SPECIFICATIONS		UNIT	JOB COST			PRICE	LOCAL AREA MODIFICATION				DATA BASE ITEM NO.
			MATLS	LABOR	TOTAL		MATLS	LABOR	TOTAL	PRICE	
STAIRS REPAIR	REPLACE BALUSTERS IN EXISTING STAIRWAY FROM FLOOR TO FLOOR, OPEN ONE SIDE, TWO BALUSTERS PER RISER										
	STOCK	PER RIS'R	10.00	13.00	23.00	34.50					.324
	SPECIAL MILLED	PER RIS'R	14.00	15.50	29.50	44.25					.325
	INSTALL NEW HANDRAIL, ONE NEWEL POST AND BALUSTERS ON EXISTING STAIRWAY, TWO BALUSTERS PER RISER	EA PLUS	53.00	45.00	98.00	147.00					.326
		PER RIS'R	14.00	15.00	29.00	43.50					.327
	REPLACE PINE TREADS ON BASEMENT STAIRWAY -- TREADS ONLY, **NO** RISERS EA = PER TREAD	EA	4.50	7.50	12.00	18.00					.329
RADIATOR ENCLO-SURE	JOB-BUILT WITH 1 X 3 PINE STILES, BOTTOM RAIL RAISED FROM FLOOR ON FRONT AND SIDES, UP TO 12" DEEP SF = FRONT	SF	7.52	6.46	13.98	20.97					.330
BUILT-UP FALSE BEAM	INSTALL FALSE WOOD BOX BEAM AGAINST STRAIGHT CEILING, 2 X 4 BASE NAILER WITH 1 X 4 FACE AND 1 X 6 SIDES, 3/4" QUARTER ROUND AT CEILING INTERSECTION										
	CWP	LF	3.50	3.89	7.39	11.09					.331
	#2 PINE	LF	1.80	3.89	5.69	8.54					.332
	CLEAR RW	LF	2.15	3.89	6.04	9.06					.333

20

21. KITCHEN CABINETS, ECONOMY

SPECIFICATIONS		UNIT	JOB COST			PRICE	LOCAL AREA MODIFICATION				DATA BASE ITEM NO.
			MATLS	LABOR	TOTAL		MATLS	LABOR	TOTAL	PRICE	
BASE CABINETS	INSTALL PREFINISHED OR UNFINISHED KITCHEN CABINETS ON PLUMB & STRAIGHT WALLS. STAINING AND PAINTING **NOT** INCLUDED BASIC BASE CABINET INSTALLATION INCLUDES THE FOLLOWING • ONE 36" WIDE, 2 DOOR AND 2 DRAWER FRONTS, SINK BASE CABINET @ $95.00 EACH RETAIL • ONE 24" WIDE, 4 DRAWER DRAWER BASE CABINET • 1 AND 2 DOOR, 1 AND 2 DRAWER BASE CABINETS AS REQUIRED LF = TOTAL OVERALL FRONT WIDTH OF ALL BASE CABINETS	LF	61.00	15.00	76.00	114.00					.000
	ADDITIONAL BASE CABINETS AS REQUIRED EA = EACH CABINET										
DRAWER BASE	WITH 4 DRAWERS 15"	EA	141.00	29.00	170.00	255.00					.002
	18"	EA	143.00	29.00	172.00	258.00					.003
	21"	EA	147.00	29.00	176.00	264.00					.004
	24"	EA	153.00	29.00	182.00	273.00					.005
SINK OR BURNER BASE	30"	EA	99.00	29.00	128.00	192.00					.006
	36"	EA	105.00	29.00	134.00	201.00					.007
	42"	EA	115.00	29.00	144.00	216.00					.008
SINK OR BURNER FRONT	FRONT AND FLOOR (**NO** SIDES) 30"	EA	85.00	29.00	114.00	171.00					.009
	36"	EA	95.00	29.00	124.00	186.00					.010
	42"	EA	99.00	29.00	128.00	192.00					.011
	48"	EA	103.00	29.00	132.00	198.00					.012
ISLAND BASE	WITH DOORS 24"	EA	158.00	34.00	192.00	288.00					.013
	2 SIDES AND 36"	EA	208.00	34.00	242.00	363.00					.014
	DRAWERS 1 SIDE 42"	EA	232.00	34.00	266.00	399.00					.015
	48"	EA	250.00	34.00	284.00	426.00					.016
CORNER BASE	(LAZY SUSAN) WITH 36" ALONG EACH WALL	EA	147.00	35.00	182.00	273.00					.017
BROOM CABINET	**Width Height Depth** 18" 7-0 12"	EA	157.00	45.00	202.00	303.00					.018
	18" 7-0 24"	EA	175.00	45.00	220.00	330.00					.019
OVEN CABINET	**Width Height Depth** 24" 7-0 24"	EA	267.00	45.00	312.00	468.00					.020
	27" 7-0 24"	EA	287.00	45.00	332.00	498.00					.021

21

SPECIFICATIONS		UNIT	JOB COST			PRICE	LOCAL AREA MODIFICATION				DATA BASE ITEM NO.
			MATLS	LABOR	TOTAL		MATLS	LABOR	TOTAL	PRICE	
WALL CABINETS	THE BASIC WALL CABINET INSTALLATION INCLUDES THE FOLLOWING: • TWO "ABOVE APPLIANCE" WALL CABINETS 15" HIGH AND 36" WIDE @ $76.00 EACH RETAIL • 1 AND 2 DOOR WALL CABINETS 30" HIGH AS REQUIRED LF = TOTAL OVERALL FRONT WIDTH OF ALL WALL CABINETS	LF	36.00	16.00	52.00	78.00					.001
	ADDITIONAL WALL CABINETS AS REQUIRED EA = EACH CABINET										
UPPER ISLAND	WITH DOORS TWO SIDES, 24" HIGH 36"	EA	154.00	32.00	186.00	279.00					.022
	SAME AS ABOVE, 30" HIGH 24"	EA	134.00	32.00	166.00	249.00					.023
	27"	EA	142.00	32.00	174.00	261.00					.024
	30"	EA	148.00	32.00	180.00	270.00					.025
	33"	EA	158.00	32.00	190.00	285.00					.026
	36"	EA	164.00	32.00	196.00	294.00					.027
	42"	EA	178.00	32.00	210.00	315.00					.028
	48"	EA	196.00	32.00	228.00	342.00					.029
CORNER WALL	(LAZY SUSAN) WITH 24" ALONG EACH WALL	EA	126.00	32.00	158.00	237.00					.030
MICRO-WAVE WALL	MICROWAVE WALL CABINET, 24"	EA	216.00	32.00	248.00	372.00					.031

21

21. KITCHEN CABINETS, BUILDER

SPECIFICATIONS		UNIT	JOB COST			PRICE	LOCAL AREA MODIFICATION				DATA BASE ITEM NO.
			MATLS	LABOR	TOTAL		MATLS	LABOR	TOTAL	PRICE	
BASE CABINETS	INSTALL PREFINISHED OR UNFINISHED KITCHEN CABINETS ON PLUMB & STRAIGHT WALLS. STAINING AND PAINTING **NOT** INCLUDED BASIC BASE CABINET INSTALLATION INCLUDES THE FOLLOWING • ONE 36" WIDE, 2 DOOR AND 2 DRAWER FRONTS, SINK BASE CABINET @ $160.00 EACH RETAIL • ONE 24" WIDE, 4 DRAWER DRAWER BASE CABINET • 1 AND 2 DOOR, 1 AND 2 DRAWER BASE CABINETS AS REQUIRED LF = TOTAL OVERALL FRONT WIDTH OF ALL BASE CABINETS	LF	95.00	15.00	110.00	165.00					.100
	ADDITIONAL BASE CABINETS AS REQUIRED EA = EACH CABINET										
DRAWER BASE	WITH 4 DRAWERS 15"	EA	219.00	29.00	248.00	372.00					.102
	18"	EA	223.00	29.00	252.00	378.00					.103
	21"	EA	227.00	29.00	256.00	384.00					.104
	24"	EA	231.00	29.00	260.00	390.00					.105
SINK OR BURNER BASE	30"	EA	155.00	29.00	184.00	276.00					.106
	36"	EA	161.00	29.00	190.00	285.00					.107
	42"	EA	177.00	29.00	206.00	309.00					.108
SINK OR BURNER FRONT	FRONT AND FLOOR (**NO** SIDES) 30"	EA	131.00	29.00	160.00	240.00					.109
	36"	EA	145.00	29.00	174.00	261.00					.110
	42"	EA	155.00	29.00	184.00	276.00					.111
	48"	EA	161.00	29.00	190.00	285.00					.112
ISLAND BASE	WITH DOORS 24" 2 SIDES AND 36" DRAWERS 1 SIDE 42" 48"	EA	242.00	34.00	276.00	414.00					.113
	36"	EA	338.00	34.00	372.00	558.00					.114
	42"	EA	362.00	34.00	396.00	594.00					.115
	48"	EA	390.00	34.00	424.00	636.00					.116
CORNER BASE	(LAZY SUSAN) WITH 36" ALONG EACH WALL	EA	227.00	35.00	262.00	393.00					.117
BROOM CABINET	Width Height Depth 18" 7-0 12"	EA	241.00	45.00	286.00	429.00					.118
	18" 7-0 24"	EA	273.00	45.00	318.00	477.00					.119
OVEN CABINET	Width Height Depth 24" 7-0 24"	EA	409.00	45.00	454.00	681.00					.120
	27" 7-0 24"	EA	437.00	45.00	482.00	723.00					.121

21

SPECIFICATIONS		UNIT	JOB COST			PRICE	LOCAL AREA MODIFICATION				DATA BASE ITEM NO.
			MATLS	LABOR	TOTAL		MATLS	LABOR	TOTAL	PRICE	
WALL CABINETS	THE BASIC WALL CABINET INSTALLATION INCLUDES THE FOLLOWING: • TWO "ABOVE APPLIANCE" WALL CABINETS 15" HIGH AND 36" WIDE @ $115.00 EACH RETAIL • 1 AND 2 DOOR WALL CABINETS 30" HIGH AS REQUIRED LF = TOTAL OVERALL FRONT WIDTH OF ALL WALL CABINETS	LF	57.00	15.00	72.00	108.00					.101
	ADDITIONAL WALL CABINETS AS REQUIRED EA = EACH CABINET										
UPPER ISLAND	WITH DOORS TWO SIDES, 24" HIGH 36"	EA	232.00	32.00	264.00	396.00					.122
	SAME AS ABOVE, 30" HIGH 24"	EA	208.00	32.00	240.00	360.00					.123
	27"	EA	218.00	32.00	250.00	375.00					.124
	30"	EA	232.00	32.00	264.00	396.00					.125
	33"	EA	242.00	32.00	274.00	411.00					.126
	36"	EA	254.00	32.00	286.00	429.00					.127
	42"	EA	274.00	32.00	306.00	459.00					.128
	48"	EA	300.00	32.00	332.00	498.00					.129
CORNER WALL	(LAZY SUSAN) WITH 24" ALONG EACH WALL	EA	158.00	32.00	190.00	285.00					.130
MICRO-WAVE WALL	MICROWAVE WALL CABINET, 24"	EA	270.00	32.00	302.00	453.00					.131

21

21. KITCHEN CABINETS, PREMIUM

SPECIFICATIONS			UNIT	JOB COST			PRICE	LOCAL AREA MODIFICATION				DATA BASE ITEM NO.
				MATLS	LABOR	TOTAL		MATLS	LABOR	TOTAL	PRICE	
BASE CABINETS	INSTALL PREFINISHED OR UNFINISHED KITCHEN CABINETS ON PLUMB & STRAIGHT WALLS. STAINING AND PAINTING **NOT** INCLUDED BASIC BASE CABINET INSTALLATION INCLUDES THE FOLLOWING • ONE 36" WIDE, 2 DOOR AND 2 DRAWER FRONTS, SINK BASE CABINET @ $230.00 EACH RETAIL • ONE 24" WIDE, 4 DRAWER DRAWER BASE CABINET • 1 AND 2 DOOR, 1 AND 2 DRAWER BASE CABINETS AS REQUIRED LF = TOTAL OVERALL FRONT WIDTH OF ALL BASE CABINETS		LF	132.00	16.00	148.00	222.00					.200
	ADDITIONAL BASE CABINETS AS REQUIRED EA = EACH CABINET											
DRAWER BASE	WITH 4 DRAWERS	15"	EA	314.00	34.00	348.00	522.00					.202
		18"	EA	320.00	34.00	354.00	531.00					.203
		21"	EA	322.00	34.00	356.00	534.00					.204
		24"	EA	334.00	34.00	368.00	552.00					.205
SINK OR BURNER BASE		30"	EA	220.00	34.00	254.00	381.00					.206
		36"	EA	228.00	34.00	262.00	393.00					.207
		42"	EA	254.00	34.00	288.00	432.00					.208
SINK OR BURNER FRONT	FRONT AND FLOOR (**NO** SIDES)	30"	EA	186.00	34.00	220.00	330.00					.209
		36"	EA	208.00	34.00	242.00	363.00					.210
		42"	EA	220.00	34.00	254.00	381.00					.211
		48"	EA	226.00	34.00	260.00	390.00					.212
ISLAND BASE	WITH DOORS 2 SIDES AND DRAWERS 1 SIDE	24"	EA	408.00	34.00	442.00	663.00					.213
		36"	EA	450.00	34.00	484.00	726.00					.214
		42"	EA	490.00	34.00	524.00	786.00					.215
		48"	EA	532.00	34.00	566.00	849.00					.216
CORNER BASE	(LAZY SUSAN) WITH 36" ALONG EACH WALL		EA	320.00	38.00	358.00	537.00					.217
BROOM CABINET	*Width* 18" / *Height* 7-0 / *Depth* 12"		EA	364.00	48.00	412.00	618.00					.218
	Width 18" / *Height* 7-0 / *Depth* 24"		EA	400.00	48.00	448.00	672.00					.219
OVEN CABINET	*Width* 24" / *Height* 7-0 / *Depth* 24"		EA	546.00	48.00	594.00	891.00					.220
	Width 27" / *Height* 7-0 / *Depth* 24"		EA	580.00	48.00	628.00	942.00					.221

21

SPECIFICATIONS		UNIT	JOB COST			PRICE	LOCAL AREA MODIFICATION				DATA BASE ITEM NO.
			MATLS	LABOR	TOTAL		MATLS	LABOR	TOTAL	PRICE	
WALL CABINETS	THE BASIC WALL CABINET INSTALLATION INCLUDES THE FOLLOWING: • TWO "ABOVE APPLIANCE" WALL CABINETS 15" HIGH AND 36" WIDE @ $170.00 EACH RETAIL • 1 AND 2 DOOR WALL CABINETS 30" HIGH AS REQUIRED LF = TOTAL OVERALL FRONT WIDTH OF ALL WALL CABINETS	LF	80.00	16.00	96.00	144.00					.201
UPPER ISLAND	ADDITIONAL WALL CABINETS AS REQUIRED EA = EACH CABINET WITH DOORS TWO SIDES, 24" HIGH 36"	EA	324.00	36.00	360.00	540.00					.222
	SAME AS ABOVE, 30" HIGH 24"	EA	286.00	36.00	322.00	483.00					.223
	27"	EA	300.00	36.00	336.00	504.00					.224
	30"	EA	314.00	36.00	350.00	525.00					.225
	33"	EA	330.00	36.00	366.00	549.00					.226
	36"	EA	352.00	36.00	388.00	582.00					.227
	42"	EA	382.00	36.00	418.00	627.00					.228
	48"	EA	410.00	36.00	446.00	669.00					.229
CORNER WALL	(LAZY SUSAN) WITH 24" ALONG EACH WALL	EA	222.00	36.00	258.00	387.00					.230
MICRO-WAVE WALL	MICROWAVE WALL CABINET, 24"	EA	390.00	36.00	426.00	639.00					.231

21

21. KITCHEN CABINETS, TOP QUALITY

SPECIFICATIONS		UNIT	JOB COST			PRICE	LOCAL AREA MODIFICATION				DATA BASE ITEM NO.
			MATLS	LABOR	TOTAL		MATLS	LABOR	TOTAL	PRICE	
BASE CABINETS	INSTALL PREFINISHED OR UNFINISHED KITCHEN CABINETS ON PLUMB & STRAIGHT WALLS. STAINING AND PAINTING **NOT** INCLUDED BASIC BASE CABINET INSTALLATION INCLUDES THE FOLLOWING • ONE 36" WIDE, 2 DOOR AND 2 DRAWER FRONTS, SINK BASE CABINET @ $300.00 EACH RETAIL • ONE 24" WIDE, 4 DRAWER DRAWER BASE CABINET • 1 AND 2 DOOR, 1 AND 2 DRAWER BASE CABINETS AS REQUIRED LF = TOTAL OVERALL FRONT WIDTH OF ALL BASE CABINETS	LF	170.00	16.00	186.00	279.00					.300
	ADDITIONAL BASE CABINETS AS REQUIRED EA = EACH CABINET										
DRAWER BASE	WITH 4 DRAWERS 15"	EA	400.00	34.00	434.00	651.00					.302
	18"	EA	414.00	34.00	448.00	672.00					.303
	21"	EA	418.00	34.00	452.00	678.00					.304
	24"	EA	434.00	34.00	468.00	702.00					.305
SINK OR BURNER BASE	30"	EA	284.00	34.00	318.00	477.00					.306
	36"	EA	294.00	34.00	328.00	492.00					.307
	42"	EA	328.00	34.00	362.00	543.00					.308
SINK OR BURNER FRONT	FRONT AND FLOOR (**NO** SIDES) 30"	EA	242.00	34.00	276.00	414.00					.309
	36"	EA	272.00	34.00	306.00	459.00					.310
	42"	EA	284.00	34.00	318.00	477.00					.311
	48"	EA	294.00	34.00	328.00	492.00					.312
ISLAND BASE	WITH DOORS 2 SIDES AND DRAWERS 1 SIDE 24"	EA	530.00	34.00	564.00	846.00					.313
	36"	EA	582.00	34.00	616.00	924.00					.314
	42"	EA	636.00	34.00	670.00	1,005.00					.315
	48"	EA	694.00	34.00	728.00	1,092.00					.316
CORNER BASE	(LAZY SUSAN) WITH 36" ALONG EACH WALL	EA	416.00	38.00	454.00	681.00					.317
BROOM CABINET	*Width Height Depth* 18" 7-0 12"	EA	604.00	48.00	652.00	978.00					.318
	18" 7-0 24"	EA	646.00	48.00	694.00	1,041.00					.319
OVEN CABINET	*Width Height Depth* 24" 7-0 24"	EA	708.00	48.00	756.00	1,134.00					.320
	27" 7-0 24"	EA	750.00	48.00	798.00	1,197.00					.321

21

SPECIFICATIONS		UNIT	JOB COST			PRICE	LOCAL AREA MODIFICATION				DATA BASE ITEM NO.
			MATLS	LABOR	TOTAL		MATLS	LABOR	TOTAL	PRICE	
WALL CABINETS	THE BASIC WALL CABINET INSTALLATION INCLUDES THE FOLLOWING: • TWO "ABOVE APPLIANCE" WALL CABINETS 15" HIGH AND 36" WIDE @ $220.00 EACH RETAIL • 1 AND 2 DOOR WALL CABINETS 30" HIGH AS REQUIRED LF = TOTAL OVERALL FRONT WIDTH OF ALL WALL CABINETS	LF	104.00	16.00	120.00	180.00					.301
	ADDITIONAL WALL CABINETS AS REQUIRED EA = EACH CABINET										
UPPER ISLAND	WITH DOORS TWO SIDES, 24" HIGH 36"	EA	420.00	36.00	456.00	684.00					.322
	SAME AS ABOVE, 30" HIGH 24"	EA	368.00	36.00	404.00	606.00					.323
	27"	EA	388.00	36.00	424.00	636.00					.324
	30"	EA	406.00	36.00	442.00	663.00					.325
	33"	EA	430.00	36.00	466.00	699.00					.326
	36"	EA	456.00	36.00	492.00	738.00					.327
	42"	EA	494.00	36.00	530.00	795.00					.328
	48"	EA	530.00	36.00	566.00	849.00					.329
CORNER WALL	(LAZY SUSAN) WITH 24" ALONG EACH WALL	EA	284.00	36.00	320.00	480.00					.330
MICRO-WAVE WALL	MICROWAVE WALL CABINET, 24"	EA	490.00	36.00	526.00	789.00					.331

21

21. KITCHEN FIXTURES & APPLIANCES -- NEW & REPLACEMENT

SPECIFICATIONS	UNIT	JOB COST			PRICE	LOCAL AREA MODIFICATION				DATA BASE ITEM NO.
		MATLS	LABOR	TOTAL		MATLS	LABOR	TOTAL	PRICE	
THE MATERIALS COSTS BELOW INCLUDE ONLY THE FIXTURES, APPLIANCES AND FITTINGS, AND ARE FOR NEW INSTALLATIONS THAT ARE ALREADY ROUGHED IN, OR WHERE EXISTING FIXTURES AND APPLIANCES HAVE BEEN REMOVED. THE LABOR COSTS INCLUDE ONLY THE COST OF SETTING IN PLACE THE FIXTURES AND APPLIANCES BY CARPENTERS ON THE JOB. FOR PLUMBING AND ELECTRICAL ROUGH IN AND INSTALLATION COSTS, SEE SECTIONS 14 AND 16.										
REFRIGERATOR — REFRIGERATOR										
ECONOMY	EA	350.00	46.00	396.00	594.00					.400
BUILDER	EA	650.00	50.00	700.00	1,050.00					.401
PREMIUM	EA	1000.00	60.00	1,060.00	1,590.00					.402
CLOTHES DRYER — ELECTRIC CLOTHES DRYER										
ECONOMY	EA	280.00	36.00	316.00	474.00					.403
BUILDER	EA	410.00	36.00	446.00	669.00					.404
PREMIUM	EA	566.00	36.00	602.00	903.00					.405
ELECTRIC COOKTOP, OVEN AND RANGE — ELECTRIC COOKTOP WITH DUCTED DOWNDRAFT 35" X 20"	EA	250.00	40.00	290.00	435.00					.406
WALL OVEN SINGLE	EA	232.00	30.00	262.00	393.00					.407
DOUBLE	EA	278.00	40.00	318.00	477.00					.408
FREE-STANDING ELECTRIC RANGE AND OVEN										
ECONOMY	EA	320.00	50.00	370.00	555.00					.409
BUILDER	EA	430.00	50.00	480.00	720.00					.410
PREMIUM	EA	700.00	60.00	760.00	1,140.00					.411
MICROWAVE — MICROWAVE OVEN										
ECONOMY	EA	176.00	20.00	196.00	294.00					.412
BUILDER	EA	200.00	20.00	220.00	330.00					.413
PREMIUM	EA	250.00	20.00	270.00	405.00					.414
TRASH COMPACTOR — UNDER-COUNTER COMPACTOR										
ECONOMY	EA	356.00	40.00	396.00	594.00					.415
BUILDER	EA	450.00	40.00	490.00	735.00					.416
PREMIUM	EA	526.00	50.00	576.00	864.00					.417
FOR INSTANT HOT, SEE PAGE 206										

21

SPECIFICATIONS		UNIT	JOB COST			PRICE	LOCAL AREA MODIFICATION				DATA BASE ITEM NO.
			MATLS	LABOR	TOTAL		MATLS	LABOR	TOTAL	PRICE	
BULKHEAD, DROP, SOFFIT	DRYWALL, FROM CEILING TO TOP OF WALL CABINET, INCLUDING FRAMING AS REQUIRED	LF	1.80	9.50	11.30	16.95					.418
COUNTER-TOP	WITH PLASTIC, METAL OR SELF EDGE AND 4" BACK-SPLASH, SHOP-BUILT BY OTHERS										
	ECONOMY	LF	7.00	7.74	14.74	22.11					.419
	BUILDER	LF	14.50	7.74	22.24	33.36					.420
	PREMIUM	LF	30.00	7.74	37.74	56.61					.421
	SOLID SURFACE COUNTER-TOP, 4" BACKSPLASH										
	1/2"	LF	76.00	24.00	100.00	150.00					.422
	3/4"	LF	113.00	29.00	142.00	213.00					.423
CUTTING BOARD	1-1/2" X 18" X 25" MAPLE LAMINATED, INSTALLED IN COUNTERTOP ABOVE	EA	81.00	79.00	160.00	240.00					.424
CERAMIC TILE	CERAMIC TILE COUNTERTOP 4-1/4 " X 4-1/4", 6" X 6", AND 8" X 8", INCLUDES TRIM, SET IN MUD	SF	2.98	9.45	12.43	18.65					.425
	SAME AS ABOVE, SET IN THIN SET OR MASTIC	SF	1.98	6.30	8.28	12.42					.426
KITCHEN CABINET REFACING	REFACE EXISTING KITCHEN CABINETS WITH PLASTIC LAMINATE DONE IN PLACE, INCLUDING NEW DOOR HARDWARE										
	WALL CABINET DOOR	EA	26.00	12.00	38.00	57.00					.427
	BASE CABINET DOOR	EA	32.00	18.00	50.00	75.00					.428
	WALL CABINET FACE	EA	10.00	16.00	26.00	39.00					.429
	BASE CABINET FACE	EA	14.00	18.00	32.00	48.00					.430
	REFACE SIDE OF CABINET WITH PLASTIC LAMINATE										
	WALL CABINET	EA	10.00	14.00	24.00	36.00					.431
	BASE CABINET	EA	14.00	18.00	32.00	48.00					.432

21

22. BATHROOM CABINETS & ACCESSORIES

SPECIFICATIONS		UNIT	JOB COST			PRICE	LOCAL AREA MODIFICATION				DATA BASE ITEM NO.
			MATLS	LABOR	TOTAL		MATLS	LABOR	TOTAL	PRICE	
MEDICINE CABINET	RECESSED CABINET WITH HINGED WINDOW GLASS MIRROR DOOR, OVERALL SIZE 16" X 22"	EA	29.00	21.00	50.00	75.00					.000
	SAME AS ABOVE WITH PLATE GLASS MIRROR DOOR	EA	43.00	21.00	64.00	96.00					.001
	RECESSED CABINET WITH HINGED PLATE GLASS MIRROR DOOR, 2 FLUORESCENT SIDELIGHTS, OVERALL SIZE 22" X 22"	EA	73.00	15.00	88.00	132.00					.002
	RECESSED CABINET WITH OVERHEAD FLUORESCENT FIXTURE, 2 MIRROR DOORS, OVERALL SIZE 27" X 23"	EA	62.00	16.00	78.00	117.00					.003
SURFACE MOUNT CABINET	SURFACE MOUNT CABINET WITH 2 SLIDING MIRROR DOORS, OVERALL SIZE 24" X 16"	EA	46.00	16.00	62.00	93.00					.004
	SAME AS ABOVE WITH OVERHEAD 4-BULB FIXTURE **Overall Size**										
	24" X 23"	EA	69.00	17.00	86.00	129.00					.005
	28" X 23"	EA	71.00	17.00	88.00	132.00					.006
	36" X 23"	EA	85.00	17.00	102.00	153.00					.007
MIRROR	WALL MIRROR 1/4" THICK WITH POLISHED EDGES, ATTACHED TO WALL WITH CLIPS OR ADHESIVE	SF	5.10	4.70	9.80	14.70					.008
CHROME ACCESSORIES	POLISHED CHROME BATH ACCESSORIES: TOWEL BAR, TOOTHBRUSH AND GLASS HOLDER, PAPER HOLDER, SOAP DISH	SET	41.00	23.00	64.00	96.00					.009
SHOWER ROD	CHROME SHOWER CURTAIN ROD, 5'	EA	30.00	10.00	40.00	60.00					.010
GRAB BAR	STAINLESS STEEL GRAB BAR, 1-1/4" DIAMETER, STRAIGHT, WITH ANCHOR PLATES										
	24"	EA	28.00	12.00	40.00	60.00					.011
	36"	EA	30.00	12.00	42.00	63.00					.012
	NOTE: FOR CERAMIC BATH ACCESSORIES, SEE P. 230										

22

SPECIFICATIONS		UNIT	JOB COST			PRICE	LOCAL AREA MODIFICATION				DATA BASE ITEM NO.
			MATLS	LABOR	TOTAL		MATLS	LABOR	TOTAL	PRICE	
FOLDING DOORS	TUB ENCLOSURE, ALUMINUM FRAME, PLASTIC PANELS, ROLLER GLIDES TOP AND BOTTOM, FOLD TO 12" FOR UP TO 66" OPENING, 58" HIGH	EA	110.00	54.00	164.00	246.00					.013
	SHOWER ENCLOSURE, SAME SPECIFICATIONS AS ABOVE, FOR OPENING 30" TO 66" WIDE AND 68" HIGH	EA	97.00	53.00	150.00	225.00					.014
BY-PASS DOOR	TUB ENCLOSURE, ALUMINUM FRAME, 2 PLASTIC PANELS 58" HIGH FOR UP TO 60" OPENING	EA	88.00	54.00	142.00	213.00					.015
	SAME AS ABOVE WITH TEMPERED GLASS	EA	124.00	54.00	178.00	267.00					.016
	SHOWER ENCLOSURE WITH 2 TEMPERED GLASS DOORS 70" HIGH FOR UP TO 48" OPENING	EA	155.00	53.00	208.00	312.00					.017
HINGED DOOR	ALUMINUM FRAME, ONE HINGED TEMPERED GLASS DOOR 64" HIGH FOR 23" TO 28" OPENINGS	EA	118.00	54.00	172.00	258.00					.018
	SAME AS ABOVE WITH ONE STATIONARY PANEL AND ONE HINGED DOOR FOR 48" OPENING	EA	200.00	58.00	258.00	387.00					.019
NEO-ANGLE GLASS SHOWER ENCLOSURE	ALUMINUM FRAME, 24" SIDE PANELS, 27" DOOR, 72" HIGH, OBSCURE GLASS, INSTALLED	EA	195.00	47.00	242.00	363.00					.020
	SAME AS ABOVE, WITH BRIGHT BRASS FINISH	EA	235.00	47.00	282.00	423.00					.021
TUB WALL KIT	TUB SURROUND 3, 4 OR 5 PIECE VINYL INSTEAD OF CERAMIC TILE, NO DOOR INCLUDED										
	ECONOMY	EA	30.00	78.00	108.00	162.00					.022
	BUILDER	EA	100.00	78.00	178.00	267.00					.023
	PREMIUM	EA	250.00	78.00	328.00	492.00					.024

22

22. SHUTTERS, METAL FIREPLACE

SPECIFICATIONS		UNIT	JOB COST			PRICE	LOCAL AREA MODIFICATION				DATA BASE ITEM NO.
			MATLS	LABOR	TOTAL		MATLS	LABOR	TOTAL	PRICE	
DOOR SHUTTERS	EXTERIOR DOOR SHUTTERS WITH STATIONARY SLATS, INSTALLED WITHOUT HINGES										
	WOOD	PAIR	55.00	15.00	70.00	105.00					.100
	PLASTIC	PAIR	55.00	15.00	70.00	105.00					.101
	ALUMINUM	PAIR	61.00	15.00	76.00	114.00					.102
WINDOW SHUTTERS	EXTERIOR WINDOW SHUTTERS WITH STATIONARY SLATS, INSTALLED WITHOUT HINGES										
	WOOD	PAIR	38.00	14.00	52.00	78.00					.103
	PLASTIC	PAIR	38.00	14.00	52.00	78.00					.104
	ALUMINUM	PAIR	46.00	14.00	60.00	90.00					.105
INTERIOR SHUTTERS	INTERIOR MOVABLE SHUTTER INSTALLED (HINGED), FOUR PANELS TO EACH SET, 3/4" PINE, PRE-ASSEMBLED WITH KNOBS, HINGES, GLUED JOINTS										
	Width **Height**										
	25" TO 29" 17" TO 21"	SET	38.00	14.00	52.00	78.00					.106
	33" TO 37"	SET	50.00	14.00	64.00	96.00					.107
	29" TO 33" 17" TO 21"	SET	52.00	14.00	66.00	99.00					.108
	33" TO 37"	SET	66.00	14.00	80.00	120.00					.109
	33" TO 37" 17" TO 21"	SET	56.00	14.00	70.00	105.00					.110
	33" TO 37"	SET	66.00	14.00	80.00	120.00					.111
	37" TO 41" 17" TO 21"	SET	62.00	14.00	76.00	114.00					.112
	33" TO 37"	SET	74.00	14.00	88.00	132.00					.113
ZERO CLEARANCE FIREPLACE	PRE-FABRICATED METAL WOOD BURNING FIREPLACE AND CHIMNEY, INCLUDING EXTERIOR CHIMNEY HOUSING, 4-5 X 5-0, 13 FEET MEASURING FROM FLOOR TO TOP OF CHIMNEY HOUSING	EA	863.00	215.00	1,078.00	1,617.00					.114
	ADDITIONAL CHIMNEY ABOVE 13 FEET	LF	16.00	9.00	25.00	37.50					.114
	COLORED FIREPLACE AND CHIMNEY **ADD**	EA	70.00	--	70.00	105.00					.115

22

SPECIFICATIONS		UNIT	JOB COST			PRICE	LOCAL AREA MODIFICATION				DATA BASE ITEM NO.
			MATLS	LABOR	TOTAL		MATLS	LABOR	TOTAL	PRICE	
BURGLAR ALARM	CONTROL BOX STANDARD	EA	270.00	400.00	670.00	1,005.00					.117
	PREMIUM	EA	430.00	600.00	1,030.00	1,545.00					.118
	SENSOR HARD WIRED (WIRES IN WALLS)	EA	24.00	24.00	48.00	72.00					.119
	EXPOSED WIRES	EA	30.00	18.00	48.00	72.00					.120
	RADIO TRANS-MITTER	EA	58.00	30.00	88.00	132.00					.121
ACCES-SORIES	OUTSIDE HORN	EA	80.00	60.00	140.00	210.00					.122
	EMERGENCY POCKET TRANSMITTER	EA	71.00	11.00	82.00	123.00					.123
	AUTOMATIC MESSAGE TRANSMITTER	EA	480.00	30.00	510.00	765.00					.124
	FREEZE ALERT	EA	75.00	15.00	90.00	135.00					.125
	MOTION DETECTOR	EA	220.00	80.00	300.00	450.00					.126
	REMOTE ALARM	EA	60.00	20.00	80.00	120.00					.127
FIRE EXTIN-GUISHER	INSTALL	EA	50.00	--	50.00	75.00					.128
	RE-TEST	EA	--	20.00	20.00	30.00					.129
SECURITY BARS	SECURITY BARS INSTALLED ON EXISTING WINDOW SF = WINDOW	SF	7.50	4.50	12.00	18.00					.130
	HINGED DOOR BARS WITH DEADBOLT	EA	106.00	90.00	196.00	294.00					.131

22

22. WOOD FENCING

SPECIFICATIONS	UNIT	JOB COST			PRICE	LOCAL AREA MODIFICATION				DATA BASE ITEM NO.
		MATLS	LABOR	TOTAL		MATLS	LABOR	TOTAL	PRICE	
WOOD FENCING COSTS BELOW INCLUDE 4" X 4" POSTS 8'-0" O.C., SET 2" INTO 6" GRAVEL BED AT BOTTOM OF 36" DEEP POST HOLE AND TAMPED DIRT AND GRAVEL TO TOP OF POST HOLE. IF, INSTEAD OF TAMPED DIRT AND GRAVEL, POSTS ARE SET IN GRAVEL BED WITH DRY CONCRETE PLACED ON GRAVEL BED TO TOP OR NEAR TOP OF POST HOLE **ADD** LF = FENCING	LF	.40	.20	.60	.90					.238

BASKET-WEAVE

SPECIFICATIONS	UNIT	MATLS	LABOR	TOTAL	PRICE					DATA BASE ITEM NO.
3/4" X 6" REDWOOD, INCLUDING 4" X 4" LINE POSTS, CORNER AND END POSTS *Height*										
48"	LF	9.19	4.24	13.43	20.15					.200
60"	LF	12.75	5.22	17.97	26.96					.201
SAME AS ABOVE, PLUS 12" CRISS-CROSS ALONG TOP										
48"	LF	11.75	5.22	16.97	25.46					.202
60"	LF	14.44	6.14	20.58	30.87					.203
2 GATEPOSTS AND 42" WIDE GATE										
48"	EA	62.00	48.00	110.00	165.00					.204
60"	EA	72.00	62.00	134.00	201.00					.205

PICKET

SPECIFICATIONS	UNIT	MATLS	LABOR	TOTAL	PRICE					DATA BASE ITEM NO.
1" X 4" CEDAR OR REDWOOD PICKETS 6" O.C., 4" X 4" POSTS, 2-RAIL FENCE										
36"	LF	5.75	3.03	8.78	13.17					.206
48"	LF	6.50	3.03	9.53	14.30					.207
60"	LF	7.95	3.03	10.98	16.47					.208
2 GATEPOSTS AND 42" WIDE GATE										
36"	EA	66.00	30.00	96.00	144.00					.209
48"	EA	67.00	33.00	100.00	150.00					.210
60"	EA	69.00	37.00	106.00	159.00					.211

STOCKADE

SPECIFICATIONS	UNIT	MATLS	LABOR	TOTAL	PRICE					DATA BASE ITEM NO.
1" X 3" CEDAR PICKETS WITH 4" X 4" OR 6" ROUND POSTS, (3) 2" X 3" RAILS										
72"	LF	9.75	4.36	14.11	21.17					.212
2 GATEPOSTS AND 42" WIDE GATE, 72"	EA	102.00	54.00	156.00	234.00					.213

PRIVACY

SPECIFICATIONS	UNIT	MATLS	LABOR	TOTAL	PRICE					DATA BASE ITEM NO.
PRESSURE TREATED 1 X 6, (3) 2 X 4 RAILS, 4 X 4 POSTS, 8' OC										
72"	LF	5.60	4.20	9.80	14.70					.234
2 GATEPOSTS AND 36" WIDE GATE, 72"	EA	72.00	56.00	128.00	192.00					.235

22

SPECIFICATIONS		UNIT	JOB COST			PRICE	LOCAL AREA MODIFICATION				DATA BASE ITEM NO.
			MATLS	LABOR	TOTAL		MATLS	LABOR	TOTAL	PRICE	
SHADOW BOX	OR "ALTERNATING BOARD" PRESSURE TREATED 1 X 6, (3) 2 X 4 RAILS, 4 X 4 POSTS, 8' OC										
	72"	LF	6.50	5.01	11.51	17.27					.236
	2 GATEPOSTS AND 36" WIDE GATE, 72"	EA	84.00	68.00	152.00	228.00					.237
SPLIT RAIL	SPLIT RAIL FENCE WITH 10 FEET BETWEEN LINE POSTS										
	Height										
	2 RAIL 36"	LF	2.88	3.58	6.46	9.69					.214
	3 RAIL 42"	LF	3.81	3.89	7.70	11.55					.215
RANCH ROUND (DUXBURY)	INCLUDING FULL ROUND PINE RAILS AND POSTS, 8 FEET BETWEEN LINE POSTS										
	36"	LF	3.25	2.85	6.10	9.15					.216
	42"	LF	4.00	3.15	7.15	10.73					.217
	2 GATEPOSTS AND 42" WIDE GATE										
	36"	EA	59.00	17.00	76.00	114.00					.218
CHAIN LINK	11 GA., GALVANIZED STEEL OR 11-1/2 GA., VINYL COVERED FENCING										
	36"	LF	2.94	2.55	5.49	8.24					.219
	42"	LF	3.89	2.55	6.44	9.66					.220
	48"	LF	4.20	2.55	6.75	10.13					.221
	60"	LF	5.00	2.73	7.73	11.60					.222
	72"	LF	5.67	2.73	8.40	12.60					.223
	2 GATEPOSTS AND 42" WIDE GATE										
	36"	EA	79.00	37.00	116.00	174.00					.224
	42"	EA	83.00	37.00	120.00	180.00					.225
	48"	EA	87.00	37.00	124.00	186.00					.226
	60"	EA	97.00	37.00	134.00	201.00					.227
	72"	EA	103.00	37.00	140.00	210.00					.228
	2 GATEPOSTS AND 12 FOOT WIDE (DOUBLE) GATE										
	36"	EA	139.00	57.00	196.00	294.00					.229
	42"	EA	141.00	57.00	198.00	297.00					.230
	48"	EA	155.00	57.00	212.00	318.00					.231
	60"	EA	170.00	60.00	230.00	345.00					.232
	72"	EA	188.00	60.00	248.00	372.00					.233

22

23. HARDWOOD FLOORS

SPECIFICATIONS		UNIT	JOB COST			PRICE	LOCAL AREA MODIFICATION				DATA BASE ITEM NO.
			MATLS	LABOR	TOTAL		MATLS	LABOR	TOTAL	PRICE	
HARD-WOOD	LAY FLOOR WITH T&G AND END-MATCHED OAK FLOORING, 25/32" X 2-1/4"										
	CLEAR	SF	5.29	1.27	6.56	9.84					.000
	SELECT	SF	4.13	1.27	5.40	8.10					.001
	#1 COMMON	SF	3.51	1.27	4.78	7.17					.002
PRE-FINISHED HARD-WOOD	PREFINISHED OAK, 25/32 X 2-1/4	SF	5.79	1.38	7.17	10.76					.003
	RANCH PLANK WITH SCREWS AND PLUGS	SF	7.77	1.61	9.38	14.07					.004
HARD-WOOD TOOTHED IN	STRIP OAK FLOORING TOOTHED IN TO EXISTING FLOOR, 25/32" X 2-1/4" **ADD** EA = EACH TOOTH-IN JOINT	EA	--	16.00	16.00	24.00					.005
PARQUET	OAK LAMINATED, SET IN MASTIC OVER CONCRETE OR PLYWOOD DECK, T&G, PRE-FINISHED, CLEAR										
	12" X 12" X 1/2"	SF	4.50	1.53	6.03	9.05					.006
FLOOR FINISHING	SAND AND FINISH NEW FLOOR										
	NATURAL	SF	.38	.80	1.18	1.77					.007
	DARK STAIN	SF	.58	1.27	1.85	2.78					.008
	BLEACHED	SF	.65	1.52	2.17	3.26					.009
	SAND AND FINISH OLD FLOOR										
	NATURAL	SF	.46	.98	1.44	2.16					.010
	DARK STAIN	SF	.72	1.55	2.27	3.41					.011
	BLEACHED	SF	.80	1.79	2.59	3.89					.012

23

SPECIFICATIONS		UNIT	JOB COST			PRICE	LOCAL AREA MODIFICATION				DATA BASE ITEM NO.
			MATLS	LABOR	TOTAL		MATLS	LABOR	TOTAL	PRICE	
UNDER-LAYMENT	UNDERLAYMENT FOR RESILIENT FLOORS NAILED OVER EXISTING FLOOR										
	PLYWOOD 3/8"	SF	.48	.38	.86	1.29					.100
	1/2"	SF	.56	.38	.94	1.41					.101
	5/8"	SF	.68	.38	1.06	1.59					.102
	3/4"	SF	.78	.38	1.16	1.74					.103
	PARTICLEBOARD 3/8"	SF	.30	.38	.68	1.02					.104
	1/2"	SF	.33	.38	.71	1.07					.105
	5/8"	SF	.36	.38	.74	1.11					.106
	3/4"	SF	.45	.38	.83	1.25					.107
ASPHALT TILE	LAID IN ADHESIVE OVER SMOOTH SURFACE, 9" X 9"										
	B GRADE	SF	.81	.66	1.47	2.21					.108
	C GRADE	SF	.97	.66	1.63	2.45					.109
	D GRADE	SF	1.18	.66	1.84	2.76					.110
VINYL COMPOSITION	LAID IN ADHESIVE OVER SMOOTH SURFACE, 9" X 9"										
	1/16"	SF	.65	.56	1.21	1.82					.111
	3/32"	SF	.90	.56	1.46	2.19					.112
	1/8"	SF	1.89	.56	2.45	3.68					.113
	SAME AS ABOVE, 12" X 12"										
	1/16"	SF	.62	.50	1.12	1.68					.114
	3/32"	SF	.87	.50	1.37	2.06					.115
	1/8"	SF	1.85	.50	2.35	3.53					.116
VINYL TILE	LAID IN ADHESIVE OVER SMOOTH SURFACE, 9" X 9"										
	1/16"	SF	1.20	.56	1.76	2.64					.117
	3/32"	SF	2.00	.56	2.56	3.84					.118
	1/8"	SF	3.00	.56	3.56	5.34					.119
	SAME AS ABOVE, 12" X 12"										
	1/16"	SF	1.20	.50	1.70	2.55					.120
	1/8"	SF	3.00	.50	3.50	5.25					.121
	PREMIUM	SF	3.50	.50	4.00	6.00					.122
SHEET VINYL	LAID IN ADHESIVE OVER SMOOTH SURFACE										
	ECONOMY	SF	1.20	.50	1.70	2.55					.123
	MEDIUM	SF	2.04	.50	2.54	3.81					.124
	PREMIUM	SF	3.74	.50	4.24	6.36					.125
BASE	VINYL BASE, INCLUDING CORNERS										
	2-1/4"	LF	.40	.65	1.05	1.58					.126
	4"	LF	.47	.65	1.12	1.68					.127
	6"	LF	.55	.68	1.23	1.85					.128

23

23. TILE FLOOR

SPECIFICATIONS		UNIT	JOB COST			PRICE	LOCAL AREA MODIFICATION				DATA BASE ITEM NO.
			MATLS	LABOR	TOTAL		MATLS	LABOR	TOTAL	PRICE	
UNDER-LAYMENT	WATER RESISTANT 1/2" THICK BACKER BOARD UNDERLAYMENT FOR CERAMIC TILE FLOOR, IN-CLUDING MORTAR, JOINT TAPE AND NAILS	SF	2.02	.74	2.76	4.14					.200
CERAMIC TILE	CERAMIC TILE FLOOR, 4-1/4" X 4-1/4", 6" X 6" AND 8" X 8", INCLUDING COVE, BASE OR BULLNOSE TRIM, SET IN MUD	SF	3.31	9.45	12.76	19.14					.201
	SAME AS ABOVE, SET IN THIN SET OR MASTIC	SF	2.21	5.88	8.09	12.14					.202
QUARRY TILE	QUARRY TILE 1/2" OR 3/4" THICK, 4" X 4", 6" X 6", 8" X 8" AND 9" X 9"	SF	2.21	5.88	8.09	12.14					.203
PAVER TILE	PAVER TILE 1" THICK, 4" X 4", 6" X 6"	SF	3.31	5.88	9.19	13.79					.204
MARBLE	MARBLE TILE 3/8" THICK, 12" X 12"	SF	6.62	8.40	15.02	22.53					.205
SLATE	SLATE TILE, 3/8" THICK, 12" X 12"	SF	4.04	6.04	10.08	15.12					.206

23

SPECIFICATIONS		UNIT	JOB COST			PRICE	LOCAL AREA MODIFICATION				DATA BASE ITEM NO.
			MATLS	LABOR	TOTAL		MATLS	LABOR	TOTAL	PRICE	
CARPETING	INDOOR-OUTDOOR CARPETING ON WOOD OR CONCRETE										
	ECONOMY	SF	.55	.30	.85	1.28					.207
	MEDIUM	SF	.80	.30	1.10	1.65					.208
	PREMIUM	SF	1.07	.30	1.37	2.06					.209
	INTERIOR CARPETING, INCLUDING PADDING										
	@ $12/YD	SF	1.70	.30	2.00	3.00					.210
	@ $15/YD	SF	2.20	.31	2.51	3.77					.211
	@ $20/YD	SF	2.93	.33	3.26	4.89					.212
	@ $25/YD	SF	3.67	.35	4.02	6.03					.213
	STAIR CARPETING ADD	Per Riser	--	5.00	5.00	7.50					.214
REMOVE CARPET	REMOVE EXISTING CARPETING FROM FLOOR										
	BONDED	SF	--	.25	.25	.38					.215
	TACKLESS	SF	--	.07	.07	.11					.216
	CARPET PAD	SF	--	.05	.05	.08					.217
CLEAN	CLEAN EXISTING CARPETING (IN PLACE) WITH STEAM	SF	--	.40	.40	.60					.218

23

24. PAINTING PREPARATION, EXTERIOR

SPECIFICATIONS		UNIT	JOB COST			PRICE	LOCAL AREA MODIFICATION				DATA BASE ITEM NO.
			MATLS	LABOR	TOTAL		MATLS	LABOR	TOTAL	PRICE	
PAINT REMOVAL	REMOVE EXISTING PAINT OR VARNISH WITH CHEMICALS, 1 OR 2 LAYERS PER COAT OF PAINT REMOVER										
	EXTERIOR WOOD SIDING	SF	.12	.92	1.04	1.56					.000
	EXTERIOR TRIM	LF	.12	.77	.89	1.34					.001
	REMOVE NON LEAD BASED PAINT BY LIGHT SAND BLASTING	SF	.14	.78	.92	1.38					.002
	REMOVE LOOSE OR BLISTERED PAINT WITH WIRE BRUSH OR PAINT SCRAPER	SF	--	.13	.13	.20					.003
BURNING OFF	EXTERIOR SIDING, NON LEAD BASED PAINT										
	AVERAGE	SF	--	.99	.99	1.49					.004
	EXTENSIVE	SF	--	1.37	1.37	2.06					.005
	EXTERIOR TRIM, NON LEAD BASED PAINT	LF	--	.75	.75	1.13					.006
SANDING AND FILLING	SAND AND FILL EXTERIOR WOOD SIDING AND TRIM										
	AVERAGE	SF	--	.16	.16	.24					.007
	EXTENSIVE	SF	--	.20	.20	.30					.008
	SAND AND FILL EXTERIOR TRIM ONLY										
	AVERAGE	LF	--	.20	.20	.30					.009
	EXTENSIVE	LF	--	.27	.27	.41					.010
REGLAZE	REGLAZE SASH	LF	--	.75	.75	1.13					.011
ACID WASH	ACID WASH GALVANIZED GUTTER AND DOWNSPOUT	LF	--	.27	.27	.41					.012
CAULK	CAULK TRIM OR SIDING JOINTS	LF	.02	.40	.42	.63					.013
WASHING	POWER WASH WITH TSP IN PREPARATION FOR EXTERIOR PAINTING	SF	--	.13	.13	.20					.014
	EACH ADDITIONAL STORY IF LADDER OR SCAFFOLDING IS REQUIRED **ADD**	SF	--	30%	20%	20%					.015

24

SPECIFICATIONS	UNIT	JOB COST			PRICE	LOCAL AREA MODIFICATION				DATA BASE ITEM NO.
		MATLS	LABOR	TOTAL		MATLS	LABOR	TOTAL	PRICE	
SIDING NOTE: PAINTING COSTS ON EVERY PAGE INCLUDE SET-UP AND CLEAN-UP TIME AND PROTECTION OF ITEMS NOT TO BE PAINTED. ALL WORK SHOWN IS AT ONE-STORY LEVEL EXCEPT AS NOTED.										
WOOD, PLYWOOD, ASBESTOS AND COMPOSITION SIDING, PAINT WITH BRUSH										
1 COAT ON PAINTED SURF.	SF	.06	.27	.33	.50					.100
PRIME AND 1 COAT	SF	.11	.47	.58	.87					.101
PRIME AND 2 COATS	SF	.16	.67	.83	1.25					.102
SAME AS ABOVE, WITH ROLLER										
1 COAT ON PAINTED SURF.	SF	.06	.23	.29	.44					.103
PRIME AND 1 COAT	SF	.11	.34	.45	.68					.104
PRIME AND 2 COATS	SF	.16	.47	.63	.95					.105
SAME AS ABOVE, WITH SPRAY GUN										
1 COAT ON PAINTED SURF.	SF	.08	.09	.17	.26					.106
PRIME AND 1 COAT	SF	.15	.17	.32	.48					.107
PRIME AND 2 COATS	SF	.22	.26	.48	.72					.108
SAME AS ABOVE, STAINED, WITH BRUSH										
1 COAT	SF	.06	.24	.30	.45					.109
2 COATS	SF	.11	.42	.53	.80					.110
ALUMINUM SIDING ALUMINUM SIDING, PAINT WITH BRUSH										
1 COAT	SF	.06	.15	.21	.32					.111
2 COATS	SF	.11	.22	.33	.50					.112
SAME AS ABOVE, WITH SPRAY GUN										
1 COAT	SF	.08	.09	.17	.26					.113
2 COATS	SF	.15	.17	.32	.48					.114
FOR EACH ADDITIONAL STORY UP, PER COAT **ADD**	SF	--	30%	20%	20%					.115

24

24. SHINGLES, STUCCO, MASONRY PAINTING

SPECIFICATIONS		UNIT	JOB COST			PRICE	LOCAL AREA MODIFICATION				DATA BASE ITEM NO.
			MATLS	LABOR	TOTAL		MATLS	LABOR	TOTAL	PRICE	
WOOD SHINGLES OR ROUGH SIDING	CEDAR SHINGLES, PAINT WITH BRUSH										
	1 COAT ON PAINTED SURF.	SF	.08	.31	.39	.59					.116
	PRIME AND 1 COAT	SF	.14	.52	.66	.99					.117
	PRIME AND 2 COATS	SF	.20	.75	.95	1.43					.118
	CEDAR SHINGLES, WITH SPRAY GUN										
	1 COAT ON PAINTED SURF.	SF	.10	.09	.19	.29					.119
	PRIME AND 1 COAT	SF	.14	.17	.31	.47					.120
	PRIME AND 2 COATS	SF	.20	.26	.46	.69					.121
	CEDAR SHINGLES, STAINED WITH BRUSH										
	1 COAT	SF	.06	.31	.37	.56					.122
	2 COATS	SF	.10	.52	.62	.93					.123
	CEDAR SHINGLES, STAINED WITH SPRAY GUN										
	1 COAT	SF	.07	.09	.16	.24					.124
	2 COATS	SF	.13	.17	.30	.45					.125
SMOOTH BRICK OR STUCCO	WITH ROLLER										
	1 COAT ON PAINTED SURF.	SF	.06	.30	.36	.54					.126
	PRIME AND 1 COAT	SF	.11	.47	.58	.87					.127
	PRIME AND 2 COATS	SF	.16	.64	.80	1.20					.128
	WITH SPRAY GUN										
	1 COAT ON PAINTED SURF.	SF	.08	.09	.17	.26					.129
	PRIME AND 1 COAT	SF	.14	.17	.31	.47					.130
	PRIME AND 2 COATS	SF	.21	.26	.47	.71					.131
POROUS BRICK OR STUCCO	WITH ROLLER										
	1 COAT ON PAINTED SURF.	SF	.10	.34	.44	.66					.132
	PRIME AND 1 COAT	SF	.13	.62	.75	1.13					.133
	PRIME AND 2 COATS	SF	.15	.88	1.03	1.55					.134
	WITH SPRAY GUN										
	1 COAT ON PAINTED SURF.	SF	.08	.10	.18	.27					.135
	PRIME AND 1 COAT	SF	.15	.19	.34	.51					.136
	PRIME AND 2 COATS	SF	.22	.30	.52	.78					.137

24

SPECIFICATIONS		UNIT	JOB COST			PRICE	LOCAL AREA MODIFICATION				DATA BASE ITEM NO.
			MATLS	LABOR	TOTAL		MATLS	LABOR	TOTAL	PRICE	
MOULDING AND TRIM	DOOR OR WINDOW MOULD-ING OR TRIM (IF SAME COLOR AND FINISH AS WALL, INCLUDE IN WALL ESTIMATE), WITH BRUSH										
	1 COAT ON PAINTED SURF.	LF	.04	.23	.27	.41					.200
	PRIME AND 1 COAT	LF	.06	.42	.48	.72					.201
	PRIME AND 2 COATS	LF	.08	.62	.70	1.05					.202
FASCIA OR RAKE	UP TO 12" FASCIA OR RAKE WITH BRUSH, ONE FACE & EDGE										
	1 COAT ON PAINTED SURF.	LF	.06	.32	.38	.57					.203
	PRIME AND 1 COAT	LF	.08	.60	.68	1.02					.204
	PRIME AND 2 COATS	LF	.10	.76	.86	1.29					.205
SOFFIT	UP TO 12" SOFFIT, WITH BRUSH										
	1 COAT ON PAINTED SURF.	LF	.06	.40	.46	.69					.206
	PRIME AND 1 COAT	LF	.11	.74	.85	1.28					.207
	PRIME AND 2 COATS	LF	.16	1.08	1.24	1.86					.208
	SOFFIT, WITH ROLLER										
	1 COAT ON PAINTED SURF.	SF	.06	.25	.31	.47					.209
	PRIME AND 1 COAT	SF	.11	.42	.53	.80					.210
	PRIME AND 2 COATS	SF	.16	.55	.71	1.07					.211
	SOFFIT, WITH SPRAY GUN										
	1 COAT ON PAINTED SURF.	SF	.08	.09	.17	.26					.212
	PRIME AND 1 COAT	SF	.14	.17	.31	.47					.213
	PRIME AND 2 COATS	SF	.20	.26	.46	.69					.214
EAVES, OPEN WITH RAFTERS	WITH BRUSH, ALL SIDES OF RAFTERS, INTERIOR OF FAS-CIA AND UNDERSIDE OF ROOF SHEATHING										
	1 COAT ON PAINTED SURF.	SF	.06	.43	.49	.74					.215
	PRIME AND 1 COAT	SF	.11	.86	.97	1.46					.216
	PRIME AND 2 COATS	SF	.16	1.30	1.46	2.19					.217
	WITH SPRAY GUN										
	1 COAT ON PAINTED SURF.	SF	.08	.11	.19	.29					.218
	PRIME AND 1 COAT	SF	.15	.20	.35	.53					.219
	PRIME AND 2 COATS	SF	.22	.28	.50	.75					.220

24

24. DECK, PORCH PAINTING

SPECIFICATIONS		UNIT	JOB COST			PRICE	LOCAL AREA MODIFICATION				DATA BASE ITEM NO.
			MATLS	LABOR	TOTAL		MATLS	LABOR	TOTAL	PRICE	
WOOD DECK OR PORCH FLOOR, TOP SIDE	**PAINT WITH BRUSH**										
	1 COAT ON PAINTED SURF.	SF	.08	.26	.34	.51					.221
	PRIME AND 1 COAT	SF	.10	.42	.52	.78					.222
	PRIME AND 2 COATS	SF	.12	.58	.70	1.05					.223
	PAINT WITH ROLLER										
	1 COAT ON PAINTED SURF.	SF	.10	.24	.34	.51					.224
	PRIME AND 1 COAT	SF	.13	.37	.50	.75					.225
	PRIME AND 2 COATS	SF	.15	.50	.65	.98					.226
	PAINT WITH SPRAY GUN										
	1 COAT ON PAINTED SURF.	SF	.12	.10	.22	.33					.227
	PRIME AND 1 COAT	SF	.16	.20	.36	.54					.228
	PRIME AND 2 COATS	SF	.18	.30	.48	.72					.229
	STAIN AND/OR CLEAR FINISH WITH BRUSH 1 COAT	SF	.06	.27	.33	.50					.230
	2 COATS	SF	.12	.44	.56	.84					.231
	3 COATS	SF	.18	.60	.78	1.17					.232
	STAIN AND/OR CLEAR FINISH WITH ROLLER 1 COAT	SF	.06	.25	.31	.47					.233
	2 COATS	SF	.12	.38	.50	.75					.234
	3 COATS	SF	.18	.52	.70	1.05					.235
	STAIN AND/OR CLEAR FINISH W/SPRAY GUN 1 COAT	SF	.08	.11	.19	.29					.236
	2 COATS	SF	.16	.20	.36	.54					.237
	3 COATS	SF	.24	.28	.52	.78					.238
PORCH CEILING	**PAINT WOOD PORCH CEILING WITH BRUSH**										
	1 COAT ON PAINTED SURF.	SF	.06	.27	.33	.50					.239
	PRIME AND 1 COAT	SF	.11	.44	.55	.83					.240
	PRIME AND 2 COATS	SF	.16	.59	.75	1.13					.241
	PAINT WITH ROLLER										
	1 COAT ON PAINTED SURF.	SF	.06	.25	.31	.47					.242
	PRIME AND 1 COAT	SF	.11	.42	.53	.80					.243
	PRIME AND 2 COATS	SF	.16	.55	.71	1.07					.244
	PAINT WITH SPRAY GUN										
	1 COAT ON PAINTED SURF.	SF	.08	.10	.18	.27					.245
	PRIME AND 1 COAT	SF	.14	.20	.34	.51					.246
	PRIME AND 2 COATS	SF	.22	.29	.51	.77					.247
CONCRETE PORCH	**PAINT TOP SURFACE AND SIDES OF CONCRETE PORCH SLAB WITH ROLLER** 1 COAT	SF	.10	.34	.44	.66					.248
	2 COATS	SF	.13	.62	.75	1.13					.249
	PAINT WITH SPRAY GUN 1 COAT	SF	.10	.10	.20	.30					.250
	2 COATS	SF	.13	.20	.33	.50					.251

24

SPECIFICATIONS	UNIT	JOB COST			PRICE	LOCAL AREA MODIFICATION				DATA BASE ITEM NO.
		MATLS	LABOR	TOTAL		MATLS	LABOR	TOTAL	PRICE	
PORCH OR DECK RAILING — SF = HEIGHT OF RAILING MULTIPLIED BY LENGTH OF RAILING, **ONE SIDE ONLY**										
COSTS SHOWN INCLUDE PAINTING ALL SIDES OF RAILINGS AND BALUSTERS, WITH BRUSH 1 COAT	SF	.09	.65	.74	1.11					.300
2 COATS	SF	.17	1.13	1.30	1.95					.301
PAINT WITH SPRAY GUN 1 COAT	SF	.13	.26	.39	.59					.302
2 COATS	SF	.24	.51	.75	1.13					.303
STAIN AND/OR CLEAR FINISH WITH BRUSH 1 COAT	SF	.06	.69	.75	1.13					.304
2 COATS	SF	.12	1.18	1.30	1.95					.305
3 COATS	SF	.18	1.66	1.84	2.76					.306
STAIN AND/OR CLEAR FINISH WITH SPRAY GUN 1 COAT	SF	.08	.26	.34	.51					.307
2 COATS	SF	.16	.51	.67	1.01					.308
3 COATS	SF	.24	.77	1.01	1.52					.309
WOOD STEPS — SF = TO OBTAIN SF AREA, *MEASURE TOP SIDE ONLY* OF STEPS										
PAINT ALL SURFACES (TOP, BOTTOM, SIDES) OF STEPS AND STRINGERS, PAINT WITH BRUSH 1 COAT ON PAINTED SURF.	SF	.06	.64	.70	1.05					.310
PRIME AND 1 COAT	SF	.11	1.05	1.16	1.74					.311
PRIME AND 2 COATS	SF	.16	1.57	1.73	2.60					.312
PAINT WITH SPRAY GUN 1 COAT ON PAINTED SURF.	SF	.08	.25	.33	.50					.313
PRIME AND 1 COAT	SF	.14	.49	.63	.95					.314
PRIME AND 2 COATS	SF	.22	.74	.96	1.44					.315
STAIN AND/OR CLEAR FINISH WITH BRUSH 1 COAT	SF	.06	.69	.75	1.13					.316
2 COATS	SF	.12	1.18	1.30	1.95					.314
3 COATS	SF	.18	1.66	1.84	2.76					.318
STAIN AND/OR CLEAR FINISH WITH SPRAY GUN 1 COAT	SF	.08	.26	.34	.51					.319
2 COATS	SF	.16	.51	.67	1.01					.320
3 COATS	SF	.24	.77	1.01	1.52					.321

24

24. EXTERIOR, FENCE PAINTING

SPECIFICATIONS		UNIT	JOB COST			PRICE	LOCAL AREA MODIFICATION				DATA BASE ITEM NO.
			MATLS	LABOR	TOTAL		MATLS	LABOR	TOTAL	PRICE	
WINDOW SCREEN	PAINT WOOD WINDOW SCREEN, ALL SIDES										
	1 COAT	EA	.70	3.95	4.65	6.98					.322
	2 COATS	EA	1.15	6.90	8.05	12.08					.323
STORM SASH	PAINT STORM SASH ALL SIDES										
	1 COAT	EA	.70	5.30	6.00	9.00					.324
	2 COATS	EA	1.15	9.00	1.15	15.23					.325
BLINDS AND SHUTTERS	WOOD OR METAL, LOUVERED, BOTH SIDES AND ALL EDGES, WITH BRUSH										
	1 COAT ON PAINTED SURF.	EA	1.00	13.50	14.50	21.75					.326
	PRIME AND 1 COAT	EA	1.50	22.00	23.50	35.25					.327
	PRIME AND 2 COATS	EA	2.00	3.00	32.00	48.00					.328
	WITH SPRAY GUN										
	1 COAT ON PAINTED SURF.	EA	1.00	6.00	7.00	10.50					.329
	PRIME AND 1 COAT	EA	1.50	9.50	11.00	16.50					.330
	PRIME AND 2 COATS	EA	2.00	12.00	14.00	21.00					.331
FENCE, PLAIN	PLAIN BOARD FENCE, ALL SIDES										
	BRUSH 1 COAT	SF	.12	.54	.66	.99					.332
	2 COATS	SF	.18	.80	.98	1.47					.333
	SPRAY 1 COAT	SF	.14	.09	.23	.35					.334
	2 COATS	SF	.20	.17	.37	.56					.335
	SF = HEIGHT OF FENCE MULTIPLIED BY LENGTH OF FENCE, **ONE FACE ONLY**										
PICKET	PICKET FENCE, ALL SIDES										
	BRUSH 1 COAT	SF	.05	.74	.79	1.19					.336
	2 COATS	SF	.08	1.29	1.37	2.06					.337
	SPRAY 1 COAT	SF	.06	.09	.15	.23					.338
	2 COATS	SF	.10	.17	.27	.41					.339
CHAIN LINK	CHAIN LINK FENCE, ALL SIDES										
	BRUSH 1 COAT	SF	.04	.57	.61	.92					.340
	2 COATS	SF	.07	.98	1.05	1.58					.341
	ROLLER 1 COAT	SF	.05	.25	.30	.45					342
	2 COATS	SF	.09	.43	.52	.78					.343
	CHAIN LINK FENCE, ALL SIDES WITH SPRAY GUN										
	1 COAT	SF	.10	.09	.19	.29					.344
	2 COATS	SF	.12	.17	.29	.44					.345

24

SPECIFICATIONS		UNIT	JOB COST			PRICE	LOCAL AREA MODIFICATION				DATA BASE ITEM NO.
			MATLS	LABOR	TOTAL		MATLS	LABOR	TOTAL	PRICE	
GUTTER AND/OR DOWN-SPOUT	METAL GUTTER AND/OR DOWNSPOUT, PAINT WITH BRUSH										
	1 COAT ON PAINTED SURF.	LF	.06	.32	.38	.57					.400
	PRIME AND 1 COAT	LF	.11	.43	.54	.81					.401
	PRIME AND 2 COATS	LF	.16	.55	.71	1.07					.402
	WOOD GUTTER, 4" X 5", WITH BRUSH										
	1 COAT ON PAINTED SURF.	LF	.10	.34	.44	.66					.403
	PRIME AND 1 COAT	LF	.14	.58	.72	1.08					.404
	PRIME AND 2 COATS	LF	.18	.76	.94	1.41					.405
LATTICE WORK	PAINT ALL SIDES OF LATTICE WORK WITH BRUSH										
	SF = HEIGHT OF LATTICE WORK MULTIPLIED BY LENGTH, **ONE FACE ONLY**										
	1 COAT	SF	.06	.50	.56	.84					.406
	2 COATS	SF	.11	.86	.97	1.46					.407
	SAME AS ABOVE, WITH SPRAY GUN										
	1 COAT	SF	.08	.09	.17	.26					.408
	2 COATS	SF	.12	.17	.29	.44					.409
COLUMNS AND PILASTERS	WOOD COLUMNS AND PILASTERS, PAINT WITH BRUSH										
	1 COAT	SF	.06	.26	.32	.48					.410
	2 COATS	SF	.11	.44	.55	.83					.411
	WOOD COLUMNS AND PILASTERS, PAINT WITH SPRAY GUN										
	1 COAT	SF	.08	.09	.17	.26					.412
	2 COATS	SF	.15	.17	.32	.48					.413

24

24. DOOR PAINTING

SPECIFICATIONS		UNIT	JOB COST			PRICE	LOCAL AREA MODIFICATION				DATA BASE ITEM NO.
			MATLS	LABOR	TOTAL		MATLS	LABOR	TOTAL	PRICE	
FLUSH ENTRANCE DOOR	PAINT ONE SIDE ONLY, INCLUDING JAMB AND CASING, WITH BRUSH										
	1 COAT ON PAINTED SURF.	EA	1.40	11.60	13.00	19.50					.414
	PRIME AND 1 COAT	EA	2.30	19.20	21.50	32.25					.415
	PRIME AND 2 COATS	EA	3.20	26.80	30.00	45.00					.416
	WITH ROLLER										
	1 COAT ON PAINTED SURF.	EA	1.40	6.20	7.60	11.40					.417
	PRIME AND 1 COAT	EA	2.30	10.10	12.40	18.60					.418
	PRIME AND 2 COATS	EA	3.20	13.80	17.00	25.50					.419
	STAIN AND/OR CLEAR FINISH WITH BRUSH										
	1 COAT	EA	1.50	15.00	16.50	24.75					.420
	2 COATS	EA	2.80	25.00	27.80	41.70					.421
	3 COATS	EA	4.00	35.00	39.00	58.50					.422
2 TO 6 PANEL ENTRANCE DOOR	PAINT ONE SIDE ONLY, INCLUDING JAMB & CASINGS, WITH BRUSH										
	1 COAT ON PAINTED SURF.	EA	1.60	14.40	16.00	24.00					.423
	PRIME AND 1 COAT	EA	2.50	24.50	27.00	40.50					.424
	PRIME AND 2 COATS	EA	3.50	33.00	36.50	54.75					.425
	STAIN AND/OR CLEAR FINISH WITH BRUSH										
	1 COAT	EA	1.70	19.00	20.70	31.05					.426
	2 COATS	EA	2.60	32.20	34.80	52.20					.427
	3 COATS	EA	3.70	43.00	46.70	70.05					.428

24

SPECIFICATIONS		UNIT	JOB COST			PRICE	LOCAL AREA MODIFICATION				DATA BASE ITEM NO.
			MATLS	LABOR	TOTAL		MATLS	LABOR	TOTAL	PRICE	
2 TO 6 LIGHT DOOR	PAINT ONE SIDE ONLY, INCLUDING JAMB AND CAS-ING, WITH BRUSH										
	1 COAT ON PAINTED SURF.	EA	1.30	18.50	19.80	29.70					.429
	PRIME AND 1 COAT	EA	2.10	28.90	31.00	46.50					.430
	PRIME AND 2 COATS	EA	2.80	39.50	42.30	63.45					.431
	STAIN AND/OR CLEAR FINISH, WITH BRUSH										
	1 COAT	EA	1.30	23.70	25.00	37.50					.432
	2 COATS	EA	2.20	37.80	40.00	60.00					.433
	3 COATS	EA	3.00	51.50	54.50	81.75					.434
9 TO 15 LIGHT FRENCH DOOR	ONE SIDE ONLY, INCLUDING JAMB AND CASING, WITH BRUSH										
	1 COAT ON PAINTED SURF.	EA	1.40	21.40	22.80	34.20					.435
	PRIME AND 1 COAT	EA	2.30	35.40	37.70	56.55					.436
	PRIME AND 2 COATS	EA	3.20	48.80	52.00	78.00					.437
	STAIN AND/OR CLEAR FINISH, WITH BRUSH										
	1 COAT	EA	1.40	28.10	29.50	44.25					.438
	2 COATS	EA	2.30	47.70	50.00	75.00					.439
	3 COATS	EA	3.20	63.80	67.00	100.50					.440
COLONIAL ENTRANCE	FOR SINGLE DOOR, PAINT WITH BRUSH										
	1 COAT ON PAINTED SURF.	EA	2.50	24.50	27.00	40.50					.441
	PRIME AND 1 COAT	EA	3.50	4.50	8.00	12.00					.442
	PRIME AND 2 COATS	EA	4.50	55.50	60.00	90.00					.443
	FOR SINGLE DOOR WITH TWO SIDELIGHTS										
	1 COAT ON PAINTED SURF.	EA	3.00	37.00	40.00	60.00					.444
	PRIME AND 1 COAT	EA	4.00	52.00	56.00	84.00					.445
	PRIME AND 2 COATS	EA	5.00	67.00	72.00	108.00					.446

24

24. DOOR PAINTING

SPECIFICATIONS		UNIT	JOB COST			PRICE	LOCAL AREA MODIFICATION				DATA BASE ITEM NO.
			MATLS	LABOR	TOTAL		MATLS	LABOR	TOTAL	PRICE	
GLIDING PATIO DOOR	PAINT ONE SIDE ONLY, IN-CLUDING EXTERIOR JAMB AND CASINGS, WITH BRUSH, TWO PANELS										
	1 COAT ON PAINTED SURF.	EA	1.30	14.00	15.30	22.95					.500
	PRIME AND 1 COAT	EA	2.30	23.50	25.80	38.70					.501
	PRIME AND 2 COATS	EA	3.30	33.00	36.30	54.45					.502
	STAIN AND/OR CLEAR FINISH WITH BRUSH 1 COAT	EA	1.40	18.00	19.40	29.10					.503
	2 COATS	EA	2.40	3.60	33.00	49.50					.504
	3 COATS	EA	3.50	43.00	46.50	69.75					.505
	SAME AS ABOVE, THREE PANELS WITH BRUSH										
	1 COAT ON PAINTED SURF.	EA	1.50	17.00	18.50	27.75					.506
	PRIME AND 1 COAT	EA	2.60	28.60	31.20	46.80					.507
	PRIME AND 2 COATS	EA	3.70	4.60	44.30	66.45					.508
	STAIN AND/OR CLEAR FINISH WITH BRUSH 1 COAT	EA	1.60	22.00	23.60	35.40					.509
	2 COATS	EA	2.75	36.60	39.35	59.03					.510
	3 COATS	EA	3.90	53.00	56.90	85.35					.511
GARAGE DOOR	BOTH SIDES, INCLUDING JAMBS AND CASINGS, PAINT WITH BRUSH, 9-0 X 7-0										
	1 COAT ON PAINTED SURF.	EA	8.60	46.30	54.90	82.35					.512
	PRIME AND 1 COAT	EA	15.80	78.60	94.40	141.60					.513
	PRIME AND 2 COATS	EA	23.00	102.00	125.00	187.50					.514
	SAME AS ABOVE, WITH SPRAY GUN										
	1 COAT ON PAINTED SURF.	EA	12.10	12.40	24.50	36.75					.515
	PRIME AND 1 COAT	EA	22.20	2.80	43.00	64.50					.516
	PRIME AND 2 COATS	EA	32.25	27.30	59.55	89.33					.517
	BOTH SIDES, INCLUDING JAMBS AND CASINGS, PAINT WITH BRUSH, 16-0 X 7-0										
	1 COAT ON PAINTED SURF.	EA	15.40	82.00	97.40	146.10					.518
	PRIME AND 1 COAT	EA	28.20	139.00	167.20	250.80					.519
	PRIME AND 2 COATS	EA	41.00	182.00	223.00	334.50					.520
	SAME AS ABOVE, WITH SPRAY GUN										
	1 COAT ON PAINTED SURF.	EA	21.50	22.00	43.50	65.25					.521
	PRIME AND 1 COAT	EA	39.40	37.00	76.40	114.60					.522
	PRIME AND 2 COATS	EA	57.30	48.50	105.80	158.70					.523

24

SPECIFICATIONS		UNIT	JOB COST			PRICE	LOCAL AREA MODIFICATION				DATA BASE ITEM NO.
			MATLS	LABOR	TOTAL		MATLS	LABOR	TOTAL	PRICE	
PAINT REMOVAL	REMOVE EXISTING PAINT OR VARNISH WITH CHEMICALS, 1 OR 2 LAYERS PER COAT OF PAINT REMOVER										
	INTERIOR TRIM	LF	.12	.69	.81	1.22					.016
	PANELING	SF	.12	.85	.97	1.46					.017
STRIP CABINETS	REMOVE EXISTING PAINT OR VARNISH WITH CHEMICALS										
	BOOKSHELVES OR CABINETS (INCLUDING BACKING)	SF	.12	1.05	1.17	1.76					.018
	CABINET DOORS (2 SIDES)	SF	.15	1.55	1.70	2.55					.019
	SF = FRONT SQUARE FOOTAGE OF SHELVES, CABINETS OR DOORS										
BURNING OFF	PLAIN SURFACES, NON-LEAD BASED PAINT										
	AVERAGE	SF	--	.87	.87	1.31					.020
	EXTENSIVE	SF	--	1.13	1.13	1.70					.021
	INTERIOR TRIM	LF	--	.77	.77	1.16					.022
	DE-GLOSS WALLS OR TRIM	SF	.01	.20	.21	.32					.023
SANDING AND FILLING	FILL HOLES AND SMALL CRACKS IN WALLS OR CEILING										
	AVERAGE	SF	--	.07	.07	.11					.024
	EXTENSIVE	SF	--	.12	.12	.18					.025
	SAND AND FILL TRIM	LF	.01	.16	.17	.26					.026
	CAULK TRIM	LF	.02	.37	.39	.59					.027
HARDWARE	CABINET HARDWARE REMOVAL AND REPLACEMENT	EA	--	1.00	1.00	1.50					.028
	LOCKSET REMOVAL AND REPLACEMENT	EA	--	7.00	7.00	10.50					.029
WASHING	WASH INTERIOR WITH TSP BEFORE REPAINTING										
	SMOOTH WALLS	SF	--	.11	.11	.17					.030
	TEXTURED WALLS	SF	--	.13	.13	.20					.031
	DOOR JAMB, CASING AND TRIM	EA	--	5.00	5.00	7.50					.032
	WINDOW, INCLUDING ALL TRIM	EA	--	8.00	8.00	12.00					.033
	MOULDINGS	LF	--	.14	.14	.21					.034

24

24. DOOR PAINTING, INTERIOR

SPECIFICATIONS		UNIT	JOB COST			PRICE	LOCAL AREA MODIFICATION				DATA BASE ITEM NO.
			MATLS	LABOR	TOTAL		MATLS	LABOR	TOTAL	PRICE	
INTERIOR DOOR	PAINT BOTH SIDES OF FLUSH OR PANEL DOOR INCLUDING CASING & JAMB, WITH BRUSH										
	1 COAT ON PAINTED SURF.	EA	1.20	12.80	14.00	21.00					.524
	PRIME AND 1 COAT	EA	2.00	23.00	25.00	37.50					.525
	PRIME AND 2 COATS	EA	2.80	33.20	36.00	54.00					.526
	WITH ROLLER										
	1 COAT ON PAINTED SURF.	EA	1.20	6.80	8.00	12.00					.527
	PRIME AND 1 COAT	EA	2.00	12.00	14.00	21.00					.528
	PRIME AND 2 COATS	EA	2.80	17.20	20.00	30.00					.529
	STAIN AND/OR CLEAR FINISH, WITH BRUSH										
	1 COAT	EA	1.26	16.74	18.00	27.00					.530
	2 COATS	EA	2.10	29.90	32.00	48.00					.531
	3 COATS	EA	2.95	43.05	46.00	69.00					.532
LOUVER DOOR	PAINT BOTH SIDES OF LOUVER DOOR, INCLUDING CASING AND JAMB, WITH BRUSH										
	1 COAT ON PAINTED SURF.	EA	2.70	35.30	38.00	57.00					.533
	PRIME AND 1 COAT	EA	5.00	63.00	68.00	102.00					.534
	PRIME AND 2 COATS	EA	7.30	90.70	98.00	147.00					.535
	STAIN AND/OR CLEAR FINISH, WITH BRUSH										
	1 COAT	EA	3.00	46.00	49.00	73.50					.536
	2 COATS	EA	5.25	82.25	87.50	131.25					.537
	3 COATS	EA	7.67	118.33	126.00	189.00					.538
BIFOLD DOORS	PAINT BOTH SIDES OF BIFOLD UNIT UP TO 5'-0" WIDE OPENING, INCL. CASING AND JAMB, WITH BRUSH										
	1 COAT ON PAINTED SURF.	EA	4.00	27.00	31.00	46.50					.539
	PRIME AND 1 COAT	EA	6.00	46.00	52.00	78.00					.540
	PRIME AND 2 COATS	EA	8.00	66.00	74.00	111.00					.541
	STAIN AND/OR CLEAR FINISH, WITH BRUSH										
	1 COAT	EA	4.20	34.80	39.00	58.50					.542
	2 COATS	EA	6.30	59.70	66.00	99.00					.543
	3 COATS	EA	8.40	85.60	94.00	141.00					.544

24

SPECIFICATIONS		UNIT	JOB COST			PRICE	LOCAL AREA MODIFICATION				DATA BASE ITEM NO.
			MATLS	LABOR	TOTAL		MATLS	LABOR	TOTAL	PRICE	
WINDOW	PAINT ONE SIDE ONLY, WOOD OR METAL DOUBLE HUNG, SLIDING, CASEMENT, AWNING OR PICTURE WINDOW, INCLUDING ALL TRIM ONE SIDE, **1-8 TO 3-0 WIDTH**										
	1 COAT ON PAINTED SURF.	EA	1.15	9.55	10.70	16.05					.600
		PLUS LITE	--	.47	.47	.71					.601
	PRIME AND 1 COAT	EA	1.80	16.60	18.40	27.60					.602
		PLUS LITE	--	.78	.78	1.17					.603
	PRIME AND 2 COATS	EA	2.50	24.00	26.50	39.75					.604
	LITE = EACH LITE; COUNT THE LITES AND MULTIPLY BY COSTS	PLUS LITE	--	1.10	1.10	1.65					.605
	3-4 TO 7-0 WIDTH 1 COAT ON PAINTED SURF.	EA	1.50	11.70	13.20	19.80					.606
		PLUS LITE	--	.47	.47	.71					.607
	PRIME AND 1 COAT	EA	2.10	20.90	23.00	34.50					.608
		PLUS LITE	--	.78	.78	1.17					.609
	PRIME AND 2 COATS	EA	2.80	29.20	32.00	48.00					.610
		PLUS LITE	--	1.10	1.10	1.65					.611
	STAIN AND/OR CLEAR FINISH, **1-8 TO 3-0 WIDTH** 1 COAT	EA	1.20	12.50	13.70	20.55					.612
		PLUS LITE	--	.62	.62	.93					.613
	2 COATS	EA	1.90	21.00	22.90	34.35					.614
		PLUS LITE	--	1.11	1.11	1.67					.615
	3 COATS	EA	2.60	31.40	34.00	51.00					.616
		PLUS LITE	--	1.53	1.53	2.30					.617
	3-4 TO 7-0 WIDTH 1 COAT	EA	1.60	21.00	22.60	33.90					.618
		PLUS LITE	--	.62	.62	.93					.619
	2 COATS	EA	2.30	36.70	39.00	58.50					.620
		PLUS LITE	--	1.04	1.04	1.56					.621
	3 COATS	EA	2.90	52.00	54.90	82.35					.622
		PLUS LITE	--	1.46	1.46	2.19					.623
	PER ADDITIONAL STORY, IF LADDER OR SCAFFOLDING IS REQUIRED **ADD** EA = EACH OPENING	EA	--	45%	40%	40%					.624
GRILLES	DIVIDED LIGHT GRILLE PAINT OR STAIN, 1 COAT	EA	.70	5.50	6.20	9.30					.625
	PAINT OR STAIN, 2 COATS	EA	1.10	10.30	11.40	17.10					.626

24

24. INTERIOR PAINTING

SPECIFICATIONS		UNIT	JOB COST			PRICE	LOCAL AREA MODIFICATION				DATA BASE ITEM NO.
			MATLS	LABOR	TOTAL		MATLS	LABOR	TOTAL	PRICE	
WALL	SMOOTH FINISH PLASTER OR PLASTERBOARD, W/ROLLER										
	1 COAT ON PAINTED SURF.	SF	.06	.15	.21	.32					.700
	PRIME AND 1 COAT	SF	.11	.24	.35	.53					.701
	PRIME AND 2 COATS	SF	.16	.32	.48	.72					.702
	SAME AS ABOVE, WITH SPRAY GUN										
	1 COAT ON PAINTED SURF.	SF	.08	.09	.17	.26					.703
	PRIME AND 1 COAT	SF	.15	.17	.32	.48					.704
	PRIME AND 2 COATS	SF	.22	.26	.48	.72					.705
PANELING	PAINT WOOD PANELLED WALLS, WITH BRUSH										
	1 COAT ON PAINTED SURF.	SF	.06	.31	.37	.56					.706
	PRIME AND 1 COAT	SF	.11	.49	.60	.90					.707
	PRIME AND 2 COATS	SF	.16	.67	.83	1.25					.708
	PAINT WITH ROLLER										
	1 COAT ON PAINTED SURF.	SF	.06	.16	.22	.33					.709
	PRIME AND 1 COAT	SF	.11	.27	.38	.57					.710
	PRIME AND 2 COATS	SF	.16	.37	.53	.80					.711
	STAIN AND/OR CLEAR FINISH, ON PANELLED WALLS WITH ROLLER										
	1 COAT	SF	.06	.22	.28	.42					.712
	2 COATS	SF	.11	.35	.46	.69					.713
	3 COATS	SF	.16	.49	.65	.98					.714
MOULDING	BASE, CEILING MOULDING, CHAIR RAIL OR DOOR OR WINDOW TRIM UP TO 6" (IF SAME COLOR AND FINISH AS WALL OR CEILING, INCLUDE IN WALL OR CEILING ESTIMATE)										
	1 COAT ON PAINTED SURF.	LF	.04	.24	.28	.42					.715
	PRIME AND 1 COAT	LF	.07	.42	.49	.74					.716
	PRIME AND 2 COATS	LF	.10	.59	.69	1.04					.717
	SAME AS ABOVE WITH STAIN AND/OR CLEAR FINISH										
	1 COAT	LF	.05	.25	.30	.45					.718
	2 COATS	LF	.08	.44	.52	.78					.719
	3 COATS	LF	.11	.62	.73	1.10					.720
STAIR HANDRAIL	PAINT HANDRAIL, BALUSTERS AND NEWEL POST, ONE FLIGHT										
	1 COAT ON PAINTED SURF.	EA	2.20	41.80	44.00	66.00					.721
	PRIME AND 1 COAT	EA	4.00	78.00	82.00	123.00					.722
	PRIME AND 2 COATS	EA	6.00	106.00	112.00	168.00					.723

24

SPECIFICATIONS		UNIT	JOB COST			PRICE	LOCAL AREA MODIFICATION				DATA BASE ITEM NO.
			MATLS	LABOR	TOTAL		MATLS	LABOR	TOTAL	PRICE	
SMOOTH CEILING	SMOOTH FINISH PLASTER OR PLASTERBOARD, W/ROLLER										
	1 COAT ON PAINTED SURF.	SF	.06	.19	.25	.38					.724
	PRIME AND 1 COAT	SF	.11	.31	.42	.63					.725
	PRIME AND 2 COATS	SF	.16	.43	.59	.89					.726
	SAME AS ABOVE, WITH SPRAY GUN										
	1 COAT ON PAINTED SURF.	SF	.08	.09	.17	.26					.727
	PRIME AND 1 COAT	SF	.15	.17	.32	.48					.728
	PRIME AND 2 COATS	SF	.22	.26	.48	.72					.729
TEXTURED CEILING	TEXTURED CEILING, WITH ROLLER										
	1 COAT ON PAINTED SURF.	SF	.06	.17	.23	.35					.730
	PRIME AND 1 COAT	SF	.11	.28	.39	.59					.731
	PRIME AND 2 COATS	SF	.16	.41	.57	.86					.732
	SAME AS ABOVE, WITH SPRAY GUN										
	1 COAT ON PAINTED SURF.	SF	.08	.09	.17	.26					.733
	PRIME AND 1 COAT	SF	.15	.17	.32	.48					.734
	PRIME AND 2 COATS	SF	.22	.26	.48	.72					.735
WOOD CEILING	WOOD TONGUE & GROOVE CEILING, WITH BRUSH										
	1 COAT ON PAINTED SURF.	SF	.06	.34	.40	.60					.736
	PRIME AND 1 COAT	SF	.11	.55	.66	.99					.737
	PRIME AND 2 COATS	SF	.16	.74	.90	1.35					.738
	SAME AS ABOVE, WITH ROLLER										
	1 COAT ON PAINTED SURF.	SF	.06	.19	.25	.38					.739
	PRIME AND 1 COAT	SF	.11	.31	.42	.63					.740
	PRIME AND 2 COATS	SF	.16	.43	.59	.89					.741
	SAME AS ABOVE, WITH SPRAY GUN										
	1 COAT ON PAINTED SURF.	SF	.08	.09	.17	.26					.742
	PRIME AND 1 COAT	SF	.15	.17	.32	.48					.743
	PRIME AND 2 COATS	SF	.22	.26	.48	.72					.744
CATHEDRAL CEILING	SINGLE OR DOUBLE PITCH CEILING										
	Highest Point Above Floor										
	6/12 SLOPE 12'-0" **ADD**	SF	15%	45%	37%	37%					.745
	16'-0" **ADD**	SF	15%	75%	58%	58%					.746
	20'-0" **ADD**	SF	15%	120%	90%	90%					.747
	12/12 SLOPE 12'-0" **ADD**	SF	45%	75%	67%	67%					.748
	16'-0" **ADD**	SF	45%	100%	85%	85%					.749
	20'-0" **ADD**	SF	45%	150%	121%	121%					.750
	SF = PLAN AREA, **NOT** ACTUAL AREA COVERED										

24

24. INTERIOR PAINTING

SPECIFICATIONS		UNIT	JOB COST			PRICE	LOCAL AREA MODIFICATION				DATA BASE ITEM NO.
			MATLS	LABOR	TOTAL		MATLS	LABOR	TOTAL	PRICE	
CASED OPENING	DOOR JAMBS AND TWO SIDES CASING, PAINT WITH BRUSH, 3'-0" OPENING										
	1 COAT ON PAINTED SURF.	EA	2.10	11.40	13.50	20.25					.800
	PRIME AND 1 COAT	EA	3.30	19.20	22.50	33.75					.801
	PRIME AND 2 COATS	EA	4.50	25.00	29.50	44.25					.802
	SAME AS ABOVE, 6'-0" OPENING										
	1 COAT ON PAINTED SURF.	EA	2.60	15.00	17.60	26.40					.803
	PRIME AND 1 COAT	EA	4.00	26.00	30.00	45.00					.804
	PRIME AND 2 COATS	EA	5.40	34.00	39.40	59.10					.805
	STAIN AND/OR CLEAR FINISH ON 3'-0" OPENING										
	1 COAT	EA	2.20	14.80	17.00	25.50					.806
	2 COATS	EA	3.50	25.50	29.00	43.50					.807
	3 COATS	EA	4.75	33.25	38.00	57.00					.808
	SAME AS ABOVE ON 6'-0" OPENING										
	1 COAT	EA	2.75	19.25	22.00	33.00					.809
	2 COATS	EA	4.20	33.00	37.20	55.80					.810
	3 COATS	EA	5.70	43.00	48.70	73.05					.811
RADIATOR	NOTE: TO OBTAIN SQUARE FOOT DIMENSION, MULTIPLY HEIGHT OF RADIATOR BY LENGTH OF RADIATOR, **THE FACE ONLY.** COSTS INCLUDE PAINTING THE ENTIRE RADIATOR WITH BRUSH										
	1 COAT	SF	.20	2.50	2.70	4.05					.812
	2 COATS	SF	.40	4.89	5.29	7.94					.813
	SAME AS ABOVE, WITH SPRAY GUN										
	1 COAT	SF	.30	.70	1.00	1.50					.814
	2 COATS	SF	.60	1.39	1.99	2.99					.815
SHELVES	OPEN SHELVES, **NO** BACKING OR VERTICAL SUPPORTS, TOP, BOTTOM AND EDGES, PAINT WITH BRUSH, UP TO 12" SHELVES										
	1 COAT ON PAINTED SURF.	LF	.13	.41	.54	.81					.816
	PRIME AND 1 COAT	LF	.23	.77	1.00	1.50					.817
	PRIME AND 2 COATS	LF	.33	1.14	1.47	2.21					.818
	STAIN AND/OR CLEAR FINISH WITH BRUSH										
	1 COAT	LF	.14	.43	.57	.86					.819
	2 COATS	LF	.24	.81	1.05	1.58					.820
	3 COATS	LF	.35	1.20	1.55	2.33					.821

24

SPECIFICATIONS		UNIT	JOB COST			PRICE	LOCAL AREA MODIFICATION				DATA BASE ITEM NO.
			MATLS	LABOR	TOTAL		MATLS	LABOR	TOTAL	PRICE	
	NOTE: TO OBTAIN SQUARE FOOT DIMENSIONS, MULTIPLY HEIGHT OF BOOKCASE, CABINET OR DOOR BY THE WIDTH, **ONE FACE ONLY**										
	COSTS SHOWN INCLUDE PAINTING ALL SIDES AND EDGES OF SHELVES & TRIM AND ONE SIDE OF BACKING										
BOOKCASE AND CABINETS	PAINT WITH BRUSH										
	SHELVES 12" DEEP										
	1 COAT ON PAINTED SURF.	SF	.24	1.50	1.74	2.61					.822
	PRIME AND 1 COAT	SF	.44	2.57	3.01	4.52					.823
	PRIME AND 2 COATS	SF	.64	3.32	3.96	5.94					.824
	SHELVES 24" DEEP										
	1 COAT ON PAINTED SURF.	SF	.48	3.00	3.48	5.22					.825
	PRIME AND 1 COAT	SF	.88	5.14	6.02	9.03					.826
	PRIME AND 2 COATS	SF	1.28	6.65	7.93	11.90					.827
	SAME AS ABOVE WITH STAIN AND/OR CLEAR FINISH *SHELVES 12" DEEP*										
	1 COAT	SF	.28	1.94	2.22	3.33					.828
	2 COATS	SF	.52	3.32	3.84	5.76					.829
	3 COATS	SF	.76	4.28	5.04	7.56					.830
	SHELVES 24" DEEP										
	1 COAT	SF	.56	3.90	4.46	6.69					.831
	2 COATS	SF	1.04	6.70	7.74	11.61					.832
	3 COATS	SF	1.52	8.57	10.09	15.14					.833
CABINET DOORS	TWO SIDES AND ALL EDGES OF CABINET DOORS, PAINT WITH BRUSH										
	1 COAT ON PAINTED SURF.	SF	.11	1.45	1.56	2.34					.834
	PRIME AND 1 COAT	SF	.21	2.46	2.67	4.01					.835
	PRIME AND 2 COATS	SF	.30	3.32	3.62	5.43					.836
	SAME AS ABOVE WITH STAIN AND/OR CLEAR FINISH										
	1 COAT	SF	.12	1.52	1.64	2.46					.837
	2 COATS	SF	.22	2.59	2.81	4.22					.838
	3 COATS	SF	.32	3.49	3.81	5.72					.839
CLOTHES CLOSET	PAINT SHELF AND TRIM ONLY										
	1 COAT ON PAINTED SURF.	LF	.20	.80	1.00	1.50					.840
	PRIME AND 1 COAT	LF	.35	1.42	1.77	2.66					.841
	PRIME AND 2 COATS	LF	.50	2.10	2.60	3.90					.842
	LF = WIDTH OF CLOSET										
LINEN CLOSET	24" DEEP SHELVES, 12" O.C. VERTICALLY MEASURED, PAINT SHELVES & TRIM ONLY										
	1 COAT ON PAINTED SURF.	LF	2.80	12.20	15.00	22.50					.843
	PRIME AND 1 COAT	LF	4.90	21.00	25.90	38.85					.844
	PRIME AND 2 COATS	LF	7.00	30.00	37.00	55.50					.845
	LF = WIDTH OF CLOSET										

24

SPECIFICATIONS		UNIT	JOB COST			PRICE	LOCAL AREA MODIFICATION				DATA BASE ITEM NO.
			MATLS	LABOR	TOTAL		MATLS	LABOR	TOTAL	PRICE	
WALL-PAPER	WALLPAPER APPLIED TO WALLS AND/OR CEILING, CEILING UP TO 8 FEET ABOVE FLOOR										
	Retail Cost of Wallpaper										
	$11 PER ROLL	SF	.44	.40	.84	1.26					.900
	$14 PER ROLL	SF	.56	.47	1.03	1.55					.901
	$17 PER ROLL	SF	.68	.55	1.23	1.85					.902
	$20 PER ROLL	SF	.80	.62	1.42	2.13					.903
	$25 PER ROLL	SF	1.00	.70	1.70	2.55					.904
	$30 PER ROLL	SF	1.20	.77	1.97	2.96					.905
VINYL	SAME AS ABOVE, WITH VINYL WALL COVERING										
	LIGHT	SF	.50	.40	.90	1.35					.906
	MEDIUM	SF	.72	.40	1.12	1.68					.907
	HEAVY	SF	.85	.52	1.37	2.06					.908
GRASS CLOTH	SAME AS ABOVE, WITH GRASS CLOTH										
	ECONOMY	SF	.65	.75	1.40	2.10					.909
	MEDIUM	SF	.95	.79	1.74	2.61					.910
	PREMIUM	SF	1.55	.87	2.42	3.63					.911
HIGH CEILING	FOR HANGING PAPER IN ROOMS WITH CEILING MORE THAN 8 FEET ABOVE FLOOR (UP TO 12 FEET), ADD TO ABOVE COSTS **ADD**	SF	--	20%	10%	10%					.912
OVER OLD PAPER	FOR HANGING PAPER OVER OLD LAYER(S) OF WALL-PAPER **ADD**	SF	--	20%	10%	10%					.913
DIFFICULT JOB	FOR DIFFICULT JOB WITH MORE THAN AVERAGE AMOUNT OF CUTTING **ADD**	SF	--	20%	10%	10%					.914
SMALL JOB	FOR HANGING PAPER IN SMALL ROOF WITH LESS THAN 200 SF OF WALL AREA **ADD**	SF	--	10%	5%	5%					.915
REMOVE WALL-PAPER	REMOVE WALLPAPER FROM PLASTER OR DRYWALL WITH STEAMING EQUIPMENT										
	ONE LAYER	SF	--	.23	.23	.35					.916
	SEVERAL LAYERS	SF	--	.36	.36	.54					.917
	WASH OFF GLUE AFTER RE-MOVING PAPER FROM WALL	SF	--	.13	.13	.20					.918

24

| SPECIFICATIONS | UNIT | JOB COST | | | PRICE | LOCAL AREA MODIFICATION | | | | DATA BASE ITEM NO. |
		MATLS	LABOR	TOTAL		MATLS	LABOR	TOTAL	PRICE	
	NOTE: COSTS SHOWN BELOW INCLUDE REMOVAL OF RUBBISH FROM PREMISES TO DUMPING GROUND, BUT **NO** TEAR-OUT COSTS AND **NO** DUMPSTER OR DUMPING FEES										
ADDITION	INCLUDING AVG. AMOUNT OF TEAR-OUT DEBRIS SF = ROOM ADDITION	EA PLUS SF	-- --	124.00 .26	124.00 .26	186.00 .39					.000 .001
BATHROOM AND KITCHEN	INCLUDING REMOVAL OF OLD TILE, PLUMBING FIXTURES, CABINETS AND APPLIANCES (DOES **NOT** INCLUDE DISCONNECTING FIXTURES) EA = ONE LOAD OF RUBBISH (5 YARDS)	EA	--	82.00	82.00	123.00					.002
DORMERS	INCLUDES REMOVAL OF OLD ROOFING AND FRAMING MATERIALS FROM PREMISES SF = SIZE OF NEW DORMER	EA PLUS SF	-- --	248.00 .26	248.00 .26	372.00 .39					.003 .004
GARAGE OR CARPORT	REMOVAL OF SCRAP LUMBER, MASONRY AND TEAR-OUT DEBRIS EA = EACH JOB	EA	--	124.00	124.00	186.00					.005
PORCH	REMOVAL OF SCRAP LUMBER AND MASONRY	EA	--	70.00	70.00	105.00					.006
BASEMENT FAMILY ROOM	REMOVAL OF SCRAP LUMBER AND TEAR-OUT DEBRIS SF = FAMILY ROOF	EA PLUS SF	-- --	70.00 .24	70.00 .24	105.00 .36					.007 .008
PORCH OR CARPORT ENCLOSURE	REMOVAL OF SCREENING, POSTS, AND OTHER TEAR-OUT DEBRIS	EA	--	106.00	106.00	159.00					.009

25

25. CLEAN-UP AND HAULING

SPECIFICATIONS		UNIT	JOB COST			PRICE	LOCAL AREA MODIFICATION				DATA BASE ITEM NO.
			MATLS	LABOR	TOTAL		MATLS	LABOR	TOTAL	PRICE	
DEMOLI-TION	REMOVAL OF DEBRIS FROM DEMOLITION WORK, LOADING MASONRY, PLASTER, LUMBER AND OTHER TEAR-OUT DEBRIS – LOAD TRUCK FROM BUILDING BY HAND AND HAUL TO DUMPING GROUND WITHIN 5 MILES DUMPING FEES AND DUMP-STER COST **NOT** INCLUDED SF = FLOOR AREA OF BUILDING	EA PLUS SF	-- --	364.00 .36	364.00 .36	546.00 .54					.010 .011
	SAME AS ABOVE, CHUTE RUBBISH FROM BUILDING AND HAUL	EA PLUS SF	-- --	364.00 .24	364.00 .24	546.00 .36					.010 .012
DIRT	SHOVEL DIRT BY HAND INTO TRUCK FROM PILES AND HAUL TO DUMPING GROUND UP TO 5 MILES EA = 5 CU. YARDS OF DIRT (ONE TRUCKLOAD)	EA	--	178.00	178.00	267.00					.013
ROOFING	FOR REMOVAL OF ROOF COVERING FROM ROOF, SEE PAGE 124										
	REMOVAL OF SCRAP LUMBER AND HAULING UP TO 5 MILES	EA PLUS SF	-- --	56.00 .12	56.00 .12	84.00 .18					.014 .015
MAID SERVICE	COMPLETE CLEANING OF INTERIOR OF AREA REMOD-ELED, INCLUDING WINDOWS INSIDE AND OUT	EA PLUS SF	-- --	50.00 .18	50.00 .18	75.00 .27					.016 .017

25

INDEX

INDEX

INDEX

INDEX

INDEX

INDEX

INDEX

NOW IS THE TIME
TO COMPUTERIZE YOUR ESTIMATING!

HomeTech can say with confidence that RIGHT NOW is the best time for you to computerize your estimating! HomeTech has taken its 30 years of experience in unit cost estimating and its 8 years in the software business and developed the best estimating software available. *HOMETECH QUICKEST* is not a glorified spreadsheet system or a generic program adapted for our use, but a complete system developed exclusively for HomeTech. *HOMETECH QUICKEST* is a high–tech program utilizing the latest technology in pull–down menus and a fully interactive database. And most importantly, *HOMETECH QUICKEST* is the only program on the market built around the HomeTech Remodeling & Renovation Cost Estimator!

For only $495, *HOMETECH QUICKEST* comes complete with the software program and database, quarterly local modifiers, complete written instructions, toll–free telephone support and the HomeTech Remodeling & Renovation Cost Estimator — that's everything you need to successfully computerize your estimating! Our program is so user friendly and our database so complete that you'll be up and running in no time. And if you ever need help, the HomeTech experts are only a phone call away.

So call 1–800–638–8292 to place your order today, risk–free! Our 30–day 100% money–back guarantee will protect your investment while our 90–day unlimited toll–free telephone support will ensure your success!